T0074371

Yong Tang
Xiaoping Ye
Na Tang

Temporal Information Processing Technology and Its Applications

Yong Tang
Xiaoping Ye
Na Tang

Temporal Information Processing Technology and Its Applications

With 129 figures

Editors
Prof. Yong Tang
Computer School
South China Normal University
Guangzhou 510631, P.R China
Email: issty@mail.sysu.edu.cn

Xiaoping Ye
Computer School
South China Normal University
Guangzhou 510631, P.R China
Email: mcsyxp@mail.sysu.edu.cn

Na Tang
Computer School
South China Normal University
Guangzhou 510631, P.R China
Email: isstn@mail.sysu.edu.cn

ISBN 978-7-302-22390-0
Tsinghua University Press, Beijing

ISBN 978-3-642-14958-0 e-ISBN 978-3-642-14959-7
Springer Heidelberg Dordrecht London New York

Library of Congress Control Number: 2010931751

Cover design: Frido Steinen-Broo, EStudio Calamar, Spain

Printed on acid-free paper

Springer is part of Springer Science+Business Media (www.springer.com)

Preface

Time is a natural attribute of everything. With the explosive growth of computer and network systems, temporal information has received extensive attention in both academia and industry. It plays an increasingly important role in the new generation information systems and also a key role in some applications. The use of temporal information modeling and processing technology in these applications can make them more useful and more convenient.

Temporal database and application problems have been mentioned during the 1970s. The groundbreaking study in this area was conducted by J. Ben Zvi, who proposed the bitemporal concept and a temporal database model in his dissertation, submitted to the University of California, Los Angeles, in 1982. In subsequent years, the temporal database theory research has grown vigorously and hundreds of temporal models have been proposed. James Clifford, Christian S. Jensen, Richard T. Snodgrass and Andreas Steiner made important contributions to temporal database models, theory and technology. In the recent years, along with information technology that can meet the increasing requirement for new applications, the temporal database theory and application technologies have made remarkable progress. However, there are many problems in temporal information processing, e.g., weakness in temporal calculus theory, low efficiency of temporal storage and access, complex temporal information processing and lack of the software development tools. There are three main trends in temporal technologies: model standardization, middleware development and application diversification.

We began to pay special attention to research on temporal database when we undertook the software application project: Intelligent Decision Support System of Salary (SIDSS), in 1998. The main concept behind SIDSS is that an employee's wage is paid according to information related to the employee and to the policies of the salary management department. SIDSS is a typical temporal system, in which the employee information that influences his or her salary is the typical temporal data and the salary policies that can be changed by the management department which also are time-varying knowledge. In the SIDSS, we used a temporal database model to design the employee database. We proposed a rule-based temporal knowledge model to represent the time-varying salary policies, and implemented a reasoning mechanism to realize the employee's salary determination and change based on the employee's temporal information and the salary policies based on temporal rule knowledge. The SIDSS has been used by more than ten thousands of agencies and millions of employees since 2000. In the past decade, we have undertaken more than 20 research projects on temporal database systems and associated software that involved the temporal data/knowledge base middleware, extension temporal to workflow, XML and role-based cooperative software, and others. Remarkably, TempDB 2.01, a temporal database management middleware developed by us, has been downloaded by hundreds of users from more than 10 countries since its release on July 2008.

This book is a collection of main study and research results that we obtained in the past years. There are five parts in this book. Part Ⅰ gives the time models, basic types of time data and their calculation methods. In Part Ⅱ, we introduce the basic concepts of temporal database. We then discuss the complex semantics of temporal variables and the basic problems of temporal database. In Part Ⅱ, we introduce database systems based on temporal information, such as spatio-temporal database systems and temporal XML database. Part Ⅲ discusses some temporal index technologies and proposes some new temporal indexing methods, such as bitemporal index, spatio-temporal data indexes and temporal XML index. These are key problems in the implementation of temporal database systems and applications. Part Ⅳ introduces the basic concepts of Temporal Database Management Systems (TDBMS) and the main techniques for the implementation of TempDB 2.1. Part Ⅴ discusses some temporal application technologies, such as temporal knowledge representation and reasoning mechanism, temporal extension to workflow and role management systems. In Part Ⅴ, as a case study, we introduce temporal data model, temporal knowledge reasoning and their implementations on a typical temporal application of SIDSS.

The aim of this book is to provide a basic understanding on calculation methods of time data and on temporal information modeling, as well as the implementation technologies of temporal applications. Researchers, graduate students and information technology professionals who are interested in the information systems and software involving temporal attributes, will find a starting point and a reference for their study, research and development from this book.

We would like to acknowledge all of the researchers in the area of temporal database modeling and applications. Based on their publications, their influence on this book is profound. We would like to thank all of the colleagues and students in the Co-soft Research and Development Center at Sun Yat-Sen University, China. We wish to express thanks to them for their valuable suggestions, discussions and assistance. We would also like to thank Dr. Jing Xiao, Dr. Gansen Zhao and Mr. Aaron Tang for their valuable suggestions and kind help.

The work described in this book was supported by The National Natural Science Foundation of China under Grant Nos. 60373081, 60673135, 60736020, and 60970044; The Natural Science Foundation of Guangdong Province of China under Grant Nos. 4105503, 7003721, 5003348, and 9151027501000054; and the Program for New Century Excellent Talents at the University of China under Grant No. NCET-04-0805.

Yong Tang, Xiaoping Ye and Na Tang
April, 2009

Contents

List of Figures and Tables

Part Ⅰ Temporal Models and Calculation Methods

- From Time Data to Temporal Information
- Time Calculation and Temporal Logic Method
- Temporal Extension of Relational Algebra

1 From Time Data to Temporal Information

Yong Tang[1,2], Zewu Peng[2], Dongning Liu[2,3], and Wenshen Zhang[2]

[1] Computer School, South China Normal University, Guangzhou 510631, P.R. China
[2] Department of Computer Science, Sun Yat-sen University, Guangzhou 510275, P.R .China
[3] Faculty of Computer, Guangdong University of Technology, Guangzhou 510006, P.R. China

Abstract Time data is one of the basic data types in database systems. There are two modes of using time data in applications, one is the explicit mode and the other is the implicit mode. In the second mode of application, time attributes of information need to be handled. In this chapter, we introduce three basic types of time data, i.e., point, interval and span. Subsequently, we propose the concepts of temporal information, temporal database and temporal systems, and introduce the basic concepts and core technologies of temporal database. We also analyze the origin and development of temporal information processing technologies and divide the evolution in this research field into three phases. Finally, we analyze the current situation in temporal research field and propose some trends of temporal information technologies.

Keywords *time data, temporal information, temporal database, temporal system, basic concept, evolution, trends*

1.1 Application Requirement

Time exists everywhere in the world. Its attributes are applied in many areas, such as e-commerce, e-government, global information system, and the stock market. However, some applications process time attribute in the same way as they would process a common attribute. For example, web sites can record logon time of users, but simply regard them as a normal attribute like number or character data type. We call these temporal applications implicit applications. There are other temporal applications that require special time processing mechanisms to manage time attributes. We call these applications explicit applications. The following examples are some explicit applications:

- Temporal application of e-government

  Government provides various kinds of important information that may only be

issty@mail.sysu.edu.cn

valid between specific time intervals. These types of information have to be handled on-time to reduce the corresponding loss. For instance, government publishes a notice of invitation for bids with a time limit. When the notice expires, the e-government system should be able to remove the notice automatically.

However, current information processing technologies applied in e-government are derived from processing methods of e-commerce or conventional applications, which always ignore the out-of-date information. Hence, there may be some negative effects. On one hand, people may be confused by the overdue information and waste effort on it. On the other hand, the e-government system fails to achieve the goal of conveying valid information. Therefore, designing temporal e-government systems is necessary to replace the current e-government systems.

- Temporal application of workflow

Workflow is a depiction of a sequence of operations, such as the work of a person, work of a simple or complex mechanism, work of a group of persons and work of an organization of staff or machines. Workflow management aims to implement the automation of business processes in the environment of computer application. In order to coordinate each member's job, a temporal attribute is added to workflow management.

The operational process of workflow is closely bound with time. For example, in the research field of integral temporal relation, time sequential relation is usually studied by Time Petri Net to analyze the earliest completion time of the entire workflow, the earliest start time, the latest start time, valid time and completion time of each part of workflow. There exist other forms of temporal attributes in a workflow, such as temporal attribute of local work item. In short, the requirement of temporal workflow has become necessary in the new generation of workflow.

- Temporal application of data warehouse

Data warehouse is a repository of an organization's electronically stored data. The methods to retrieve and analyze temporal data are considered an essential component of a data warehousing system because there are large amounts of temporal data in a data warehouse that can be mined to get useful information. For example, in an electronic commerce system we can analyze the fact that customers usually buy milk with coffee in time interval 7:00 – 9:00, but we cannot infer the same fact for time interval 7:00 – 21:00. Hence, temporal data mining is important and the requirement of temporal data in data warehouse is also necessary.

1.2 What Is Time Data

All of the above-mentioned applications make use of data related with time, such as the logon time, the time interval [7:00, 9:00] in a data warehouse. We refer to the types of data associated with time as **time data**. Time data can be represented by **time elements**. As the basis of time data, time elements are significant for denoting time attributes of records accurately. As we can see, time elements contain several concrete forms, such as discrete **time point**, **time interval**, **time span** and even **time set**.

As for time point, its time-varying characteristic is discrete, and every time point corresponds to the value of the attribute (Becher et al. 2000). Time interval (Allen 1983), the attribute of which possesses continuous characteristic, is recorded with a continuous interval to save a lot of space. Time span means the length of a period of time. When time attribute is complex and hard to describe, the time set may be a good representation to time data. We will interpret each of them in detail.

1.2.1 Time Point

First of all, **time point system** is given as follows.

Let the whole system $I = \langle P, \leqslant_t \rangle, P = \{p_1, p_2, \cdots, p_n\}$ be a finite set for time point. "$\leqslant_t$" represents the time sequence in set P, $p_i \leqslant_t p_{i+1}$ represents the fact that p_i does not appear behind p_{i+1}, that is, p_i happens before p_{i+1} or p_i and p_{i+1} happens at the same time. Then we can get:

$\forall p_i \in P$, $p_i \leqslant_t p_i$, so the relation $\leqslant_t$ is reflexive in P;

$\forall p_i, p_j \in P$, if $p_i \leqslant_t p_j$ and $p_j \leqslant_t p_i$, then $p_i = p_j$; the relation $\leqslant_t$ is symmetric in P;

$\forall p_i, p_j, p_k \in P$, if $p_i \leqslant_t p_j$ and $p_j \leqslant_t p_k$, then $p_i \leqslant_t p_k$, so relation $\leqslant_t$ is transitive in P.

Hence, I is a partial order set. Meanwhile, $\forall p_i, p_j \in P$, we can get $p_i \leqslant_t p_j$ or $p_j \leqslant_t p_i \in I$, which means I is also a total order set.

Secondly, we discuss time element based on point in conventional database. There are also some attributes with time feature in conventional relational database, such as "date of birth". The management of these attributes is similar to the management of normal attributes like integer, real number and so on. Time, in this way, is called user-defined time.

In the following example, we can see that the operation of user-defined time is the same as the operation of a normal attribute.

Table 1.1 shows that a conventional relational database contains an attribute of birthday (denoted as D.O.B), which is a user-defined time. When we want to get the information of persons who were born before 1965, we just need to compare the values of time attributes in the database to 19650000. The third tuple meets the requirement for which the value of time attribute is 19640920.

Moreover, we may need to calculate the value of a time attribute in a database for a special application. For example, if we want to find the tuples whose D.O.B. is 10 days before October 1st, we can verify whether the last 4 digits of

Table 1.1 Employees' information with time point attribute

Working number	Name	D.O.B	Academic title	Salary (Yuan)
019504478	Li	19720218	Assistant	2000
019504479	Zhang	19680118	Associate professor	3000
019504480	Wang	19640920	Professor	4000

the attribute value of D.O.B. plus 10 is equal to 1001. Obviously, no item meets the requirement.

As we can see, in the conventional relational database, there is little difference between the operations of time attributes and normal attributes.

1.2.2 Time Interval

Time interval (Philippens et al. 2009) is referred to as a segment of time, which is described by starting point and ending point, such as 1995 – 1996. According to the end points of time interval, we have four representations in Fig. 1.1.

	Intervals	Definition of interval	Graph representation
(1)	$[p_i, p_j]$	$p_i \leqslant t \leqslant p_j$	●……●
(2)	$[p_i, p_j)$	$p_i \leqslant t < p_j$	●……○
(3)	$(p_i, p_j]$	$p_i < t \leqslant p_j$	○……●
(4)	(p_i, p_j)	$p_i < t < p_j$	○……○

p_i, p_j: represents two separate time points
[ ] : represents closed interval
( ): represents open interval

Figure 1.1 Representation of time intervals

Time interval is able to describe time as powerful as time point. For example, we can get every time point by letting $p_i = p_j$, making sure that the time point is equal to p_i or p_j.

From Table 1.2, we know that if the attribute values of ending time points change more frequently, more tuples would be created. The difference between old tuples and new tuples are just the time-varying attributes. However, the database with time-varying attributes is not a relational database, because if several key "working number" items have the same attribute value, then the changed database does not agree with the property of relational database. Therefore, we can conclude that the characteristics of database changes a lot when time attribute is added to it.

Table 1.2 Employees' information with time interval attribute

Working number	Name	Academic title	Salary (Yuan)	Start-to-end
019504478	Li	Assistant	2000	[2005-03, 2008-09]
019504478	Li	Associate professor	3000	[2005-09, *Now*]
019504479	Zhang	Associate professor	3000	[2005-09, *Now*]
019504480	Wang	Professor	4000	[2005-09, *Now*]

Note that the time value "*Now*" is a time variable, which usually means the current time.

1.2.3 Time Span

Time span, which means an unbroken period of time, represents the length of time, such as one year, 30 days and 24 hours. It is usually denoted as an integral in database systems. Although time span represents a period of time in a similar way as time interval, it contains neither start point nor end point.

1.2.4 Complex Time Data

The basic representations of time element are time point and time interval, which can represent most time attributes. However, several time attributes can be represented in neither the single-time point approach nor single-time interval approach. For example, the departure time of Table 1.3 is represented by a set of time points like {6:00, 9:00, 12:00, 15:00, 18:00, 21:00}. The item "working years" of Table 1.4 shows that the working years of Li in the department computer engineering is the union of [1989, 1998] and [2000, 2003], which is a set of time intervals. Here, time elements in both tables are recorded by **time sets**.

Table 1.3 Departure information of bus with time point set

Bus number	Station of departure	Departure time
25642	Kangle Village	6:00, 9:00, 12:00, 15:00, 18:00, 21:00
25643	Tianhe Square	7:00, 10:00, 13:00, 16:00, 19:00, 22:00

Table 1.4 Employees' information with time interval set

Name	Department	Working years
Li	Dept. of Computer Engineering	[1989,1998], [2000,2008]
Wang	Dept. of Electronics	[1989,1996], [2000,2008]

1.3 Temporal Information, Temporal Database and Temporal System

1.3.1 What Is Temporal Information

The concept of time data is mainly used to obtain information related with time. The type of time-related information that possesses time-varying attributes is called temporal information. Temporal information is what we really want. Let us see how **temporal information** (Tang et al. 2003) appears. From Table 1.5, we can see that Alice's academic title changed at different time intervals. Alice was a lecturer from 2002 to 2005 and an associate professor from 2005 to 2008. Such information is temporal information.

Table 1.5 Temporal information of academic title

Name	Academic title
Alice	lecturer [2002, 2005], associate professor [2005, 2008]
Bob	associate professor [2005, 2008]
Charlie	professor [2005, 2008]

1.3.2 Temporal Database

Temporal information is represented by **temporal data** in a database system. Temporal data is data reflecting temporal information in a computer system. Like other types of data, temporal data are usually stored in a database system. Hence, we call a database system storing temporal data a **temporal database system** (Tang 2004). Different from a conventional database system, a temporal database system records and manages temporal information.

A conventional database system regards user-defined time as the only type of time. However, a temporal database system extends the time concept with valid time and transaction time. Valid time is a period of time during which the fact is true in the modeled reality. A database fact is stored in a database at some point in time and after it is stored, it is current until logically deleted. Transaction time means the time when the fact is current in the database and may be retrieved. In terms of the ability to manipulate different types of times, we divide temporal databases into the following types.

- Snapshot database: there is only one type of user-defined time in a snapshot database. It just reflects the current state of a database system and shares a similar property with a conventional database system.
- Rollback database: transaction time is applicable in rollback database. Rollback

database focuses on storing the station of transactions committing and the station before change in a temporal database.

- Historical database: valid time is adopted in historical database. Historical database records and manages the events at valid time point and changing states.
- Bitemporal database: valid time and transaction time are both supported in bitemporal database. Bitemporal database incorporates the characteristics of three other temporal databases and status of all times of changing database.

In the research field, there are several aspects of core technologies of temporal databases as follows:

- When temporal attribute is added to relational database, temporal calculus changes a lot compared to conventional relational calculus and becomes a key technology of temporal database system.
- **Temporal data models** are an essential component of a temporal database system. Current temporal data models are able to cover most of temporal problems and become mature. The famous instance of temporal data model is bitemporal conceptual data model (BCDM), which focuses on the implementation of semantics of temporal data.
- Temporal query languages and temporal database management systems are other kernel parts of a temporal database system. Most of current temporal query languages are based on conventional query language, and they extend temporal query ability to meet the requirements of a temporal database system. For instance, TempSQL, a temporal query language, is based on the framework of conventional query language SQL and adds temporal semantics in it.

1.3.3 Temporal System

However, temporal database is just a carrier for recording temporal data. The goal of proposing temporal database is to handle temporal information in information systems. As an important part of an information system, temporal database system is mainly used to record and manage temporal data. The information system with a temporal database is called a temporal system, which treats temporal database as the most important technology.

1.4 Origin and Development of Temporal Information Technologies

After understanding the concepts of temporal information, we introduce the evolved phases of temporal information technology.

Concerning the origin and development of temporal information technology, the publication of several books on temporal database were treated as the landmark of the temporal database research field. Tansel, Clifford, Gadia, Jajodia, Segev, and Snodgrass co-edited *Temporal databases: theory, design, and implementation* (Tansel et al. 1993) to collect almost all the previous important results on temporal database in 1993. Chinese researcher Tang Changjie reviewed the first 20 years' development of temporal database technology (Tang 1999a, 1999b). However, since temporal database research has always been developing, there exist many new results. Through consulting the literature, especially the latest developments, we divide the origin and development of temporal information technology into three phases (Tang 1999a, 1999b): the founding phase (before 1982), the development phase (1983 – 1993), and the application phase (since 1994).

1.4.1 Founding Phase

The feature "temporal" is an attribute of the information in nature, which created interest as early as the 1970s on the application of temporal information. In 1970, Wiederhold and Fries made an attempt at dealing with temporal information in the development of the health care system. In 1977, Kahn Ketal's paper "*Mechanizing temporal knowledge*" (Kahn and Gorry 1977), published in *Artificial Intelligence,* reflects the basic research work of the early stage on temporal information. At the early 1980s, the maturity of database technologies and the development of high-speed large-capacity storage devices had made the presence of temporal database technology possible. The doctoral theses of Ben-Zvi from University of California, Los Angeles (Zvi 1982), and Clifford from New York University (Clifford 1982) in 1982, are two landmark results of temporal database technology during founding phase.

In the years 1979 – 1982, Zvi made a systematic study on temporal database technology and completed his doctoral thesis titled *The time relational model* in 1982. The main contributions of the study include:

- proposing temporal database model;
- making N1NF (non-1NF) of the TDB (temporal database), referring to the time interval as field values, breaking through the limitation that the database field values can only be a number or string;
- introducing the concept of bitemporal, which uses valid time as the managed object in the database life cycle and transaction time as the history of the database itself;
- introducing a temporal index structure.

The above concepts and ideas, which mark the formation of temporal database, are still in use now.

In the same year, Clifford published a series of papers and completed his doctoral thesis, *A logical framework for the temporal semantics and natural language querying of historical database*. His main contributions include a pioneering work on historical database and the proposal of the historical relational data model (HRDM) model. He studied the technical details on plugging temporal information into relationships, tuples, and field values, introduced historical relational models and historical relational algebra, and treated the traditional relational database as a special case of a historical database, that is, a reduced point (*Now*, *Now*) in historical database.

Hence, it can be thought that the concept of temporal database was formed in 1982.

1.4.2 Development Phase

After 1982, the research on temporal database began to make remarkable progress. The subsequent decade saw computer academia give increasing attention to temporal database, and thousands of papers on temporal database were published. For example, EI CompendexWeb Search indexed more than 2000 papers associated with temporal database during the years from 1982 to 1994. Hundreds of **temporal database models** and **temporal information processing** methods were proposed. During this period, a large number of researchers from important international universities and research institutions developed several specialized temporal database research groups. The most prominent were groups at New York University, Iowa State University, University of California at Berkeley, University of Southern California, University of Arizona, and some specialized laboratories at HP and Bell companies.

The most important outcome of this phase was the publication of *Temporal databases: theory, design, and implementation*, which was co-edited by Tansel, Clifford, Gadia, Jajodia, Segev, Snodgrass in 1993. It was said to be the first monograph on temporal database in the world. The book summarized the previous temporal database technology research comprehensively and listed more than 500 papers on TDB during the two decades, the authors of which came from United States, Europe, China, Japan, India, Switzerland, Turkey and so on.

The significant contributions during this period include theoretical research on temporal database and the building of temporal data models.

1.4.3 Application Phase

After the book *Temporal databases: theory, design, and implementation* was published, the academic focus on the direction of temporal information technology became consistent. Although the evolution from early understanding based on relational algebra, to the subsequent calculus-based view, datalog-based view and object-oriented view, was remarkable, researchers failed to find a major new breakthrough on temporal models. After 1994, people began to study how to make the temporal data model standard and apply it in practice.

In the "standardization" process of the temporal data model, Snodgrass proposed bitemporal data model TSQL2 in 1994, which is the temporal expansion of the standard language SQL-92. He even proposed integrating the relevant structures of TSQL2 into SQL3 standards. In 1996, the SQL/Temporal program on adding valid time was submitted to ISO in May and the scheme on adding transaction time was submitted to ANSI Committee in October. However, the process of standardization for temporal data models still needs to explore. *Temporal database management* (Jensen 2000) co-edited by Jensen, Clifford, and Snodgrass in 2000 was the important literature during this stage.

TimeDB 2.0 launched by the Swiss TimeConsult company in 1998 is the representation of the results in the production process of temporal database. TimeDB is a bitemporal relational database system (Consult 2005) that supports the platform for Java/JDBC. It is based on the SQL query language and supports temporal query language ATSQL2, whose basic model derives from R. T. Snodgrass' TSQL2, combined with ChronoLog model raised by Michael Böhlen, University of Aalborg, Denmark, and Bitemporal ChronoSQL model proposed by Andreas Steiner from Switzerland. However, TimeDB is not a perfect product and needs to improve.

To obtain a complete temporal data model software component in China, TempDB (Tang 2008) was developed by the temporal data processing component research and development team of the database and co-soft lab led by Professor Yong Tang, Sun Yat-sen University. TempDB, a temporal data component, is based on the relational database system MySQL, uses the JAVA language, and possesses some outstanding characteristics such as high portability and ease of deployment.

Another important feature of this period was the application of temporal information. In the 1980s, due to the rapid development of computer-related technology, especially network and multimedia technology, many demands and applications of temporal information were implemented. The main application areas included geographic information systems, agricultural information systems, telecommunications information systems, e-commerce, intelligent decision support systems, data warehouse and data mining, in particular temporal **spatio-temporal information technology** and temporal applications of multimedia information technology. However, the theory of temporal database model was not mature enough. The applications of temporal database during this period were just implemented by using the traditional database technology and related applications of technology with some concepts of temporal information management and operations.

1.5 Current Situation, Problems and Trends

1.5.1 Current Situation

Increasing literature on temporal information processing technology was published in recent years. We used EI CompendexWeb[1] and several related web sites to retrieve the papers with temporal keywords. There were 182 papers whose abstracts include temporal database and 904 papers whose abstracts present temporal information in the 1980s. There were 1033 and 4454 papers, respectively, of the above-mentioned types in 1990s. There were 2531 and 14886 papers, respectively, from 2000 until now.

Furthermore, some web sites (Wu et al. 1998) introduced researchers' information on temporal database and classified papers on temporal database into the following types: ① models, ② database designs, ③ query languages, ④ constraints, ⑤ time granularities, ⑥ access methods, ⑦ real time database, ⑧ sequence database, ⑨ data mining, ⑩ implements, ⑪ concurrency control, and ⑫ others. Many of these subfields have made remarkable achievements.

Temporal data models are usually the extensions of traditional relational databases and traditional relational databases are regarded as their special cases. The main approaches of the extensions include adding a calculus, such as AFTER, BEFORE and OVERLAP to TDB, or expanding regular operations to temporal operations, such as temporal selection, temporal projection and temporal connections. More than two-dozen temporal data models were proposed, including several well-known data models. For example, Ariav proposed a temporally oriented data model (Ariav 1986) with the temporal isomorphism assumption, in which there is a tight correspondence between the database and the temporally concurrent reality. The POSTGRES data model (Rowe and Stonebraker 1987), proposed in 1987, supports degenerate bitemporal relations. In 1988, a multi-dimensional data model is presented, which is in turn restricted to a two-dimensional data model with valid and transaction time as the dimensions (Gadia and Yeung 1988). Recently, spatio-temporal data model has become the hot issue. Chen proposed 2-D spatio-temporal data model for moving objects in Network in 2006 (Chen et al. 2006), Zacharouh presented a k-anonymity model for spatio-temporal data in 2007 (Zacharouh et al. 2007) and Salguero applied spatio-temporal data model for data warehousing in 2008 (Salguero et al. 2008).

There are three main types of database management systems: relational, object-oriented and object-relational DBMS. Current research and application of temporal relational databases is a hot issue. TimeDB and TempDB are two products of temporal database. TimeDB is considered as quite a successful temporal

[1] http://www.engineeringvillage2.org.cn/

database prototype system throughout the world, but it still has some defects. TempDB was developed by the temporal data processing component research and development team of the database and co-soft lab, which is led by Professor Yong Tang, Sun Yat-sen University. TempDB, a temporal data component, is based on the relational database system MySQL, uses JAVA language to develop, which characterizes it with high portability and ease in deployment.

At present, most temporal data query languages are expansions of the current popular query languages such as SQL and Quel. There are many important results on temporal extension of conventional query language. Table 1.6 (Ozsoyglu and Snodgrass 1995) lists several major temporal relational query languages.

Table 1.6 Temporal query languages

Temporal query language	Citation	Underlying data model	Basic query language
HQL	(Sadeghi et al. 1988)	Sadeghi	DEAL
HQuel	(Tansel 1991)	Tansel	Quel
HSQL	(Sarda 1990)	Sarda	SQL
HTQuel	(Gadia and Yeung 1988)	Gadia-1	Quel
Legol 2.0	(Jones et al. 1979)	Jones	Relational algebra
TDM	(Segev and Shoshani 1987)	Segev	SQL
Temporal Relational Algebra	(Lorentzos and Johnson 1988)	Lor-entzos	Relational algebra
Temp SQL	(Yau and Chat 1991)	Yau	SQL
Time by Example	(Tansel and Arkun 1989)	Tansel	QBE
TQuel	(Snodgrass 1987)	Snodgrass	Quel
TSQL	(Navathe and Ahmed 1989)	Navathe	SQL
TSQL2	(Snodgrass et al. 1994)	BCDM	SQL-92

Many papers on data mining with temporal information processing were published in recent years. We use ISI citation index to get 69 papers on data mining and temporal information. Many of the latest corresponding papers are closely tied up with spatio-temporal data mining. For example, Huang Y proposed a framework for mining sequential patterns from spatio-temporal event data sets (Huang et al. 2008). Kechadi presented a visual approach for spatio-temporal data mining (Kechadi and Bertolotto 2006).

Because the number of other sorts of papers is small relatively, we are not introducing them in detail.

As we know, temporal database research fields are at the stage of study and development. On one hand, people are proposing several types of temporal database models from different points of view. On the other hand, many methods and technologies are put forward due to the requirement of practical applications and the increasing effect of the temporal information processing. The applications of these studies have contributed greatly to the development of TDB. The above conclusions reflect the research achievements in current temporal database field.

1.5.2 Existent Problems in Temporal Database Research

There are many problems existing in the various types of subfields of temporal database research.

- In the field of temporal data models, the calculative system is not complete. Moreover, there are not enough systematic theories of mathematics to support temporal relational calculi. Although temporal data models are proposed, there is no uniform temporal data model. Some applications for the standards on temporal data model are in progress. The fact that many researchers are studying the same problem based on different models is a repetitive job to some extent. Moreover, divertive situation of temporal data models may discourage the development of temporal database. Therefore, the standardization of temporal data model is necessary.
- At present, most of the current **temporal query languages** are extended from conventional query languages, such as SQL and Quel, the efficiency of which is relatively low because the interpretation of the temporal query language needs extra steps compared to normal query language. This discourages database producers from applying temporal database technology into their database products.
- The current temporal database research has made considerable progress, but it is mostly confined to the study of temporal database property. There are only a small number of research results to the temporal attribute of other information, such as knowledge base. Temporal data technology is still at the stage of data processing. The main advantages of temporal logic and inference research are symbolic calculus and inference ability. However, information processing ability is weak and is deviated from the research of temporal databases and temporal information processing. In the field of temporal knowledge and logic, the expansion of time interval and logic operation are studied in depth, but temporal knowledge model is not.
- With the rapid development of database technology, especially multimedia technology and network technology in the 1990s, the application of temporal information was broad. In geographic information systems, agricultural information systems, telecommunications information systems, e-commerce, intelligent decision support systems, data warehouse and data mining, in particular, in temporal and spatio-temporal information technology and multimedia information systems, temporal information processing was emphasized and applied unprecedentedly. However, due to lack of mature model and software products, the applications associated with temporal information are implemented by temporal data model and conventional technology. Thus, the interpretation of temporal part relies on another application, and does not the database itself.

1.5.3 Trends

Temporal information processing has become key technology for the new generation of database and information systems, especially e-government, e-commerce, data warehousing, data mining, decision support system, and so on. Many attractive subjects related to temporal information technology are becoming an important direction of development for database and information systems.

The main trends of temporal database research are as follows:

- The standardization of temporal data models is an important trend for temporal database research development. The fact that functions of many temporal data models are similar, shows the difficulty in choosing a specified temporal data model for temporal database researches and temporal applications. To unify temporal data models is required.
- Developing important temporal database products including temporal middleware will be a main research trend in temporal database research.
- The application areas of temporal information are becoming vast. For instance, temporal applications vary from temporal data handling to application of temporal knowledge, from conventional text database to spatio-temporal database and multimedia database system.

Some hot issues of temporal database research are as follows:

- Recently, the work of extending SQL with XML technology and supporting query languages are done by temporal database researchers and SQL standardization groups. For example, as a new powerful data model and query language, XML and XQuery raise the question whether they can provide a better basis for representing and querying temporal database information. Fusheng Wang's paper (Wang et al. 2008) shows that transaction-time, valid-time and bitemporal database can be effectively represented in XML and application of XQuery requires no other extension of current query language standards. However, the extension of XML based temporal data model needs an even further thorough study. Several research groups are working to apply XML technology into temporal data model. Seo-Young Noh raised an XML based approach to implement a parametric data model related to temporal database (Noh and Gadia 2005). In another paper (Noh et al. 2008), he found a native XML-based implement approach parametric temporal data model, in which he developed his own storage technology and query language rather than using existing storage technologies including relational, object-oriented, and native XML database systems. We believe that there needs to be further studies on XML-based temporal database.
- Today, there are many researchers who are concerned about the field of spatio-temporal database. It has a wide range of application areas including sailing, traveling, military and so on. In the field of spatio-temporal database, there should be many more research results on various hot topics. The discrepancies in the implementation of many spatio-temporal database systems and the

application in the real world are large, so the research on implementation of spatio-temporal database will go on. In the next few years, research of spatio-temporal technology should be hot to keep up with the upcoming areas such as mobile, wireless and sensor networks. Because sensor networks and streaming data is a big and upcoming area, it will certainly promote the development of spatio-temporal database. Spatio-temporal access method is also an important aspect of spatio-temporal database research. There are mainly two aspects of spatio-temporal access methods, one is the storage and retrieval of historical information, and the other is the prediction of future information. Because the volume of spatio-temporal data is overwhelming, a new parallel algorithm proposed to operate them is certainly a good result of temporal research.

- Other hot issues include theoretical researches such as, logic methods of temporal information, expression and inference of temporal knowledge and temporal algebra, application researches such as temporal workflow, temporal data mining and so on.

References

[1] Allen JF (**1983**) Maintaining knowledge about temporal intervals. Communication of the ACM **26**(11): 832 – 843

[2] Ariav G (**1986**) A temporally oriented data model. ACM Transactions on Database System **11**(4): 499 – 527

[3] Becher G, Clerin-Debart F, et al. (**2000**) A qualitative model for time granularity. Computational Intelligence **16**(2): 137 – 168

[4] Chen BY, Chen XL, et al. (**2006**) 2-D spatio-temporal data model for moving objects in network. Transportation studies: Sustainable Transportation, Proceedings of the 11th International Conference of Hong Kong Society for Transportation Studies: 583 – 592

[5] Clifford J (**1982**) A logical framework for the temporal semantics and natural-language querying of historical databases. PhD thesis: 263. State University of New York at Stony Brook

[6] Consult T (**2005**) TimeDB—A bitemporal relational DBMS. http://www.timeconsult.com/Software/AboutTimeDB1.0.html

[7] Gadia S, Yeung C (**1988**) A generalized model for a relational temporal database. In: Proceedings of the 1988 ACM SIGMOD international conference, 251 – 259

[8] Huang, Y, Zhang LQ, et al. (**2008**) A framework for mining sequential patterns from spatio-temporal event data sets. IEEE Transactions on Knowledge and Data Engineering **20**(4): 433 – 448

[9] Jensen CS (**2000**) Temporal database management. http://www.cs.auc.dk/~csj/Thesis/

[10] Jones S, Mason P, et al. (**1979**) Legol 2.0: A relational specification language for complex rules. Information Systems **4**(4): 293 – 305

[11] Kahn K, Gorry GA (**1977**) Mechanizing temporal knowledge. Artificial Intelligence **9**(1): 87 – 108

[12] Kechadi MT, Bertolotto M (**2006**) A visual approach for spatio-temporal data mining. In: Proceedings of the 2006 IEEE International Conference on Information Reuse and Integration(IRI 2006): 504 – 509

[13] Lorentzos NA, Johnson RG (**1988**) Extending relational algebra to manipulate temporal data. Information Systems **13**(3): 289 – 296

[14] Navathe SB, Ahmed R (**1989**) A temporal relational model and a query language. Information Sciences **49**: 147 – 175

[15] Noh SY, Gadia SK (**2005**) An XML-based framework for temporal database implementation. In: Proceedings 12th International Symposium on Temporal Representation and Reasoning, 180 – 182

[16] Noh SY, Gadia SK, et al. (**2008**) An XML-based methodology for parametric temporal database model implementation. Journal of Systems and Software **81**(6): 929 – 948

[17] Ozsoyglu G, Snodgrass RT (**1995**) Temporal and real-time databases: A survey. IEEE Transactions on Knowledge and Data Engineering **7**(4): 513 – 532

[18] Philippens MEP, Pop LAM, et al. (**2009**) Bath and shower effect in spinal cord: The effect of time interval. International Journal of Radiation Oncology Biology Physics **73**(2): 514 – 522

[19] Rowe LA, Stonebraker MR (**1987**) POSTGRES data model. Proceedings of the 13th VLDB Conference

[20] Sadeghi R, Samson WB, Deen SM (**1988**) HQL—A historical query language.BNCOD: 69 – 86

[21] Salguero A, Araque F, et al. (**2008**) Spatio-temporal ontology based model for data warehousing. New Aspects of Telecommunications and Informatics: 125 – 130

[22] Sarda N (**1990**) Extensions to SQL for historical database. IEEE Transactions on Knowledge and Data Engineering **2**(2): 220 – 230

[23] Segev A, Shoshani A (**1987**) Logical modeling of temporal data. ACM SIGMOD Int'l Conference on Management Data: 454 – 466

[24] Snodgrass RT (**1987**) The temporal query language TQuel. ACM Transactions on Database System **12**(2): 247 – 298

[25] Snodgrass RT, Ahn I, et al. (**1994**) TSQL2 language specification. ACM SIGMOD Record **23**(1): 65 – 86

[26] Tang, CJ (**1999a**) The achievements, deficiency and future work in temporal database (in Chinese). Computer Science **26**(4): 63 – 66

[27] Tang CJ (**1999b**) The background, characteristics and representative researchers of temporal databases (in Chinese). Computer Science **26**(3): 27 – 31

[28] Tang Y (**2004**) Introduction to temporal database (in Chinese). Peking University Press, Beijing, China

[29] Tang Y (**2008**) TempDB—A temporal relational DBMS. http://www.cosoft.sysu.edu.cn/TempDB/index.asp.

[30] Tang Y, Tang N, Ye XP (**2003**) Temporal Information processing research survey (in Chinese). Acta Scientiarum Natralium Universitatis Sunyatseni, 42(4): 4 – 8

[31] Tansel AU (**1991**) A historical query language. Information Sciences **53**: 101 – 133

[32] Tansel AU, Arkun ME (**1989**) Time-by-example query language for historical database. IEEE Transactions on Software Engineering **15**(4): 464 – 478

[33] Tansel AU, Clifford J, et al. (eds) (**1993**) Temporal databases: theory, design, and implementation. Benjamin-Cummings Pub Co

[34] Wang FS, Zaniolo C, et al. (**2008**) ArchIS: An XML-based approach to transaction-time temporal database systems. VLDB Journal **17**(6): 1445 – 1463

[35] Wu Y, Jajodia S, Wang XS. (**1998**) Temporal database bibliography update. Lecture Notes in Computer Science 1399, Springer, Berlin/Heidelberg: 338 – 366

[36] Yau C, Chat G (**1991**) TempSQL—A language interface to a temporal relational model. Information Science and Technology: 44 – 60

[37] Zacharouh PA, Gkoulalas-Divanis, et al. (**2007**) A *k*-anonymity model for spatio-temporal data. 2007 IEEE 23rd International Conference on Data Engineering Workshop, Vols 1 – 2: 555 – 564

[38] Zvi JB (**1982**) The time relational model. PhD thesis, Department of Computer Science, University of California at Los Angeles

2 Time Calculation and Temporal Logic Method

Dongning Liu[1,2], Yong Tang[1,3+], Shu Li[3] , and Wenchong Fang[3]

[1] Computer School, South China Normal University, Guangzhou 510631, P.R. China
[2] Faculty of Computer, Guangdong University of Technology, Guangzhou 510006, P.R. China
[3] Department of Computer Science, Sun Yat-sen University, Guangzhou 510275, P.R. China

Abstract Time data calculation is the basic functions of time based on database systems, and temporal knowledge representation is the basic method for processing temporal information. The first task of these is the method to define time model and structure. In this chapter, we first introduce time structure, property of time and the order of time set. Based on these, we introduce the non-axiomatic temporal logic. With Gabbay's work, axiomatic systems and proof methods for temporal logic have so far relatively few applications in the context of database. Therefore, the non-axiomatic systems are the pivot in the knowledge representation of temporal database. There are two kinds of time representations, point-based and interval-based, in non-axiomatic temporal logic. The former regards time as discreteness, while the latter regards time as continuum. Subsequently, we discuss calculation approach in time, along with time span and time granularity, which is another core of formal knowledge representation in temporal database. Being an important method of knowledge representation, axiomatic logic system can help us to understand the properties of time structure easily. We will introduce it at the end of this chapter.

Keywords *time model, temporal logic, Allen's method, temporal calculations, time span, time granularity*

Temporal knowledge representation is the basic and important part of temporal database. The first job of temporal knowledge representation is to find the method to define and dispose time model. There are many basic time models, such as continuous model, stepwise model, discrete model. These models are defined by the choice of time axis structure, but they are still too coarse. Time structure and time set are seemingly very easy, but in fact they have many properties that must be dealt with seriously in information system, such as order

---

+ Corresponding author: issty@mail.sysu.edu.cn

relations of time set, first order properties of time flow and their modalities (Goble 2001).

Through richly analysis, we can define and use several types of temporal logic to deal with the knowledge representation. In this field, Allen, Gabbay and Chomicki have done some work. In 1983, Allen established one type of interval-based temporal logic, which is the headstone of work related to temporal logic. In 1994, Gabbay proved that axiomatic systems and proof methods for temporal logic have so far found relatively few applications in the context of database (Gabbay et al. 1994). Based on his result, Chomicki and Saake used non-axiomatic temporal logic system to formalize temporal query language (Chomicki and Saake 1998), most of which are affirmative (Alvarez and Mossay 2006).

Besides logic, time calculation is another method for temporal knowledge representation. The time calculation is based on time span and time granularity, which are still based on time structure. Bettini and SeanWang have established a general framework for time granularity in 1998 (Bettini et al. 1998). Becher and Clerin-Debart built a qualitative model for time granularity in 2000 (Becher et al. 2000). After their work, Anurag and Sen debated the chronon based model for temporal database in 2008 (Anurag and Sen 2008). As an increasingly important information system, temporal database is desired to be more accurate. All these are guaranteed by rigorous time calculation.

In this chapter, we will introduce all these in sequence, most of which are the pivotal formalization method of temporal database.

2.1 Time Model

Time is continuous in the time axis from the point of the objective world's view. However, in computer science, we should quantize the continuous time into time point or time interval, in order to build time model. If time were recorded in its original manner, the capacity for the records would have to be very large. Moreover, most of the time, people do not need to acquire the time status of the entire real world. Hence, the solution to the problem, we can think of, is to build a proper time model, which can not only be applied to data model, but also meet the requirements for various application situations.

Based on the choice of time axis structure, we classify time models as follows.

2.1.1 Continuous Model

In **continuous model**, time is viewed as real numbers, each of which corresponds to a time point (Alvarez and Mossay 2006). In terms of the property of real number, there exist some other time points between any two time points that lie in the time axis. This type of model is able to build model for time with high accuracy. However, thanks to the operative mode of modern computer, which is based on digital logic, it is impossible to record time without loss. On many real-time control situations, such as the industrial control field, a large quantity of continually changing data is required to be recorded. In this situation, we usually record the change of data in sampling method.

2.1.2 Stepwise Model

Stepwise model views the status of data as a time function (Jochens et al. 2007). The status of the records is not changed unless the status of data on some time point is changed. Therefore, the value of each point on the time sequence matches the value of last changed data. When we query the value of current data, we need to backtrack to the last changed status. For example, Fig. 2.1 presents that there have been three changes in the identity of Li since 1997, so only three time points need to be recorded when we adopt a stepwise model. If the identity of Li is queried on the time point *Now*, the value returned is not Null, but the value "associate professor" by backtracking from time axis. This model is different from continuous model as we are not able to get the value of the point between any two points through interpolation method.

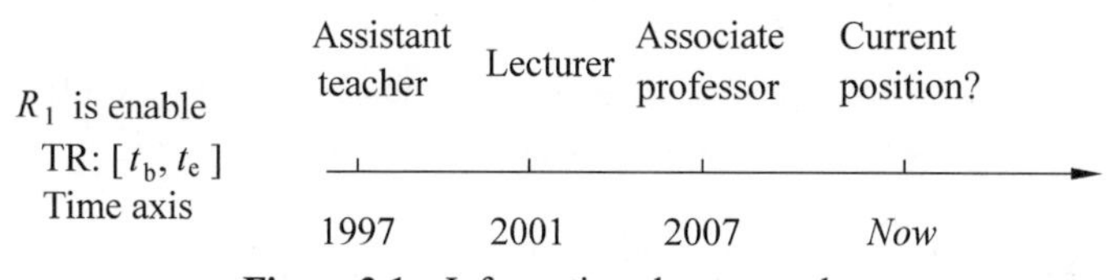

Figure 2.1 Information about a teacher

2.1.3 Discrete Model

Discrete model maps time with an integer, and there exists no time point between two adjacent time points (Rocha-Martinez 1998). The model is applicable to the practical applications in which only the key time point needs to record. For example, considering total amount of pay for the employees of a company, total amount of all employees' salaries from January to March is 190,000, 180,000 and 200,000. From statistics we can see that:

- there exists no other month between two adjacent months, such as January and February;
- we are not able to get the statistic of February through interpolation method from the statistics of January and March;

- there would be a false message if we adopt stepwise model. For example, on 10th April, we would get the false statistic that the total amount for the pay of April is $200,000, if we backtrack.

2.1.4 Non Temporal Model

There are also some statistics that are not influenced by time, such as birthplace. Therefore, their temporal properties are not considered when building temporal model. If necessary for change, the value of temporal properties will be replaced by new values.

2.2 Properties of Time Structure

Models above-mentioned are dependent on time axis structure, but these are the original stratifications, which are coarse. For continuous and discrete models, it is easy to find that the order relations of time sets are the most important theoretical basis.

2.2.1 Order Relations of Time Sets

The order relations of time sets have two representation methods, the first is **partial order**, and the second is **not partial order**.

As partial order, the structure is $T = \langle T, \leqslant \rangle$, and $T = \{t_1, t_2, \cdots, t_n\}$, which is the finite set of time points. $\leqslant$ is the time order, $t_i \leqslant t_{i+1}$ represents that t_i is not later than t_{i+1}, so $\leqslant$ is reflexive, transitive and antisymmetry.

As not partial order, the structure is $T = \langle T, < \rangle$, and $T = \{t_1, t_2, \cdots, t_n\}$, which is the finite set of time points. $<$ is the time order, $t_i < t_{i+1}$ represents that t_i is faster than t_{i+1}, so $<$ is just transitive.

In the opinion of algebra, the "$<$" seems no better than "$\leqslant$", because it does not have its own reflexive and antisymmetry. However, for these reasons, it is actually fitter for the database, which will be expatiated in the following sections.

2.2.2 First Order Properties of Time Flow

As indicated above, we simply debate the **order relations of time sets**, but it is more generic. However, the representation of time in temporal database is more. It must, especially satisfy left-serial, right-serial, transitive, non-branching and have finite intervals.

- Left serial and right serial: Except the database management system that has defined the start time and end time of transaction, time set of database must satisfy left-serial and right-serial. Left-serial represents that for any points in time axes, its preorder set is not empty. Its first order definition is as follows:

$$\forall x \exists y . Ryx$$

Right-serial represents that for any points in time axes, its post order set is not empty. Its first order definition is as follows:

$$\forall x \exists y . Rxy$$

After satisfied left-serial, we can insert tuple in front of head tuple of any table, and vice versa, as right-serial.
- Non-branching: In generic applications of temporal database, their axes are linear, while it is often branching in model checking (Wooldridge and Huget 2006), so it must satisfy non-branching in TDB. However, non-branching time is not first order definable, but modal definable. In the following text, we will give its modal definition.
- Continuous and having finite intervals: Although time is continuous intuitively, in computer science and database, it is data acquisition. Therefore, according to time in database, it is not continuous but has finite intervals, which means that it can insert finite points between two time points. Having finite interval is not first order definable but modal definable. In the following text, we will give its modal definition.
- Others properties: In temporal database, reflexive and antisymmetry are two controversial properties. They bring forward stronger order relations. As tuples' time attributes are cycles, such as an on duty schedule, antisymmetry must be excluded, so that it is reflexive.

On different ontology, reflexive and irreflexive time have different controversies (Goble 2001). Without loss of generality, we exclude them from this book as our work is just for generic temporal database.

Because we are excluding reflexive and antisymmetry, it is obvious that structure $\langle T, < \rangle$ is better than structure $\langle T, \leqslant \rangle$.

Based on these, there are several flows of time that correspond to number sets, such as $N = \langle N, < \rangle, Z = \langle Z, < \rangle, Q = \langle Q, < \rangle$ and $R = \langle R, < \rangle$. All of these structures are non-branching. $N = \langle N, < \rangle$ has properties "Having a first point", "Right-seriality" and "Having finite intervals". $Z = \langle Z, < \rangle$ has properties "Left-seriality", "Right-seriality" and "Having finite intervals". $Q = \langle Q, < \rangle$ has properties "Left-seriality", "Right-seriality" and "Density". $R = \langle R, < \rangle$ has properties "Left-seriality", "Right-seriality", "Density" and "Continuity". All these properties and their modal formulas will be introduced in the last section of this chapter.

In brief, it is very important for these structures to represent temporal knowledge. A majority of methods are based on them, such as temporal logic.

2.3 Point-Based Temporal Logic

Temporal databases provide a uniform and systematic way of dealing with historical data. Many languages have been proposed for temporal databases, such as **temporal logic**. Temporal logic combines abstract, formal semantics with the amenability to efficient implementation.

However, axiomatic systems and proof methods for temporal logic have so far found relatively few applications in the context of database (Gabbay et al. 1994). Moreover, one needs to bear in mind that for the standard linearly-ordered time domains, temporal logic is not recursively axiomatizable, so recursive axiomatizations are by necessity incomplete (Chomicki and Saake 1998). Hence, temporal logic used in temporal databases is mainly non-axiomatic systems.

There are two kinds of non-axiomatic temporal logic: **point-based** and **interval-based**. In this section, we discuss the first one. In order to emphasize the relations between non-axiomatic system and temporal logic, we must define the abstract temporal database.

Definition 2.1 (Abstract Temporal Database) (Chomicki and Saake 1998) Let $\rho = (r_1, \cdots, r_k)$ be a database schema, D a data domain and T a temporal structure, such as $\langle N, < \rangle$, $\langle Z, < \rangle$, $\langle Q, < \rangle$, $\langle R, < \rangle$.

A relation symbol R_i is a temporal extension of a relation symbol r_i, if it contains all attributes of r_i and a single additional attribute t of sort T. The sort of the attributes of R_i is $T \times \text{Darity}(r_i)$.

A timestamp temporal database is a first order structure $D \cup T \cup \{R_1, \cdots, R_k\}$, where R_i represents temporal relations: instances of the temporal extensions R_i. In addition, we require that the set $\{a : (t, a) \in R_i\}$ is finite for every $t \in T$ and $0 < i \leqslant k$.

A snapshot temporal database over D, T and ρ is a map DB: $T \rightarrow \text{DB}(D, \rho)$, where $\text{DB}(D, \rho)$ is the class of finite relational databases over D and ρ.

It is easy to see that snapshot and timestamp abstract temporal databases are merely different views of the same data and thus can represent the same class of temporal databases (Chomicki and Saake 1998).

2.3.1 Temporal Extensions Based Snapshot Model

Snapshot and timestamp models lead to two different temporal extensions of the first-order logic. The first gives rise to special temporal connectives and it is especially appealing because it encapsulates all the interactions with its temporal domain inside the temporal connectives.

Definition 2.2 (First-Order Temporal Connectives Based Snapshot) (Chomicki and Saake 1998) Let $k \geqslant 0$ and $O = t_i < t_j \mid O \wedge O \mid \neg O \mid \exists t_i . O \mid X_i$

This kind of method defines a k-ary temporal connective to be an O-formula with exactly one free variable t_0 and k free propositional variables X_i ($X_1, \cdots, X_k$). Variable t_i is the temporal context, and t_0 defines the outer temporal context of the connective that is made available to the surrounding formula, while the variables $t_1, \cdots, t_k$ define the temporal contexts for the sub-formulas substituted for the propositional variables $X_1, \cdots, X_k$. It is not difficult to find that this sort of definition with temporality is recursive.

Temporal connectives are key points of this method. We can use k-ary connectives to represent some temporal knowledge, such as U (until) and S (since) as follows:

$$X_1 \text{ until } X_2, \text{ if } \exists t_2.t_0 < t_2 \wedge X_2 \wedge \forall t_1 \ (t_0 < t_1 < t_2 \rightarrow X)$$

$$X_1 \text{ since } X_2, \text{ if } \exists t_2.t_2 < t_0 \wedge X_2 \wedge \forall t_1 \ (t_2 < t_1 < t_0 \rightarrow X)$$

There are still four basal 1-ary connectives, such as F, P, H and G. F and P are mnemonics for "future" and "past", respectively. The intended meaning of the formula "$F\varphi$" is "at some time in the future, φ holds", while "$P\varphi$" is to be read as "at some time in the past, φ holds".

Formally, H and G are mnemonics for "henceforth" and "hitherto", respectively. The intended meaning of the formula "$G\varphi$" is "at all times in the future, φ holds", while "$H\varphi$" is to be read as "at all times in the past, φ holds".

These are like the "box" and "diamond" of classical Modal Logic that $Fq = \neg G \neg q$ and $Pq = \neg H \neg q$. However, "box" and "diamond" are undirected, while modalities of temporal logic are directed by time.

It is not difficult to explain the equivalence between U and S from F and P. To make the writing style uniform, we write F as "◊", G as "□", P as "◆" and H as "■".

$$\Diamond X_1 \quad \text{iff} \quad \text{true until } X_1$$

$$\blacklozenge X_1 \quad \text{iff} \quad \text{true since } X_1$$

For a discrete linear order, we also define the ○ (next) and ● (previous) operators as:

$$\circ X_1 \quad \text{iff} \quad \exists t_1.(t_1 = t_0 + 1) \wedge X_1$$

$$\bullet X_1 \quad \text{iff} \quad \exists t_1.(t_1 = t_0 - 1) \wedge X_1$$

Now, we can define first-order temporal logic based snapshot.

Definition 2.3 (First-Order Temporal Logic Based Snapshot) (Chomicki and Saake 1998) Let Ω be a finite set of temporal connectives. First-order Temporal Logic $L(\Omega)$ over a schema ρ is defined as:

$$F = r(x_{i1}, \cdots, x_{ik}) \mid x_i = x_j \mid F \wedge F \mid \neg F \mid \omega(F_1, \cdots, F_k) \mid \exists x_i.F$$

where $r \in \rho$ and $\omega \in \Omega$.

Let us show how various temporal connectives are used to formulate the queries using an example.

Example 2.1 A communication company wants to query the working status of its employees. There are two relationship tables Workplace(Year, Name, City) and Duty(Year, Name, Post, Sale) as shown in Table 2.1a and Table 2.1b.

The following are several queries and their expressions in temporal logic:

- Find all employees who have worked only in one city

$$\blacklozenge\Diamond(\exists c.\text{Workplace}(x, c)) \wedge \neg\exists c'.(\blacklozenge\Diamond\text{Workplace}(x, c') \wedge c \neq c')$$

- Find all managers who sold switch continuously in GZ since they started work

$$\text{Workplace}(x, \text{GZ}) \text{ since } \text{Duty}(x, \text{manager}, \text{Switch})$$

- Find all employees who never spent more than two years in one place

$$\blacksquare\square(\neg\exists c.\text{Workplace}(x, c) \wedge \circ\text{Workplace}(x, c) \wedge \circ\circ\text{Workplace}(x, c))$$

Table 2.1a Relation of workplace

Workplace		
Year	**Name**	**City**
2004	Li	GZ
2005	Li	GZ
⋮	⋮	⋮
2009	Li	GZ
2007	Wang	BJ
2008	Wang	SH
2009	Wang	GZ
2006	Zhang	BJ
2007	Zhang	BJ
2008	Zhang	SH
2009	Zhang	SH

Table 2.1b Relation of duty

Duty			
Year	**Name**	**Post**	**Sale**
2004	Li	Manager	Switch
2007	Wang	Seller	Router
2006	Zhang	Seller	Server

2.3.2 Temporal Extensions Based Timestamp Model

While snapshot model encapsulates temporalities, timestamp method uses mapping. It requires the introduction of explicit attributes and quantifiers for handling time.

Definition 2.4 (First-Order Temporal Logic Based Time Stamp) (Chomicki and Saake 1998) The first-order logic $L(P)$ over a database schema ρ is defined as:

$$M = R(t_i, x_{i1}, \cdots, x_{ik}) \mid t_i < t_j \mid x_i = x_j \mid M \wedge M \mid \neg M \mid \exists\, t_i.M \mid \exists x_i.M$$

where R is the temporal extension of r for $r \in \rho$.

The query "find all employees who have worked only in one city" of Example 2.1 can be expressed as:

$$\exists c, t.\text{Workplace}(t, x, c) \wedge \neg\exists c', t'.(\text{Workplace}(t', x, c') \wedge c \neq c')$$

Obviously, while the snapshot method gives rise to special temporal connectives, the timestamp method requires the introduction of explicit attributes and quantifiers for handling time.

Finally, as mentioned necessary, in the expressive power of first-order Temporal Logic of $L(\Omega)$ and $L(P)$, Kamp has proved that $L(\Omega) \subseteq L(P)$. However, that is not the key context of this paper, so we ignore this topic.

2.4 Interval-Based Temporal Logic

In this section, we introduce the non-axiomatic temporal logic, which is interval-based and is second-order logic. While point-based method regards time as discreteness, interval-based method regards time as continuum.

Allen (1983) has developed a temporal logic in which time intervals are the primitives. He suggests thirteen possible relationships between time intervals, namely "Before", "After", "During", "Contains", "Overlaps", "Overlapped-by", "Meets", "Met-by", "Starts", "Started-by", "Finishes", "Finished-by" and "Equals", and shows that there can be a time point in point-based model where either two

Table 2.2 Allen's relationships of temporal intervals

Before (t_1, t_2)	Time interval t_1 comes before interval t_2 and they do not overlap in any way
After (t_1, t_2)	Time interval t_1 comes after interval t_2 and they do not overlap in any way
During (t_1, t_2)	Interval t_1 starts after and ends before t_2
Contains (t_1, t_2)	Interval t_1 starts before and ends after t_2
Overlaps (t_1, t_2)	Interval t_1 starts before t_2, and they overlap
Overlapped-by (t_1, t_2)	Interval t_1 starts after t_2, and they overlap
Meets (t_1, t_2)	Interval t_1 comes before interval t_2, but there is no interval between them, i.e., t_1 ends where t_2 starts
Met-by (t_1, t_2)	Interval t_1 comes after interval t_2, but there is no interval between them, i.e., t_2 ends where t_1 starts
Starts (t_1, t_2)	Time interval t_1 shares the same beginning as t_2, but ends before t_2 ends
Started-by (t_1, t_2)	Time interval t_2 shares the same beginning as t_1, but ends before t_1 ends
Finishes (t_1, t_2)	Time interval t_1 shares the same end as t_2, but begins after t_2 begins
Finished-by (t_1, t_2)	Time interval t_2 shares the same end as t_1, but begins after t_1 begins
Equals (t_1, t_2)	t_1 and t_2 are the same intervals

events could occur at the same time or no event could occur. Fig. 2.2 depicts them graphically. Allen uses these temporal relations to define events and activities.

Among these thirteen possible relationships between time intervals, there are six pairs of relationships that are inter-translatable, such as:

- Before (t_1, t_2) = After (t_2, t_1)
- During (t_1, t_2) = Contains (t_2, t_1)
- Overlaps (t_1, t_2) = Overlapped-by (t_2, t_1)
- Meets (t_1, t_2) = Met-by (t_2, t_1)
- Starts (t_1, t_2) = Started-by (t_2, t_1)
- Finishes (t_1, t_2) = Finished-by (t_2, t_1)

The relationship "Equals" has the property Equals (t_1, t_2) = Equals (t_2, t_1). Thus, only seven cut lines can depict all of the possible relationships between intervals.

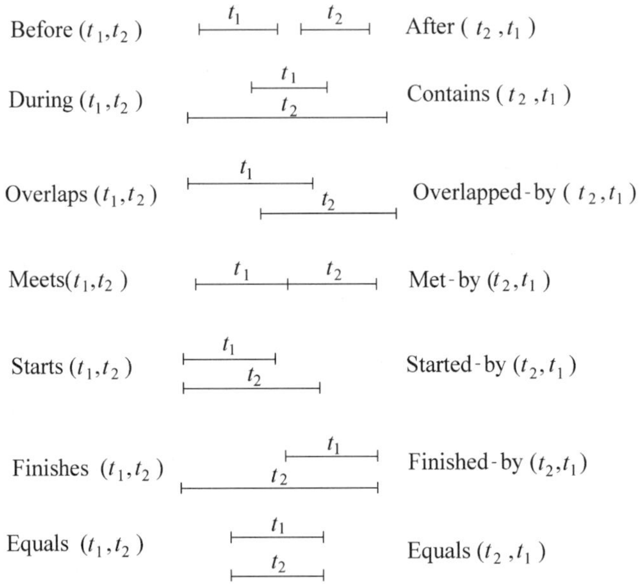

Before (t_1,t_2) t_1 t_2 After (t_2,t_1)
During (t_1,t_2) t_1 t_2 Contains (t_2,t_1)
Overlaps (t_1,t_2) t_1 t_2 Overlapped-by (t_2,t_1)
Meets(t_1,t_2) t_1 t_2 Met-by (t_2,t_1)
Starts (t_1,t_2) t_1 t_2 Started-by (t_2,t_1)
Finishes (t_1,t_2) t_1 t_2 Finished-by (t_2,t_1)
Equals (t_1,t_2) t_1 t_2 Equals (t_2,t_1)

Figure 2.2 Allen's temporal relations

2.4.1 From Interval to Point

Allen's method is interval-based, but time points can be regarded as time intervals when continuous time is zero. Therefore, based on these, we can depict the possible relationships between time interval and time point by five relationships, namely "Before", "After", "Meets (or Starts)", "Met-by (or Finishes)", "During".

Table 2.3 and Fig. 2.3 depict such five relations of temporal relationships between time intervals and points.

Table 2.3 Temporal relationships between time intervals and points

Before (p, t)	Point p starts before interval t starts
After (p, t)	Point p ends after interval t ends
Meets (p, t) Starts (p, t)	Point p and interval t start at the same time
Met-by (p, t) Finishes (p, t)	Point p and interval t end at the same time
During (p, t)	Point p starts after and ends before interval t

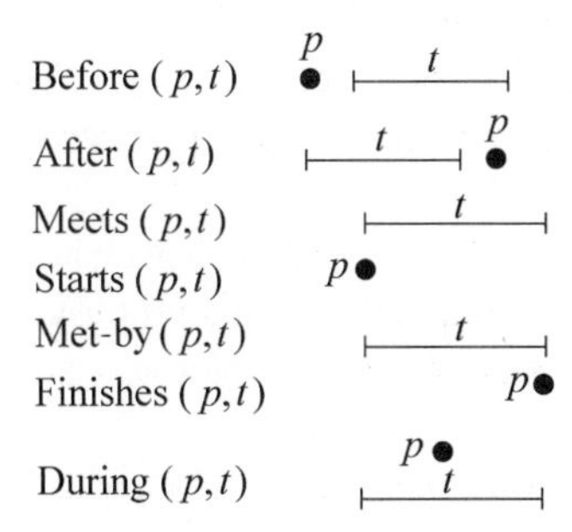

Figure 2.3 Temporal relationships between time intervals and points

2.4.2 From Point to Point

The relationship between time points is easier than former. There are just three relationships, namely "Before", "After", "Equals". Table 2.4 and Fig. 2.4 depict these relationships.

Table 2.4 Temporal relationships between time points

Before (p, q)	Point p starts before q
After (p, q)	Point p starts after q
Equals (p, q)	Point p and q start at the same time

Before (p, q) p q

After (p, q) q p

Equals (p, q) p q

Figure 2.4 Temporal relationships between time points

2.4.3 Temporal Predict

Temporal predict is the first-order predict, which expands the description of time.

For example, engineer(Li) expresses that Li is an engineer, but does not express what time Li is an engineer. Whereas, engineer(Li, t_1) expresses that Li is an engineer and the period of time when he is engineer.

Example 2.2 Li has graduated in 2007, and after he is engaged for a year, he can be an assistant engineer.

This is defined as follows:

(1) A graduate: Graduate(A, [2007, $+\infty$))

(2) The period of employment: $T_1 = [2007.\ 7, Now)$[1]

(3) The date applied for Assistant Engineer: $T_2 = [Now - 1\ \text{year}, Now)$

(4) A is an Assistant Engineer: Assistant Engineer(A, [Now,$+\infty$))

such that:

Graduate(Li, [Now, $+\infty$]) $\wedge$During(T_2,T_1) $\rightarrow$ Assistant Engineer(Li, [Now,$+\infty$))

This means that if Li is graduated, and his employment period T_1 is contained by the date applied for assistant engineer, then we can infer that Li is an assistant engineer now.

To represent property holding over an interval T, Allen needs two other predicates: Holds (p, T) and In (t, T). Holds (p, T) means that the property p holds over an interval T and In (t, T) means that the subinterval t is in interval T (Pathirage 2001).

Allen defines "In" as follows:

$$\text{In}(T_1, T_2) \Leftrightarrow (\text{During}(T_1, T_2) \vee \text{Starts}(T_1, T_2) \vee \text{Finishes}(T_1, T_2))$$

He defines "Holds" as follows:

$$\text{Holds}(p, T) \Leftrightarrow (\forall t.\ \text{In}(t, T) \Rightarrow \text{Holds}(p, t))$$

The negation of a property p is defined as

$$\text{Holds}(\text{not}(p), T) \Leftrightarrow (\forall t.\ \text{In}(t, T) \Rightarrow \neg\text{Holds}(p, t))$$

This means that if a property p holds over an interval T, it holds over all sub-intervals of T. However, Allen could not define events and actions using "Holds" predicate. Therefore, he defines another predicate "Occur". Reason for the use of another predicate is his definition of "Holds" predicate. Holds (p, T) says that property p holds over an interval T as well as all sub-intervals of T. Occur (e, t) says that the event e occurs in the time interval t.

Example 2.3 "Event E_1 occurs before event E_2" can be expressed by:

$$\text{Occur}(E_1, T_1) \wedge \text{Occur}(E_2, T_2) \wedge \text{Before}(T_1, T_2)$$

---

[1] Before we explain the semantics of *Now*, here its meaning is just current time.

However, Galton and Critical (1990) shows the weaknesses of Allen's temporal interval in representing continuous changes. According to Allen's definition of "Holds", for Holds (p, T) to be true, p must be held in all subintervals of T. If p is spasmodically moving with time, p cannot be held in all sub-intervals at the same time. Hence, if Holds (p, T) is true, then p cannot move spasmodically but has to rest over time interval T. Galton argues that in order to overcome this problem, one must combine the presentation methods of time points and time intervals.

2.5 Calculation Based on Span

In Example 2.2 (3), the date applied for Assistant Engineer can be expressed as "$T_2 = [Now - 1$ year, $Now]$". However here, "$Now - 1$ year" cannot be represented by Allen's method and other temporal interval-based logic. It must be evaluated by the calculations based time span.

Time span is a continuous period that expresses the length of time, such as "one year and three months", "thirty days", "twenty-eight hours". It has no start time and end time. Time span is more like time interval representing periods of time.

Without loss of generality, the temporal calculations based time span can be divided into the following types:

- Between point and point: Let p_1 and p_2 be TimePoint types, then:
  $p_1 - p_2$: Result is TimeSpan type, which is evaluated by integer subtraction, and expresses the span from p_1 to p_2. For example, July − May = 2 months.
- Between point and span: Let p be TimePoint type, len be TimeSpan type, then:
  - $p + len$: Result is TimePoint type, which is evaluated by integer addition, and expresses the point after p_1 len granularities. For example, May + 2 months = July.
  - $p - len$: Result is TimePoint type, which is evaluated by integer subtraction, and expresses the point before p_1 len granularities. For example, July − 2 months = May.

  After that, we can represent "$Now - 1$ year" accurately by subtraction.
- Between span and span: Although time spans are often integers, they have their own meaning. The data of TimeSpan type can be calculated by integer addition, subtraction and division. Among these, the results of addition and subtraction are still TimeSpan type, but integer type due to division. For example, let $p_1 = 1$ year, $p_2 = 1$ month. They are all TimeSpan types, then:
  - $p_1 + p_2$: Result is TimeSpan type, which is evaluated by integer addition and expresses the span of thirteen months;
  - $p_1 - p_2$: Result is TimeSpan type, which is evaluated by integer subtraction and expresses the span of thirteen months;
  - $p_1 \div p_2$: Result is integer type. It can be understood 12 months in semantic.

- Between interval to span: Let $T=[p, q]$ be TimeInterval type, *len* be TimeSpan type, then:
  - $T+len=[p+len, q+len]$. Result is TimeSpan type, which is expressed as the interval after T *len* granularities, For example, $T=$ [May, July], $len=2$ months, $T+len=$ [July, September].
  - $T-len=[p-len, q-len]$. Result is TimeSpan type, which is expressed as the interval before T *len* granularities. For example, $T=$ [May, July], $len=2$ months, $T-len=$ [March, May].
- Between interval to interval: Let $T_1=[p_1, q_1]$, $T_2=[p_2, q_2]$ be TimeInterval types, then:
  - $T_1-T_2=p_1-q_2$ $(q_2<p_1)$: Result is TimeSpan type, which is expressed as the span from T_1 to T_2. For example, $T_1=$ [July, September], $T_2=$ [March, May], $T_1-T_2=$ July $-$ May $=2$ months.
- Between span to integer: Let *len* be TimeSpan type and *i* an integer, then:
  - $len \times i$: Result is TimeSpan type, which is evaluated by integer multiplication. For example, 2 months $\times\, 2=4$ months.
  - $len \div i$: Result is TimeSpan type, which is evaluated by integer division. For example, 4 months $\div\, 2=2$ months.

2.6 Other Temporal Calculations in Common Use

Besides time span based calculations, there are other **temporal calculations** in common use.

Let p, q be TimePoint types, $T=[p, q]$ be TimeInterval type, then:

- min (p, q): Result is TimePoint type, i.e., the result is the minimum of p and q. For example, min (May, July) = May.
- max (p, q): Result is TimePoint type, i.e., the result is the maximum of p and q. For example, max (May, July) = July.
- begin (T): Result is p, which evaluates the start time of T. Such as, $T=$ [May, July], begin$(T)=$ May.
- end (T): Result is q, which evaluates the end time of T. Such as, $T=$ [May, July], end$(T)=$ July.
- length (T): Result is $q-p$ of TimeSpan type, which evaluates the span from start time to end time. Such as $T=$ [May, July], length$(T)=2$ months.

2.7 Time Granularity and Conversion Calculation

Calculation based on time span is a useful method in temporal database, but how do we define the unit of measure for it? Although the structure $R=\langle R,<\rangle$ is a flow of time, but how do we represent continuous time in computer that is discrete?

What we could do is, we could denote it by discrete form, **time granularity** (Becher et al. 2000). Time granularity (Bettini et al. 1998) is the measure of the discrete degree, yet it does not mean that the more accurate time granularity is, the better. This is because the more accurate time granularity is the more memory space it takes. Therefore, proper time granularity would be the best choice for temporal database.

2.7.1 Time Granularity and Chronon

Time granularity is the smallest time element of time data, which reflects the discrete degree of time points. However, the smallest computer-supported unit is **chronon** (Anurag and Sen 2008), which is the greatest lower bound for the choices of time granularity. Chronon determines the precision of computer system, for example, two things that happen at different times but at the same chronon cannot be identified by a computer system.

Overall, chronon is a fixed measurement but time granularity is a database defined time unit that can be changed in order to meet practical requirements.

2.7.2 State of Existence of Time Granularity

There are two types of time granularities: single granularity and dynamic granularity.

- Single granularity: Single granularity means that only one time granularity can be supported in some database systems, but granularities of different database systems may be different. For example, we can describe the information of pay by granularity day, while we adopt granularity microsecond in some electronic systems.
- Dynamic granularity: When dynamic granularity is adopted in some database systems, there exist several different time granularities in terms of different attributes.

Table 2.5 shows two relational tables that belong to a database on flights and vacations.

Table 2.5a stores the information of departure time that can be calculated by the time granularity of minute. Table 2.5b records the information of vacation whose time granularity is day and whose time element is time interval. For instance, Thanksgiving Day, shown as the second item of Table 2.5b, is a festival starting from the forth thursday of November and lasts for four days.

Although Table 2.5a and Table 2.5b belong to the same database system, their time granularities are different. In the first place, different granularities are related with the requirements of the practical application. If time granularities of departure time and holiday were the same, there would be severe consequences. In the second

Table 2.5a Flight departures

Flight No.	Departure Time	Flight No.	Departure Time
53	2007-11-20 14:38	653	2007-11-27 12:38
200	2007-11-27 14:34	658	2007-11-30 10:03

Table 2.5b Vacations

Vacation	From-Time	To-Time
Labor Day	2007-09-01	2007-09-03
Thanksgiving	2007-11-24	2007-11-28
Christmas	2007-12-24	2007-12-26

place, the system can save lots of memory space when it adopts different time granularities. The ideal temporal database is even able to adjust time granularity intelligently to describe time on a larger scale in case of limited memory resources. For instance, in order to record the long ancient history and the precise modern history, Table 2.6 describes a time axis with dynamic granularity. In Table 2.6, time granularity of the *i*-th time segment is G_i. It can be seen that G_1 = hundred years, G_4 = day and so on.

Table 2.6 presents the number of *era*, and *G*. means time granularity.

Table 2.6 Time axis of dynamic granularity

No.	0	1	2	3	4
Era	palaeogeographic stage	archeorganism	archeozoic	Ancient History	Modern History
G.	Million years	Hundred years	Ten years	Year	Day

2.7.3 Operations of Time Granularity

System can usually handle only one granularity, so the system containing more than one granularity should be able to transfer different granularities. For instance, a customer needs to go home before Thanksgiving Day. He can use the following query language to determine the proper flight.

```
Select *
    from vacations, flight_departures
    where vacation='thanksgiving' and
    flight_departures.at_time overlaps
    (vacations.from_time, vacations.to_time);
```

In the query process, the operation of temporal intersection "overlaps" is used to judge the flight that is suitable for Thanksgiving Day. However, time element involved in operation is situated in different time granularities. Due to lack of disposing of the mixed time granularities, SQL-92 will return a false prompt. Therefore, we need a tool for the conversion between different time granularities.

- Conversion between different time granularities: We introduce two functions to implement the conversion, scale and cast. The following examples interpret their usage:

$$\text{scale(2007years,days)} = \text{2007-01-01}{\sim}\text{2007-12-31years} \tag{2.1}$$

$$\text{scale(scale(}2007_{\text{years}}\text{,days),months)} = \text{2007-01}_{\text{months}}{\sim}\text{2007-12}_{\text{months}} \tag{2.2}$$

$$\text{cast(}2007_{\text{years}}\text{,days)} = \text{2007-01-01}_{\text{days}} \tag{2.3}$$

$$\text{cast(cast(}2007_{\text{years}}\text{,days),years)} = 2007_{\text{years}} \tag{2.4}$$

$$\text{scale(cast(}2007_{\text{years}}\text{,days),months)} = \text{2007-01}_{\text{years}} \tag{2.5}$$

Scale function creates a time interval and (2.1) converts the time value with granularity year into the time interval value with granularity day. Cast function converts the time value with granularity year into the time point value with granularity day.

- Comparison of semantics interpretation of time point operator between different time granularities: Suppose g, h are two time points, belonging to G, H time granularities, respectively, $\odot$ is a time point comparison operator, F is a granularity that is more precise than G and H, C is the smallest time granularity, satisfying $C \leqslant G$, $C \leqslant H$. If we operate $g \odot h$, there will be several semantics as follows:
  - A time granularity false caused by false match;
  - Semantics of left operand: the operation is performed making use of the granularity of the left operand, namely, $g \odot h = g \odot \text{scale}(h, G)$.
  - Semantics of right operand: the operation is performed making use of the granularity of the right operand, namely, $g \odot h = \text{scale}(g, H) \odot h$.
  - Precise semantics:

$$g \odot h = \begin{cases} g \odot \text{scale}(h, G) & \text{if } G \text{ is more precise than } H \\ \text{scale}(g, h) \odot h & \text{if } H \text{ is more precise than } G \\ \text{scale}(g, C) \odot \text{scale}(h, C) & \text{otherwise} \end{cases}$$

Rough semantics:

$$g \odot h = \begin{cases} \text{scale}(g, h) \odot h & \text{if } G \text{ is more precise than } H \\ g \odot \text{scale}(h, G) & \text{if } H \text{ is more precise than } G \\ \text{scale}(g, C) \odot \text{scale}(h, C) & \text{otherwise} \end{cases}$$

Different interpretations of the same sentence may bring different sequences. For instance, we can give two interpretations for "$2007\text{-}01_{months} < 2007\text{-}01\text{-}15_{days}$":

- "$2007\text{-}01_{months} < 2007\text{-}01\text{-}15_{days}$" = true.(cast function with right operand semantics, or precise semantics)
- "$2007\text{-}01_{months} < 2007\text{-}01_{months}$" = false.(cast or scale function with left operand semantics, or rough semantics)

2.7.4 Relational Chart of Time Granularity Conversion

In order to convert time granularity conveniently, we present granularity charts created by the set of converting functions. Figure 2.5 is a granularity in Gregorian calendar system.

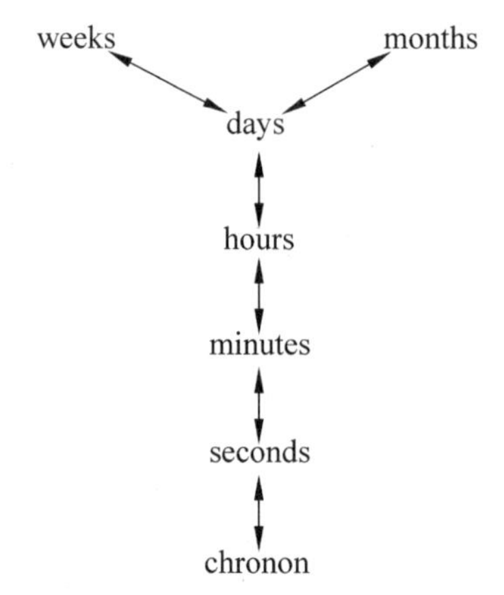

Figure 2.5 Granularity in Gregorian calendar system

The above chart represents a simplified Gregorian calendar system. Figure 2.6 shows a calendar chart containing several types of calendars to convert time point of different calendars conveniently.

2.8 Tense Logic

Although axiomatic systems and proof method for temporal logic have so far found relatively few applications in the query language modeling of temporal database by Gabby (Gabby et al. 1994), we still introduce some basic axiomatic temporal logic in this section. The primary reason is that axiomatic system can depict the order properties more precisely, and it is still an important method to present temporal knowledge. All of these can help us to understand the properties of time structure easily.

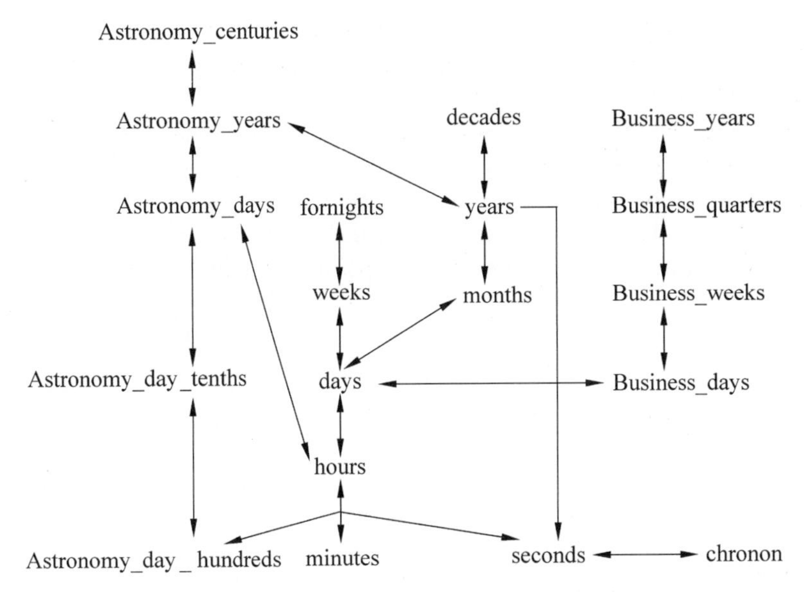

Figure 2.6 Conversion chart of multi-calendar

Hence, we will introduce **Tense Logic**, which is Prior's basic system of temporal logic (Goble 2001).

2.8.1 Syntax and Semantics of Tense Logic

Syntax and semantics of Tense Logic is an extension of classical propositional logic.

Classically propositional formulas are interpreted as truth values, such as 1 for "true" and 0 for "false". It is inductively determined by a valuation: function mapping propositional variables to truth values. Once the valuation is known, the truth value or any formula is fixed.

The first basic idea underlying temporal logic is to make valuations time dependent. More precisely, one associates a separate valuation with each point in a given flow of time. Formally, let $T = \langle T, < \rangle$ be a flow of time. A valuation on T is a mapping $\pi : (T \rightarrow (\Pi \rightarrow \{0,1\}))$. Π denotes the set of propositional variables. Therefore, an algebra model is a pair $M = \langle T, \pi \rangle$ consisting of time and a valuation.

With this definition, one can already interpret classical formulas in each point of a model in a standard way. For example, the formula $p \wedge \neg q$ is said to be true at a time point t precisely if $\pi(t)(p) = 1$ and $\pi(t)(q) = 0$.

After this, we can define modalities on $M = \langle T, \pi \rangle$. As Tense Logic, it has four modalities, F, P, H and G. F and P are mnemonics for "future" and "past", respectively. The intended meaning of the formula "$F\varphi$" is "at some time in the future, φ holds", while "$P\varphi$" is to be read as "at some time in the past, φ holds".

Formally, H and G are mnemonics for "henceforth" and "hitherto", respectively. The intended meaning of the formula "$G\varphi$" is "at all times in the future, φ holds", while "$H\varphi$" is to be read as "at all times in the past, φ holds".

Which are liked the box "□"and diamond "◊" of classical Modal Logic that $Fq = \neg G \neg q$ and $Pq = \neg H \neg q$. However, box "□"and diamond "◊" are undirected, while modalities of Tense Logic are directed by time.

According to Tarski's semantics, given the following inductive definition of the notion of truth or a formula φ at a time point, in a model $M = \langle T, \pi \rangle$:

$M, t \models q$	if $\pi(t)(q) = 1$
$M, t \models \neg\varphi$	if not $M, t \models \varphi$
$M, t \models \varphi \wedge \psi$	if $M, t \models \varphi$ and $M, t \models \psi$
$M, t \models G\varphi$	if $\forall s, t<s, M, s \models \varphi$
$M, t \models H\varphi$	if $\forall s, s<t, M, s \models \varphi$

2.8.2 Axiomatics and Properties

While analogous to K is a minimal logic system of classical model logic, Kt is the minimal one of temporal logic, and defined as the smallest class of formulas that is closed under the following axioms and derivation rules (Goble 2001):

(CT) all classical propositional tautologies

(US) if φ is a theorem, then so is $\varphi[\psi/q]$

(MP) if φ and $\varphi \to \psi$ are theorems, then so is ψ

(TG) if φ is a theorem, then so are $G\varphi$ and $H\varphi$

(CV) $q \to GPq$

$$q \to HFq$$

(DB) $G(q \to r) \to (Gq \to Gr)$

$$H(q \to r) \to (Hq \to Hr)$$

(4) $Gq \to GGq$

$$Hq \to HHq$$

Among these, (CT) contains all classical propositional tautologies; (US) is the rule of uniform substitution; (MP) is the rule of modus ponens; (TG) is the rule of temporal generalization; (CV) is the Converse axiom that ensures that the accessibility relations for the operators G and H are each other's converse; (DB) is the distribution laws of G and H, which analogues to Rule K of classical model logic; (4) is transitivity law corresponding to transitive of first order property, which is implied by $T = \langle T, < \rangle$.

It is easy to prove that Kt is sound and complete, but since this book is not a logic book, we ignore it.

Theorem 2.1 The logic Kt is sound and complete with respect to the class of all flows of time.

Like (4), there are many corresponding laws from Tense Logic to first order or not first order properties, such as:

Having a first point $\quad H\bot \vee PH\bot \quad$ (L1)

Left-seriality $\quad P \quad$ (L2)

Having a final point $\quad G\bot \vee FG\bot \quad$ (L3)

Right-seriality $\quad F \quad$ (L4)

Discreteness $\quad (F \wedge q \wedge Hq) \to FHq \quad$ (L5)

Density $\quad Fq \to FFq \quad$ (L6)

Continuity[1] $\quad (Fq \wedge \Diamond\neg q \wedge \Box(q \to Hq)) \to \Diamond((q \wedge G\neg q) \vee (\neg q \wedge Hq)) \quad$ (L7)

Having finite intervals

$$(G(Gq \to q) \to (FGq \to Gq)) \wedge (H(Hq \to q) \to (PHq \to Hq)) \quad \text{(L8)}$$

Non-branching

$$PFq \to (Pq \vee q \vee Fq) \quad FPq \to (Pq \vee q \vee Fq) \quad \text{(L9)}$$

Especially, logic $Kt + \text{L9}$ is named logic Lin, which is the basic linear system of Tense Logic, and it is sound and complete with respect to the class of linear flows of time.

Theorem 2.2 The logic **Lin** is sound and complete with respect to the class of linear flows of time.

Turning to the structures of time, now we can define the following logics (Goble 2001):

Lin.N: $\quad$ **Lin** + L1 + L4 + L8

Lin.Z: $\quad$ **Lin** + L2 + L4 + L8

Lin.Q: $\quad$ **Lin** + L2 + L4 + L6

Lin.R: $\quad$ **Lin** + L2 + L4 + L6 + L7

Theorem 2.3 The logic **Lin.N**, **Lin.Z**, **Lin.Q** and **Lin.R** are sound and complete axiomatizations of the set of validities of the flows of time $N = \langle N, < \rangle$, $Z = \langle Z, < \rangle$, $Q = \langle Q, < \rangle$, and $R = \langle R, < \rangle$, respectively.

Based on this, we can now comprehend the first order properties of time flow from modality and axiomatic systems, which is presented in Section 2.2. It is obvious that logic **Lin.Z** seems to be the most appropriate axiomatic system to

---

[1] Here, $\Diamond q$ abbreviates $Pq \vee q \vee Fq$, and $\Box q$ abbreviates $Hq \wedge q \wedge Gq$.

represent the knowledge of information system and temporal database, but regrettably, it is not recursively axiomatizable.

References

[1] Allen JF (**1983**) Maintaining knowledge about temporal intervals. Communication of the ACM 26(11): 832 – 843

[2] Alvarez J, Mossay P (**2006**) Estimation of a continuous spatio-temporal population model. Journal of Geographical Systems **8**(3): 307 – 316

[3] Anurag D, Sen AK (**2008**) The chronon based model for temporal databases. Database Systems for Advanced Applications, LNCS **4947**: 461 – 469

[4] Becher G., Clerin-Debart F, et al. (**2000**) A qualitative model for time granularity. Computational Intelligence **16**(2): 137 – 168

[5] Bettini C, SeanWang X, et al. (**1998**) A general framework for time granularity and its application to temporal reasoning. Annals of Mathematics and Artificial Intelligence **22**: 29 – 58

[6] Gabbay DM, Hodkinson I, Reynolds M (**1994**) Temporal logic: mathematical foundations and computational aspects. Oxford University Press

[7] Jochens A, Caliebe A, et al. (**2007**) Temporal behaviour of the stepwise mutation model. Annals of Human Genetics **71**: 551

[8] Chomicki J, Saake G (**1998**) Logics for databases and information systems: temporal logic information systems. Kluwer Academic Publishers, Boston/Dordrecht/London

[9] Galton A, Critical A (**1990**) Examination of Allen's theory of action and time. Artificial Intelligence 42: 159 – 188

[10] Goble L (**2001**) The blackwell guide to philosophical logic. Blackwell Publishers Inc., USA

[11] Wooldridge M, Huget MP (**2006**) Michael Fisher, Simon Parsons. Model checking for multiagent systems: the mable language and its applications. International Journal on Artificial Intelligence Tools (IJAIT) 15(2): 195 – 225

[12] Pathirage GW (**2001**) An end-user intelligible specification language and its execution. Ph.D. thesis, Department of Information Systems Design, The University of Electro-Communications, Tokyo

[13] Rocha-Martinez JM (**1998**) On some applications of a discrete-time model of failure and repair. Applied Stochastic Models and Data Analysis **14**(3): 189 – 198

3 Temporal Extension of Relational Algebra

Yong Tang[1,2], Shu Li[2,3], Dongning Liu[2], and Pan Shi[2]

[1] Computer School, South China Normal University, Guangzhou 510631, P.R. China
[2] Department of Computer Science, Sun Yat-sen University, Guangzhou 510275, P.R. China
[3] College of Computer Science and Software Engineering, Shenzhen University, Shenzhen 518060, P.R. China

Abstract Temporal relational operations can be considered as temporal extensions of regular relational operations, which are the basic contents of temporal database technology. Like traditional relational operations, temporal relational operations can also be classified into two types: temporal relational algebra and temporal relational calculus. The former takes relation as a set of tuples, while the latter uses temporal predicates to select temporal tuples, which should meet not only the requirements in terms of attributes but also temporal predicates. In this chapter, first the regular relational operations are reviewed. Subsequently, temporal relational operations based on historical relational database model (HRDM) and bitemporal conceptual data model (BCDM) are discussed. Finally, three important properties—snapshot reducibility, temporal semi-completeness, and temporal completeness are introduced.

Keywords *temporal relational algebra, temporal relational calculus, HRDM, BCDM, snapshot reducibility, temporal semi-completeness, temporal completeness*

Temporal relational operations are the basic contents of temporal database technology, which can be considered as temporal extension of regular relational operations. Like regular relational operations, temporal relational operations can be classified into two types according to their theoretic basis: **temporal relational algebra** and **temporal relational calculus**. These two kinds of temporal operations are equivalent in expressive power (Zhang 1993).

Relational operations are important theoretic basis for query languages. Whenever relational database has been extended to temporal relational database, temporal relational operations have drawn many researchers' attention.

James Clifford and Albert Croker (1987) proposed the historical relational data model (HRDM) and defined a temporal relational algebra based on it. Frank and

issty@mail.sysu.edu.cn

others (1988) proposed a temporal relational calculus in order to give a model of intelligent database. Ellen Rose and Arie Segev (1993) presented a temporal, object-oriented algebra on temporal, object-oriented data model. Shi-chao Zhang and Cheng-qi Zhang (1997) established a gap-interval based temporal tuple calculus language and the corresponding temporal relational algebra. Mehmet A. Orgun (1999) introduced a clocked temporal relational algebra that supports temporal relations based on multiple timelines. Jae Young Lee and Ramez A. Elmasri (2001) gave a temporal algebra on ER-based temporal data model. Lu-bang Wang and Yong Tang (Wang et al. 2004) gave a bitemporal relational algebra based on bitemporal conceptual data model (BCDM). Abdullah Uz Tansel and Canan Eren Atay (2006) introduced a nested bitemporal relational data model and defined algebra for this model.

HRDM and BCDM are two representative models among temporal database models. We will discuss **historical relational algebra** and **bitemporal relational algebra** in this chapter.

Before we discuss temporal algebra, we should review regular algebra and calculus.

3.1 Regular Relational Operations

Regular relational operations (Codd 1970) are the basis of temporal relational operations. The primitive operations in regular relational database include relational algebra and relational calculus. Relational algebra is essentially equivalent in expressive power to relational calculus (and thus first-order logic). This result is known as Codd's theorem (Codd 1972). Before we review regular relational algebra and regular relational calculus, some basic notions should be given here.

3.1.1 Basic Notions

Definition 3.1 Domain is the set of values allowed in an attribute.

For example, {1, 2, 3, 4, 5} and {male, female} are two different domains.

Definition 3.2 Let $D_1, D_2, \cdots, D_n$ be n sets. They may or may not be the same. Cartesian product of $D_1, D_2, \cdots, D_n$ is defined as follows:

$$D_1 \times D_2 \times \cdots \times D_n = \{(d_1, d_2, \cdots, d_n) \mid d_i \in D_i, i = 1, 2, \cdots, n\} \tag{3.1}$$

$(d_1, d_2, \cdots, d_n)$ is an n-tuple, where d_i is an element of it.

For example, $D_1 = \{1,3\}$, $D_2 = \{2,4\}$, thus the Cartesian product of D_1 and D_2 is defined as: $D_1 \times D_2 = \{(1,2),(3,2),(1,4),(3,4)\}$.

Definition 3.3 Any subset of $D_1 \times D_2 \times \cdots \times D_n$ is called a relation, which is a set of n-tuples, where n is the degree of the relation.

In the case of the example of $D_1 \times D_2$ mentioned above, {(1,2), (1,4), (3,4)} is a relation that denotes the "smaller" relationship and the degree of which is 2.

Definition 3.4 Relation schema can be defined as an ordered 3-tuple (R,U,D), where R is the name of relation, U is the set of attributes of R, D is the set of domains of attributes in U.

Relation schema is often denoted as $R(A_1, A_2, \cdots, A_n)$, where A_i is an attribute in U. A relation r on relation schema R is often denoted as $r(R)$, which is a group of mappings from U to D. The element t_i of tuple t in r is the value of t on attribute A_i, which is restricted in domain of A_i.

Relational algebra takes relations as sets of tuples, while relational calculus uses predicates to specify the selected tuples. We will describe these two types of operations.

3.1.2 Relational Algebra

Relational algebra is a group of operations such that the operands are relations. These operations include the regular set operations, such as union, intersection difference, Cartesian product and special operations defined for relations such as selection, projection, join and division.

Union, intersection and difference in relational database have an additional constraint that two relations involved must be on the same relation schema.

- Union

Let (R, U, D) be a relation schema. r_1 and r_2 are two relations on this schema, t represents any tuple in a relation. Union of r_1 and r_2 is defined as:

$$r_1 \cup r_2 = \{t \mid t \in r_1 \vee t \in r_2\} \tag{3.2}$$

- Intersection

Let r_1 and r_2 be two relations on the same relation schema, t represents any tuple. Intersection of r_1 and r_2 is defined as:

$$r_1 \cap r_2 = \{t \mid t \in r_1 \wedge t \in r_2\} \tag{3.3}$$

- Difference

Let r_1 and r_2 be two relations on the same relation schema, t represents any tuple. Difference of r_1 and r_2 is defined as:

$$r_1 - r_2 = \{t \mid t \in r_1 \wedge t \notin r_2\} \tag{3.4}$$

- Cartesian product

Let r_1 and r_2 be two relations, where the degree of r_1 is n and the degree of r_2 is m. The Cartesian product of r_1 and r_2 is a relation of $(n+m)$ degree. The first n elements of any tuple come from n-tuple of r_1 and the last m clements come from

m-tuple of r_2. $r_1 \times r_2$ can be expressed as follows:

$$r_1 \times r_2 = \{(a_1, a_2, \cdots, a_n, b_1, b_2, \cdots, b_m) \mid (a_1, a_2, \cdots, a_n) \in r_1 \wedge (b_1, b_2, \cdots, b_m) \in r_2\} \quad (3.5)$$

r_1 and r_2 do not need to be on the same schema. The attributes and numbers of tuples of r_1 and r_2 can be different.

- Selection

Selection is selecting tuples from a relation that meet the selection criteria. It can be expressed as follows:

$$\sigma_F(r) = \{t \mid t \in r \wedge F(t) = \text{“T”}\} \quad (3.6)$$

F is a boolean expression whose operators are the logical connectives and arithmetic comparison operators. The value of $F(t)$ can be true, denoted as T, or false, denoted as F. The result of selection is a new relation on the same relation schema of r.

- Projection

Selection can be seen as choosing a subset of rows, while projection is choosing a subset of columns. Projection of r can be expressed as follows:

$$\Pi_A(r) = \{t[A] \mid t \in r\} \quad (3.7)$$

A is a subset of attributes in r. $t[A]$ is a tuple containing the elements of t on attributes in A.

- Join

Let r_1 and r_2 be two relations. Join of r_1 and r_2 is choosing tuples from $r_1 \times r_2$, which meet the given conditions. It can be expressed as follows:

$$r_1 \underset{\theta}{\rhd\lhd} r_2 = \{t_1 t_2 \mid t_1 \in r_1 \wedge t_2 \in r_2 \wedge r_1(A) \theta r_2(B)\} \quad (3.8)$$

A is a subset of attributes in r_1 and B is a subset of attributes in r_2. θ is a comparison operator, such as $<, \leqslant, =, >, \geqslant, \neq$.

Natural join is the most important join operator. It requires that the values of two relations being joined should be equal on their common attributes. The common attributes only appear once in the result. Natural join can be expressed as follows:

$$r_1 \rhd\lhd r_2 = \{t_1 t_2 \mid t_1 \in r_1 \wedge t_2 \in r_2 \wedge r_1(A) = r_2(A)\} \quad (3.9)$$

- Division

Let r_1 and r_2 be two relations. The degrees of r_1 and r_2 are m and n, where m is larger than n, r_2 is not empty and there are n attributes in r_1 that are equal to n attributes of r_2. The division of r_1 by r_2 produces a new relation, whose degree is $m - n$. This can be expressed as follows:

$$r_1 \div r_2 = \{t^{(m-n)} \mid \forall t^{(n)} \in r_2, t^{(m-n)} \times t^{(n)} \in r_1\} \quad (3.10)$$

3.1.3 Relational Calculus

Besides relational algebra, relational operation can also be presented by predicates calculus, which is called relational calculus. Relational algebra takes relations as operands, while relational calculus takes tuples as operands. There are two types of relational calculus, one is **tuple relational calculus** and the other is **domain relational calculus**. These two calculi are equivalent in expressive power. (Lacroix and Pirotte 1977)

- Tuple relational calculus

Tuple relational calculus takes tuples as variables, which is normally expressed as:

$$\{t[\langle \text{attributes} \rangle] \mid P(t)\} \tag{3.11}$$

t is a tuple variable. The selection object can be tuple *t* or parts of attributes of *t*. *P*(*t*) is predicates about *t*. In fact, the formulae from (3.1) to (3.10) mentioned above, use tuple relational calculus to define relational algebra. Therefore, we can conclude that tuple relational calculus shares the same expressive power of relational algebra.

- Domain relational calculus

Domain relational calculus takes domains (or attributes) as variables, which are normally expressed as:

$$\{(x_1, x_2, \cdots, x_n) \mid P(x_1, x_2, \cdots, x_n, x_{n+1}, \cdots, x_{n+m})\} \tag{3.12}$$

$x_1, x_2, \cdots, x_n, x_{n+1}, \cdots, x_{n+m}$ are domain variables, in which $x_1, x_2, \cdots, x_n$ appear in the result. The other *m* domains do not appear in the result but appear in predicates *P*.

3.2 Relational Algebra of Historical Database

Allen (1983) introduced an interval-based temporal algebra, which takes the notion of a temporal interval as primitive. However, Allen's interval-based model is hard to represent the common sense notion of a single event with "gaps" (Morris and Al-Khatib 1991). Shi-chao Zhang (1994) proposed a model to represent temporal knowledge based on taking time interval as a set of discrete time points. This is called a point-based model. In the following sections, we will describe temporal relational algebra and calculus on point-based model.

In this section, we will introduce temporal relational algebra based on. Clifford's HRDB. HRDB is a typical temporal database model, which supports valid time but not transaction time. The time (valid time) in historical relational database is the time when the attribute of tuple is true in real world.

3.2.1 Basic Notions and Terminologies

First, we will define some basic notions.

- Chronon

A **chronon** is defined as quantum of time, the minimal time unit supported by temporal system, which cannot be divided. Chronon is often set to 0.01~1 second.

- System time domain

The start time supported by temporal system is denoted as 0. **System time domain** is a finite set of sequent discrete chronons, denoted as $Sys\_T = \{0, 1, 2, \cdots, Now, \cdots, \text{MaxSysTime}\}$.

- Time interval

Time interval is a subset of *Sys_T* whose chronons are between two chronons, a start chronon and an end chronon. For example, $\{t, t+1, t+2, \cdots, T\}$ is a time interval, which can concisely be denoted as $[t, T]$.

- Lifespan of attribute value and tuple

In historical relational database model (HRDM), valid time is denoted as **lifespan**, which is the period of time during which the tuple or attribute of tuple is true.

Table 3.1 shows how the salaries of employees change over time, where "Salary" is a time-varying attribute. From this table, we can observe that the valid time of salary of $1800 is 2000 to 2002, that is, the lifespan of salary $1800 is 2000 to 2002. Similarly, the lifespan of salary $2000 is 2003 to 2005 and the lifespan of salary $2400 is 2006 to *Now*. These lifespans can be denoted as [2000, 2002], [2003, 2005], [2006, *Now*], respectively.

Table 3.1 Changes in employees' salary

Employee number	Salary	Lifespan
929502288	[2000, 2002] $1800 [2003, 2005] $2000 [2006, *Now*] $2400	[2000, *Now*]
⋮	⋮	⋮

Besides, there is another lifespan [2000, *Now*], which is tuple's lifespan and is the period of time during which the tuple exists.

- Set operation of lifespan

Lifespans are subsets of system time domain *Sys_T*, so they can be operated by set operators such as union, intersection, difference and Not. The results of these operations are also lifespans.

3.2.2 HRDM Model

Let U be a universal set of attributes, denoted as $U = \{A_1, A_2, \cdots, A_n\}$. $DOM(A)$

denotes the domain of attribute A, which includes at least two elements. Let T be the system time domain and $D_i = DOM(A_i)$, $D = D_1 \cup D_2 \cup \cdots \cup D_n$. We can define two sets of temporal mappings, TD and TT. $TD = \{TD_1, TD_2, \cdots, TD_n\}$, where for each i, TD_i is the set of all partial temporal mappings from system time domain into attribute domain D_i, which can be expressed as $TD_i = \{f_j \mid f_j : T \rightarrow D_i\}$. TT is the set of all partial functions from system time domain into itself, i.e., $TT = \{g \mid g: T \rightarrow T\}$.

All attributes in the historical relational data model are defined over sets of partial temporal functions. That is, the value of tuple r on an attribute is a partial temporal function. The domain of an attribute is a set of temporal functions.

Let $HD = (TD \cup TT) = \{TT, TD_1, TD_2, \cdots, TD_n\}$. Among the functions in each function set in HD, some are constant-valued, i.e., they map every time in their domain into a same value. Let CD be the set derived from HD by restricting each of the function set in HD to only those functions having a constant image. That is, for each function set TD_i, (and TT) in HD, restrict TD_i (and TT) to only those functions that map their domain to a single value (Clifford and Croker 1987).

We use $f \mid D$ to denote restricting f with domain D.

Definition 3.5 A relation schema in HRDM is an ordered 4-tuple $R = \langle A, K, ALS, DOM \rangle$, where:

- $A = \{A_1, A_2, \cdots, A_n\} \subseteq U$ is the set of attributes of R. We will sometimes refer to A as the schema of R to avoid any confusion.
- $K = \{A_1, A_2, \cdots, A_m\} \subseteq A$ is the set of key attributes of R.
- ALS: $A \rightarrow 2^T$ is a function assigning a lifespan to each attribute in R. We will refer to the lifespan of attribute A in relation schema R as $ALS(A, R)$.
- DOM: $A \rightarrow HD$ is a function assigning a domain to each attribute in R, with restrictions that ① for all key attributes A_i, $DOM(A_i) \in CD$, i.e., the key attributes must all be constant-valued, and ② the domain of each of the partial functions in any $DOM(A)$ is contained within $ALS(A, R)$ (Clifford and Croker 1987).

Definition 3.6 A tuple t on schema R is an ordered 2-tuple, $t = \langle v, e \rangle$, $t.e$ is the lifespan of tuple t and $t.v$ is the value of the tuple. This is a mapping such that for any attribute $A \in R$, the value of the mapping denoted as $t.v(A)$, is a function that maps $t.e \cap ALS(A, R)$ into $DOM(A)$.

Property 3.1 Let r be a relation on schema R in HRDM. r is a finite set of such tuples that satisfy the following criteria: If t_1 and t_2 are tuples of r, for any $u \in t_1.e$ and $v \in t_2.e$, $t_1.v(K)(u) \neq t_2.v(K)(v)$, where $t.v(A)(s)$ is the value of tuple t for attribute A at time s. This means that the values of any two tuples on key attributes are different (Clifford and Croker 1987).

Property 3.2 The lifespan of relation r, denoted as $LS(r)$, is the union of lifespans of all tuples in r, i.e., if $r = \{t_1, t_2, \cdots, t_n\}$, then $LS(r) = t_1.e \cup t_2.e \cup \cdots \cup t_n.e$.

3.2.3 Historical Relational Algebra of HRDM

Like regular relations, historical relations are also sets of tuples. Thus, the standard set operations of union, intersection, set difference and Cartesian product can also be used on them.

Let relations r_1 on $R_1 = \langle A_1, K_1, ALS_1, DOM_1 \rangle$ and r_2 on $R_2 = \langle A_2, K_2, ALS_2, DOM_2 \rangle$ be two relations involved in union, intersection, set difference. They should be union-compatible, i.e., $A_1 = A_2$ and $DOM_1 = DOM_2$.

1. Union

Two relations r_1 and r_2 are union-compatible, r_1 is on model $R_1 = \langle A_1, K_1, ALS_1, DOM_1 \rangle$ and r_2 is on model $R_2 = \langle A_2, K_2, ALS_2, DOM_2 \rangle$. Union of r_1 and r_2 can be expressed as follows:

$$r_1 \cup^t r_2 = \{t \text{ on } R_3 \mid t \in r_1 \text{ or } t \in r_2 \wedge R_3 = \langle A_1, K_1, ALS_1 \cup ALS_2, DOM_1 \rangle\} \quad (3.13)$$

It should be noted that the tuples having the same values on key attributes should be merged. For example, the union of relation r_1 (shown in Table 3.2) and relation r_2 (shown in Table 3.3) is relation r_3, which is shown in Table 3.4.

Table 3.2 Relation r_1

Employee's number	Salary	Lifespan
929502288	[2000, 2002] \$1800 [2003, 2005] \$2000 [2006, *Now*] \$2400	[2000, *Now*]
929502289	[2001, 2002] \$1800 [2003, 2005] \$2100 [2006, *Now*] \$2400	[2001, *Now*]

Table 3.3 Relation r_2

Employee's number	Salary	Lifespan
929502288	[2003, 2005] \$2000 [2006, 2007] \$2400 [2008, *Now*] \$2600	[2003, *Now*]
929502289	[2003, 2005] \$2100 [2006, 2007] \$2400 [2008, *Now*] \$2500	[2003, *Now*]

From this example, we notice that the number of tuples from r_3 is not the sum of number of tuples from r_1 and r_2, because tuples of same values on key attributes are merged. Accordingly, the lifespans of those tuples are merged too.

2. Intersection

Two relations r_1 and r_2 are union-compatible. r_1 is on model $R_1 = \langle A_1, K_1, ALS_1,$

Table 3.4 Relation r_3

Employee's number	Salary	Lifespan
929502288	[2000, 2002] \$1800 [2003, 2005] \$2000 [2006, 2007] \$2400 [2008, *Now*] \$2600	[2000, *Now*]
929502289	[2001, 2002] \$1800 [2003, 2005] \$2100 [2006, 2007] \$2400 [2008, *Now*] \$2500	[2001, *Now*]

$DOM_1\rangle$ and r_2 is on model $R_2 = \langle A_2, K_2, ALS_2, DOM_2\rangle$. Intersection of r_1 and r_2 can be expressed as follows:

$$r_1 \cap^t r_2 = \{t \text{ on } R_3 \mid t \in r_1 \text{ and } t \in r_2 \wedge R_3 = \langle A_1, K_1, ALS_1 \cap ALS_2, DOM_1\rangle\} \quad (3.14)$$

It should be noted that the common parts of tuples that have the same values on key attributes are selected. For example, the intersection of relation r_1 (shown in Table 3.2) and relation r_2 (shown in Table 3.3) is relation r_4, which is shown in Table 3.5.

Table 3.5 Relation r_4

Employee's number	Salary	Lifespan
929502288	[2003, 2005] \$2000 [2006, 2007] \$2400	[2003, 2007]
929502289	[2003, 2005] \$2100 [2006, 2007] \$2400	[2003, 2007]

From this example, we notice that the lifespan of tuples in result relation may be changed.

3. Difference

Two relations r_1 and r_2 are union-compatible. r_1 is on model $R_1 = \langle A_1, K_1, ALS_1, DOM_1\rangle$ and r_2 is on model $R_2 = \langle A_2, K_2, ALS_2, DOM_2\rangle$. Difference of r_1 and r_2 can be expressed as follows:

$$r_1 -^t r_2 = \{t \text{ on } R_1 \mid t \in r_1 \text{ and } t \notin r_2\} \quad (3.15)$$

It should be noted that the common parts of tuples that have the same values on key attributes should be subtracted. For example, the difference of relation r_1 (shown in Table 3.2) and relation r_2 (shown in Table 3.3) is relation r_5, which is shown in Table 3.6.

Table 3.6 Relation r_5

Employee's number	Salary	Lifespan
929502288	[2000, 2002] $1800	[2000, 2002]
929502289	[2001, 2002] $1800	[2001, 2002]

In this example, the lifespan of tuple needs to be changed accordingly.

4. Cartesian product

Cartesian product on HRDM is an extension of regular Cartesian product. Cartesian product can be seen as an inverse operation of project. If attribute set of r_1 is A_1 and attribute set of r_2 is A_2, then it must satisfy $\Pi A_1\ (r_1 \times r_2) = r_1$ and $\Pi A_2\ (r_1 \times r_2) = r_2$, where Π denotes project operation.

Let r_1 and r_2 be two relations. r_1 is on model $R_1 = \langle A_1, K_1, ALS_1, DOM_1 \rangle$ and r_2 is on model $R_2 = \langle A_2, K_2, ALS_2, DOM_2 \rangle$. Cartesian product of r_1 and r_2 can be expressed as follows:

$$\begin{aligned} r_1 \times^t r_2 = \{t \text{ on } R_3 \mid & \exists t_1 \in r_1, \exists t_2 \in r_2, t.e = t_1.e \cup t_2.e \wedge \\ & \forall A \in R_1 \forall S \in t_1.e(t.v(A)(S) = t_1.v(A)(S)) \wedge \\ & \forall A \in R_2 \forall S \in t_2.e(t.v(A)(S) = t_2.v(A)(S)) \wedge \\ & \forall A \in R_1 \forall S \in t.e - t_1.e(t.v(A)(S) = \perp) \wedge \\ & \forall A \in R_2 \forall S \in t.e - t_2.e(t.v(A)(S) = \perp)\} \end{aligned} \tag{3.16}$$

R_3 is a new model, such that $R_3 = \langle A_1 \square A_2, K_1 \square K_2, ALS_1 \square ALS_2, DOM_1 \square DOM_2 \rangle$. $\perp$ represents empty values.

For example, Cartesian product of relation R (shown in Table 3.7) and relation S (shown in Table 3.8) is relation T, which is shown in Table 3.9.

In Table 3.9, the last four tuples include empty values denoted as "$\perp$". According to formula (3.15), tuple "a_1, b_1, [1, 2]" from R and tuple "c_2, d_2, [3, 4]" from S produced three new tuples: "a_1, b_1, c_2, d_2, [1, 4]", "a_1, b_1, $\perp$, $\perp$, [1, 2]",

Table 3.7 Relation R

A	*B*	Lifespan
a_1	b_1	[1, 2]
a_2	b_2	[3, 4]

Table 3.8 Relation S

C	*D*	Lifespan
c_1	d_1	[1, 2]
c_2	d_2	[3, 4]

Table 3.9 Relation $T = R \times^t S$

A	*B*	*C*	*D*	**Lifespan**
a_1	b_1	c_1	d_1	[1, 2]
a_1	b_1	c_2	d_2	[1, 4]
a_2	b_2	c_1	d_1	[1, 4]
a_2	b_2	c_2	d_2	[3, 4]
a_1	b_1	$\perp$	$\perp$	[1, 2]
a_2	b_2	$\perp$	$\perp$	[3, 4]
$\perp$	$\perp$	c_1	d_1	[1, 2]
$\perp$	$\perp$	c_2	d_2	[3, 4]

"$\perp$, $\perp$, c_2, d_2, [3, 4]". The first tuple is easy to understand. The second tuple is produced by subtracting [3, 4] from [1, 2]. We get lifespan [1, 2] and empty values on attributes *C* and *D* during this lifespan. Similarly, the third tuple is produced by subtracting [1, 2] from [3, 4]. We get lifespan [3, 4] and empty values on attributes *A* and *B*.

5. Selection

Because temporal attribute, valid time and lifespan are associated with HRDM, "select" operation should be extended so that it can select not only regular attribute values but also lifespan of attribute values or tuples. Before we discuss this extension, traditional Boolean expressions should be extended to temporal Boolean expressions.

Definition 3.7 (Temporal Boolean expressions) Temporal Boolean expressions are made of logic operation "$\wedge$", "$\vee$", relational expressions and set expressions. The basic form of relational expressions is $X\theta Y$, $\theta = \{>, \geqslant, <, \leqslant, =, \neq\}$, where *X* and *Y* are attribute names or given attribute values. The basic form of set expressions is $U\S V$, $\S = \{\cup, \cap, -\}$, where *U* and *V* are lifespans of relations or given lifespan values, and the value of $U\S V$ is defined as True if $U\S V \neq \varnothing$, else the value is defined as False.

"Select" operation is selecting those tuples from a relation *r* that satisfy a given selection criteria. The result of selection is denoted as $\sigma^t(r)$.

$$\sigma^t(r) = \{t \mid P(t) \wedge F(t)\} \tag{3.17}$$

P(*t*) is a regular Boolean expression, which gives predicate restrictions for tuples, while *F*(*t*) is a temporal Boolean expression, which gives temporal restrictions for tuples. Only those tuples that satisfy criteria of both *P*(*t*) and *F*(*t*) can be selected.

There are two temporal selection operators in HRDM, Select_IF and Select_WHEN, which will be explained by examples as follows.

(1) Select_IF

Select_IF is selecting qualified tuples over their entire lifespans, that is, if the selection criterion is met by a tuple *t*, then the entire tuple is returned, and its lifespan is unchanged.

For example, the result of $\text{Select\_IF}_{\text{employee's number}=929502288}(r_1)$ is shown in Table 3.10, where r_1 is the relation shown in Table 3.2.

Table 3.10 $\text{Select\_IF}_{\text{employee's number}=929502288}(r_1)$

Employee's number	Salary	Lifespan
929502288	[2000, 2002] $1800 [2003, 2005] $2000 [2006, *Now*] $2400	[2000, *Now*]

$\text{Select\_IF}_{\text{employee's number}=929502288 \wedge [1998, 1999]}(r_1) = \varnothing$, because there is no tuple in r_1 that satisfies "employee's number = 929502288" during [1998,1999].

$\text{Select\_IF}_{[2000, Now]}(r_1) = r_1$, because the two tuples in r_1 are satisfying the criterion during [2000, *Now*].

$\text{Select\_IF}_{\text{employee's number}=929502288 \wedge [2003, 2005]}(r_1)$ is the same as relation shown in Table 3.10, because the tuple in Table 3.10 satisfies "employee's number = 929502288" during [2003, 2005]. It should be pointed out that it is the entire tuple that is selected, so the lifespan of the selected tuple is still [2000, *Now*].

(2) Select_WHEN

Select_WHEN is selecting qualified tuples, ignoring all but a relevant subset of their lifespans, that is, if the selection criterion is met by a tuple *t* at some time in its lifespan, what is returned is a new tuple t'. The lifespan of this new tuple t' is exactly those points in time when the criterion is met, and whose value is the same as *t* for those points. The Select_WHEN is therefore a hybrid operation, reducing a historical relation in both the value and the temporal dimensions (Clifford and Croker 1987).

For example, the result of $\text{Select\_WHEN}_{\text{employee's number}=929502288 \wedge [2003, 2005]}(r_1)$ is shown in Table 3.11.

Table 3.11 $\text{Select\_WHEN}_{\text{employee's number}=929502288 \wedge [2003, 2005]}(r_1)$

Employee's number	Salary	Lifespan
929502288	[2003, 2005] $2000	[2003, 2005]

6. Projection

Temporal projection is similar to regular projection. We only need to change the lifespans of some tuples or tuple attributes.

Projection is a unary operation that selects some attributes from a relation to produce a new relation. Let *A* be the attribute set of relation *r*, *B* be a subset of *A*,

$t.B$ be the component of t on attributes in B. The projection of r on B is given by:

$$\Pi^t_B(r) = \{t.B \mid t \in r \wedge B \subseteq A\} \quad (3.18)$$

HRDM is an easy-to-understand model among the 13 important temporal models. It adds lifespans of attributes and tuples to RDB. It can only manage the history of object itself, but not the history of inserting, deleting and updating the object. HRDM is a compatible extension of RDM.

When lifespan is shortened to a point $[t, t]$, historical relational database becomes a snapshot database. The present state of RDB can be seen as a special case of HRDM at [*Now*, *Now*].

Furthermore, many operational rules on traditional relational database can be used in temporal relational database, such as commutative law, distributive law and associative law of selective operation.

3.3 Bitemporal Relational Algebra of BCDM

Bitemporal relational algebra is much more complex than HRDM, because it supports not only valid time but also transaction time.

Michael D. Soo, Christian S. Jensen and R. T. Snodgrass have described this algebra by means of TSQL2, which is a query language that supports both kinds of time (Soo et al. 1995). Relational algebra is proved to be both minimal and complete by mapping TSQL2 to it.

Lu-bang Wang and Yong Tang gave a bitemporal relational algebra based on BCDM in 2004 (Wang et al. 2004). In this section, we will give a detailed description for this relational algebra.

3.3.1 Basic Notions and Terminologies

Bitemporal relational algebra supports both valid time and transaction time. Let $VT = \{0, 1, \cdots, Now\}$ be the domain of valid time, $TT = \{0, 1, \cdots, UC\}$ be the domain of transaction time, where *Now* means "present time", *UC* means "until changed".

Definition 3.8 (Convex time interval) Valid time interval $I^V = [t_1, t_2]$ is said to be convex, when it satisfies: ① $\forall t \in I^V, t_1 \leqslant t \leqslant t_2$, and ② $\forall t$, if $t_1 \leqslant t \leqslant t_2$, then $t \in I^V$. Transaction time interval $I^T = [t_1, t_2)$ is said to be convex, when it satisfies: ① $\forall t \in I^T, t_1 \leqslant t \leqslant t_2$, and ② $\forall t$, if $t_1 \leqslant t \leqslant t_2$, then $t \in I^T$.

Definition 3.9 (Temporal element) Let P be the set of all intervals on VT or TT. For any n elements from P, denoted as $I_1, I_2, \cdots, I_n \in P$, where $n < +\infty$, u is the union set of $I_1, I_2, \cdots, I_n$. u is defined as a temporal element on P, denoted as $u = I_1 + I_2 + \cdots + I_n$.

Definition 3.10 (Normal temporal element) Let $u=I_1+I_2+\cdots+I_n$ be a temporal element on *VT* or *TT*. u is said to be normal if u satisfies:

(1) $I_i \cap I_j = \varnothing,\ \forall I_i, I_j \in u, i \neq j$;

(2) $I_1, I_2, \cdots, I_n$ are all convex time intervals.

Definition 3.11 Let u and v be normal temporal elements. The union, intersection and difference of u and v are defined as follows:

$$u+v=\{t \mid t \in u \vee t \in v\} \tag{3.19}$$

$$u \times v=\{t \mid t \in u \wedge t \in v\} \tag{3.20}$$

$$u-v=\{t \mid t \in u \wedge t \notin v\} \tag{3.21}$$

Definition 3.12 (Bitemporal point and bitemporal label) A bitemporal point P_{bt} is denoted as (t^t, t^v), where t^t is a transaction time point and t^v a valid time point. Bitemporal label L_{bt} is a set of bitemporal points, denoted as $\{(t^t, t^v) |\ t^t \in TT, t^v \in VT\}$.

Definition 3.13 (Normal bitemporal element) Bitemporal element is a set of two-dimensional time intervals, denoted as $\{([a,b),[c,d]) \mid a,b \in TT, 0 \leqslant a<b \leqslant UC, c,d \in VT,\ 0 \leqslant c<d \leqslant Now\}$. Bitemporal element$\{([a_1,b_1),[c_1,d_1]), ([a_2,b_2),[c_2,d_2]), \cdots,([a_n,b_n),[c_n,d_n])\}$ is said to be normal if it satisfies:

(1) If $[a_i,b_i] \cap [a_j,b_j] = \varnothing$, there is no restriction for $[c_i,d_i]$ and $[c_j,d_j]$.

(2) If $[a_i,b_i] \cap [a_j,b_j] = b_i$ or b_j, then $(c_i,d_i) \neq (c_j,d_j)$, that means $c_i \neq c_j$ or $d_i \neq d_j$.

(3) If $[a_i,b_i] \cap [a_j,b_j] \neq \varnothing$, then $[c_i,d_i] \cap [c_j,d_j] = \varnothing$.

This means, if we take two-dimensional time intervals as rectangles in two-dimensional plane, where two axes denote valid time and transaction time, respectively, then any two rectangles from a normal bitemporal element do not overlap or can be merged into one rectangle.

Figure 3.1(a) gives an example of normal bitemporal element and Fig. 3.1(b) gives an example of non-normal bitemporal element. Let x-axes denote transaction time and y-axes denote valid time and every rectangle denote a two-dimensional time interval in bitemporal element.

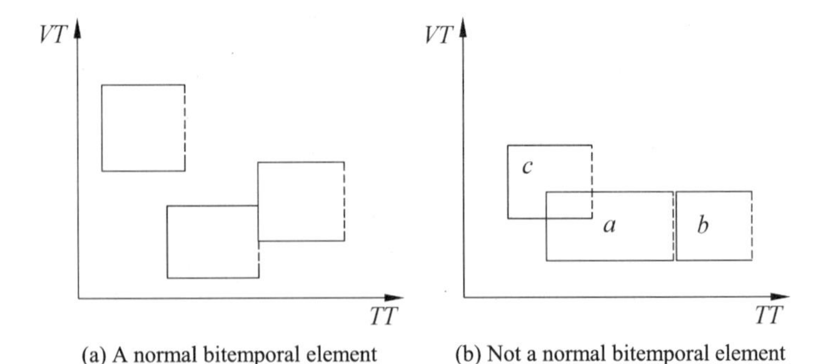

(a) A normal bitemporal element (b) Not a normal bitemporal element

Figure 3.1

Bitemporal element in Fig. 3.1(b) is not a normal bitemporal element because rectangles a, b can be merged to one rectangle and c overlaps a.

Property 3.3 Any bitemporal label can be represented by a normal bitemporal element (Wang et al. 2004).

Definition 3.14 (Consistency of bitemporal elements) Let $u = \{([a_1,b_1),[c_1,d_1]), ([a_2,b_2),[c_2,d_2]), \cdots, ([a_n,b_n),[c_n,d_n])\}$, $v = \{([a_1',b_1'),[c_1',d_1']), ([a_2',b_2'),[c_2',d_2']), \cdots, ([a_n',b_n'),[c_n',d_n'])\}$ be two normal bitemporal elements. u and v are said to be consistent if they satisfy:

$\forall(\xi, \Sigma_v) \in u$, $\forall(\xi', \Sigma_v') \in v$, if $\xi = \xi'$ then $\Sigma_v = \Sigma_v'$, where Σ_v, Σ_v' are sets of valid time corresponding to transaction time points ξ, ξ'.

The union, intersection and difference of consistent bitemporal elements u and v are defined as follows:

$$
\begin{aligned}
u \cup v = \{&([a,b),[c,d]) \mid (\exists([a_i,b_i),[c_i,d_i]) \in u,\\
&\exists([a_i',b_i'),[c_i',d_i']) \in v([a_i,b_i) \cap [a_i',b_i') \neq \varnothing)\\
&\wedge([a,b) = [a_i,b_i) \cup [a_i',b_i'),[c,d] = [c_i,d_i]))\\
&\vee(\exists([a_i,b_i),[c_i,d_i]) \in u, \forall([a_i',b_i'),[c_i',d_i']) \in v,\\
&([a_i,b_i) \cap [a_i',b_i') = \varnothing) \wedge (([a,b),[c,d]) = ([a_i,b_i),[c_i,d_i])))\\
&\vee(\exists([a_i',b_i'),[c_i',d_i']) \in v, \forall([a_i,b_i),[c_i,d_i]) \in u,\\
&([a_i',b_i') \cap [a_i,b_i) = \varnothing) \wedge (([a,b),[c,d]) = ([a_i',b_i'),[c_i',d_i'])))\}
\end{aligned}
\tag{3.22}
$$

$$
\begin{aligned}
u \cap v = \{&([a,b),[c,d]) \mid \exists([a_i,b_i),[c_i,d_i]) \in u, \exists([a_i',b_i'),[c_i',d_i']) \in v\\
&([a_i,b_i) \cap [a_i',b_i') \neq \varnothing) \wedge ([a,b) = [a_i,b_i) \cap [a_i',b_i'),[c,d] = [c_i,d_i])\}
\end{aligned}
\tag{3.23}
$$

$$
\begin{aligned}
u - v = \{&([a,b),[c,d]) \mid (\exists([a_i,b_i),[c_i,d_i]) \in u,\\
&\exists([a_i',b_i'),[c_i',d_i']) \in v([a_i,b_i) \cap [a_i',b_i') \neq \varnothing\\
&\wedge([a,b) = [a_i,b_i) - [a_i',b_i'),[c,d] = [c_i,d_i]))\\
&\vee(\exists([a_i,b_i),[c_i,d_i]) \in u, \forall([a_i',b_i'),[c_i',d_i']) \in v,\\
&([a_i,b_i) \cap [a_i',b_i') = \varnothing) \wedge (([a,b),[c,d]) = ([a_i,b_i),[c_i,d_i])))\}
\end{aligned}
\tag{3.24}
$$

For example, let $u = \{([1,3), [1,2]), ([1,3), [3,4])\}$, $v = \{([2,4), [1,2]), ([5,6), [1,2])\}$ be two consistent normal bitemporal elements, which can be expressed in Fig. 3.2(a). $u \cup v$, $u \cap v$ and $u - v$ can be expressed in Fig. 3.2(b), Fig. 3.2(c) and Fig. 3.2(d), respectively.

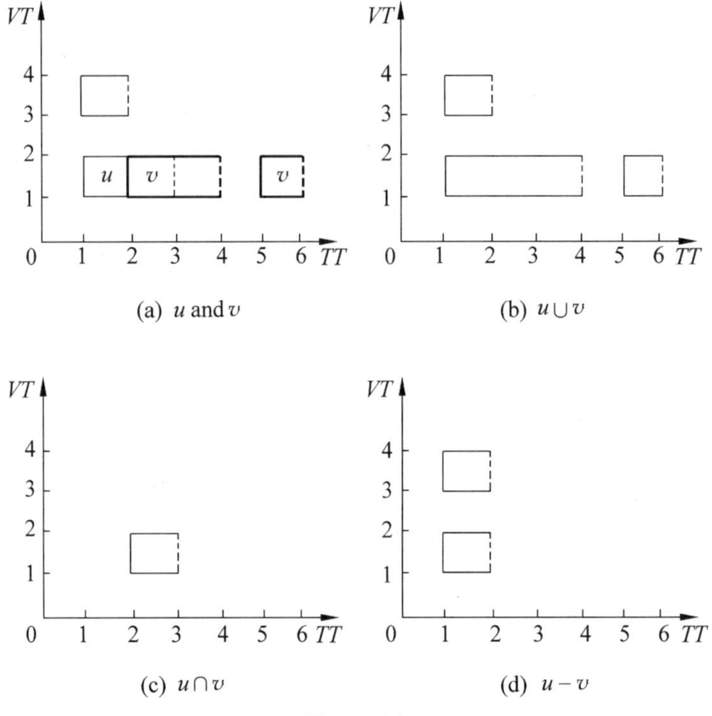

Figure 3.2

Definition 3.15 (Pure relational tuples) Removing time labels from a tuple in BCDM, we can get a pure relational tuple, which is denoted as $(a_1, a_2,\cdots, a_n)$. Relation composed of pure relational tuples is a non-temporal relation.

For example, ("John", \$1800, 2005-1, 2008-1, 2005-2, *UC*) is a bitemporal tuple, where [2005-1, 2008-1] is valid time interval, [2005-2, *UC*) is transaction time interval. If we remove bitemporal label ([2005-2, *UC*), [2005-1, 2008-1]), represented as a normal bitemporal element, from the tuple, we get a pure relational tuple ("John", \$1800).

Definition 3.16 (Bitemporal mapping) Bitemporal mapping is a mapping from normal bitemporal element to pure relational tuple, which is denoted as $\xi{:}u \rightarrow (a_1, a_2,\cdots, a_n)$. In BCDM, a bitemporal tuple is a bitemporal mapping. $\overline{\xi} = u$ is the domain of mapping. $|\xi| = (a_1, a_2,\cdots, a_n)$ is the range of mapping.

In the previous example, ("John", \$1800, 2005-1, 2008-1, 2005-2, *UC*) can be seen as a mapping from ([2005-2, *UC*), [2005-1, 2008-1]) to ("John", \$1800).

Definition 3.17 (Bitemporal mapping operation) Let ξ_1, ξ_2 be two bitemporal mappings on the same schema, $|\xi_1| = |\xi_2|$. Operators $\oplus, \otimes, \ominus$ of ξ_2, ξ_2 are defined as follows:

$$\xi_1 \oplus \xi_2 = \xi, \text{ where } |\xi| = |\xi_1|, \overline{\xi} = \overline{\xi_1} \cup \overline{\xi_2} \tag{3.25}$$

$$\xi_1 \otimes \xi_2 = \xi, \text{ where } |\xi| = |\xi_1|, \overline{\xi} = \overline{\xi_1} \cap \overline{\xi_2} \tag{3.26}$$

$$\xi_1 \ominus \xi_2 = \xi, \text{ where } |\xi| = |\xi_1|, \overline{\xi} = \overline{\xi_1} - \overline{\xi_2} \tag{3.27}$$

For example, ξ_1 = ("John", \$1800, 2005-1, 2008-1, 2005-2, *UC*), ξ_2 = ("John", \$1800, 2005-1, 2008-1, 2005-1, 2005-8), $\xi_1 \oplus \xi_2$ = ("John", \$1800, 2005-1, 2008-1, 2005-1, *UC*), $\xi_1 \otimes \xi_2$ = ("John", \$1800, 2005-1, 2008-1, 2005-2, 2005-8), $\xi_1 \ominus \xi_2$ = ("John", \$1800, 2005-1, 2008-1, 2005-8, *UC*).

3.3.2 Bitemporal Relational Algebra

Definition 3.18 (Bitemporal relation) Bitemporal relation r is an $N \times 1$ matrix composed by N bitemporal mappings, which can be denoted as $r = \begin{bmatrix} \xi_1 \\ \vdots \\ \xi_n \end{bmatrix}$, where ξ_i is a bitemporal mapping in BCDM.

In BCDM, union, intersection, difference, Cartesian product, natural join, projection, and selection of bitemporal relation are denoted as $\cup^{BT}$, $\cap^{BT}$, $-^{BT}$, $\times^{BT}$, $\rhd\lhd^{BT}$, Π^{BT}, σ^{BT}, which will be defined as follows. It should be noted that two relations under such an operation must be on the same bitemporal schema.

1. Bitemporal union ($\cup^{BT}$)

$$\begin{aligned} r = r_1 \cup^{BT} r_2 = \{\xi \mid & (\exists \xi_1 \in r_1, \exists \xi_2 \in r_2 (|\xi_1| = |\xi_2|) \wedge (\xi = \xi_1 \oplus \xi_2)) \\ & \vee (\exists \xi_1 \in r_1, \forall \xi_2 \in r_2 (|\xi_1| \neq |\xi_2|) \wedge (\xi = \xi_1)) \\ & \vee (\exists \xi_2 \in r_2, \forall \xi_1 \in r_1 (|\xi_1| \neq |\xi_2|) \wedge (\xi = \xi_2))\} \end{aligned} \tag{3.28}$$

For example, r_1 and r_2 are two bitemporal relations on the same bitemporal schema, which is shown in Table 3.12 and Table 3.13. $r_1 \cup^{BT} r_2$ is shown in Table 3.14.

Table 3.12 Relation r_1

Employee's number	Salary	Valid time	Transaction time
929502288	\$1800	[2005, 2007]	[2007, *UC*)
929502288	\$2000	[2008, *Now*]	[2008, *UC*)
929502289	\$2800	[2005, *Now*]	[2005, *UC*)

Table 3.13 Relation r_2

Employee's number	Salary	Valid time	Transaction time
929502288	\$1800	[2005, 2007]	[2005, 2008)

Table 3.14 $r_1 \cup^{BT} r_2$

Employee's number	Salary	Valid time	Transaction time
929502288	$1800	[2005, 2007]	[2005, *UC*)
929502288	$2000	[2008, *Now*]	[2008, *UC*)
929502289	$2800	[2005, *Now*]	[2005, *UC*)

2. Bitemporal intersection ($\cap^{BT}$)

$$r = r_1 \cap^{BT} r_2 = \{\xi \mid \exists \xi_1 \in r_1, \exists \xi_2 \in r_2 (|\xi_1| = |\xi_2|) \wedge (\xi = \xi_1 \otimes \xi_2)\} \qquad (3.29)$$

Consider the former relations r_1 and r_2 as shown in Table 3.12 and Table 3.13. $r_1 \cap^{BT} r_2$ is shown in Table 3.15.

Table 3.15 $r_1 \cap^{BT} r_2$

Employee's number	Salary	Valid time	Transaction time
929502288	$1800	[2005, 2007]	[2007, 2008)

3. Bitemporal difference ($-^{BT}$)

$$\begin{aligned} r = r_1 -^{BT} r_2 = \{\xi \mid & (\exists \xi_1 \in r_1, \exists \xi_2 \in r_2 (|\xi_1| = |\xi_2|) \wedge (\xi = \xi_1 \ominus \xi_2)) \\ & \vee (\exists \xi_1 \in r_1, \forall \xi_2 \in r_2 (|\xi_1| \neq |\xi_2|) \wedge (\xi = \xi_1))\} \end{aligned} \qquad (3.30)$$

Consider the former relations r_1 and r_2 as shown in Table 3.12 and Table 3.13. $r_1 -^{BT} r_2$ is shown in Table 3.16.

Table 3.16 $r_1 -^{BT} r_2$

Employee's number	Salary	Valid time	Transaction time
929502288	$1800	[2005, 2007]	[2008, *UC*)
929502288	$2000	[2008, *Now*]	[2008, *UC*)
929502289	$2800	[2005, *Now*]	[2005, *UC*)

4. Bitemporal cartesian product ($\times^{BT}$)

$$\begin{aligned} r = r_1 \times^{BT} r_2 = \{\xi \mid & (\exists \xi_1 \in r_1, \exists \xi_2 \in r_2 (\overline{\xi_1} \cap \overline{\xi_2} \neq \varnothing)) \\ & \wedge (|\xi| = (|\xi_1|, |\xi_2|) \wedge (\overline{\xi} = \overline{\xi_1} \cap \overline{\xi_1}))\} \end{aligned} \qquad (3.31)$$

5. Bitemporal natural join ($\rhd\!\lhd^{BT}$)

Let r_1, r_2 be bitemporal relations on schema R and S. r_1, r_2 can be natural joins if R and S satisfy: $R \cap S \neq \varnothing$, i.e., the set of common attributes in R and S is not empty. Let $X = R \cap S$, $A = R - X$, $B = S - X$, then

$$r = r_1 \rhd\lhd^{BT} r_2 = \{\xi \mid \exists \xi_1 \in r_1, \exists \xi_2 \in r_2 (\overline{\xi_1} \cap \overline{\xi_2} \neq \varnothing) \wedge \forall A_i \in X,$$
$$(|\xi_1|_{Ai} = |\xi_2|_{Ai}) \wedge (|\xi| = (|\xi_1|_{R-X}, |\xi_1|_X, |\xi_2|_{S-X}))\} \quad (3.32)$$

Here $|\xi|_{A_i}$ denotes the value of pure relational tuple $|\xi|$ on attribute A_i.

For example, r_1 is a bitemporal relation on schema R, r_2 is a bitemporal relation on schema S, which are shown in Table 3.17 and Table 3.18. $r_1 \rhd\lhd^{BT} r_2$ is shown in Table 3.19.

Table 3.17 Relation r_1 on schema R

Employee's number	Salary	Valid time	Transaction time
929502288	$1800	[2005, 2007]	[2008, *UC*)
929502288	$2000	[2008, *Now*]	[2008, *UC*)
929502289	$2800	[2005, *Now*]	[2005, *UC*)

Table 3.18 Relation r_2 on schema S

Employee's number	Position	Valid time	Transaction time
929502288	salesman	[2005, 2006]	[2008, *UC*)
929502288	Sales manager	[2007, *Now*]	[2008, *UC*)
929502289	HR manager	[2005, *Now*]	[2005, *UC*)

Table 3.19 $r_1 \rhd\lhd^{BT} r_2$

Employee's number	Salary	Position	Valid time	Transaction time
929502288	$1800	salesman	[2005, 2006]	[2008, *UC*)
929502288	$1800	Sales manager	[2007, 2007]	[2008, *UC*)
929502288	$2000	Sales manager	[2008, *Now*]	[2008, *UC*)
929502289	$2800	HR manager	[2005, *Now*]	[2005, *UC*)

6. Bitemporal projection (Π^{BT})

$$\Pi_X^{BT}(r) = \cup^{BT}(\Pi_X(r)) \quad (3.33)$$

For example, let r be bitemporal relation shown in Table 3.19. X= {Employee's number, salary}. If we first execute a regular project on r, we get $\Pi_X(r)$ as shown in Table 3.20.

Then, because of bitemporal union on every tuple in $\Pi_X(r)$, we get $\Pi_X^{BT}(r)$, which is the same as r_1 shown in Table 3.17. It proves that bitemporal projection is an inverse operation of bitemporal join.

Table 3.20 $\Pi_X(r)$

Employee's number	Salary	Valid time	Transaction time
929502288	$1800	[2005, 2006]	[2008, *UC*)
929502288	$1800	[2007, 2007]	[2008, *UC*)
929502288	$2000	[2008, *Now*]	[2008, *UC*)
929502289	$2800	[2005, *Now*]	[2005, *UC*)

7. Bitemporal selection (σ^{BT})

$$\sigma^{BT}_{P\wedge Q}(r) = \{\xi \mid \exists \xi_1 \in r, \mid \xi \mid = (\mid \xi_1 \mid \uparrow P) \wedge \overline{\xi} = \overline{\xi_1} \uparrow Q\} \quad (3.34)$$

Let r_1 be a bitemporal relation shown in Table 3.17. σ^{BT} (Employee's number = "92950228") $\wedge$ ([2008,*UC*), [2005,*Now*])(r_1) is shown in Table 3.21.

Table 3.21 $\sigma^{BT}_{\text{(Employee's number = "92950228")} \wedge \text{([2008,}UC\text{), [2005,}Now\text{])}}(r_1)$

Employee's number	Salary	Valid time	Transaction time
929502288	$1800	[2005, 2007]	[2008, *UC*)
929502288	$2000	[2008, *Now*]	[2008, *UC*)

3.4 Snapshot Reducibility and Temporal Completeness

In this section, we introduce three important criteria to evaluate temporal algebra or temporal query languages.

3.4.1 Snapshot Reducibility

As we already know, temporal relational algebra is an extension of regular relational algebra. Temporal relational algebra is obtained by adding temporal semantics on regular relational algebra. After this extension, we need a method using regular relational algebra to describe the semantics of temporal relational operators and temporal query languages.

R. T. Snodgrass (1991) pointed out that the semantics of temporal algebra should be consistent with the intuitive view of a snapshot relation as a two-dimensional slice of a temporal relation at a valid time t_1 and transaction time t_2.

Though Snodgrass gave the definition of snapshot reducibility on historical database model, it is also useful to rollback database and bitemporal database. The main idea of Snodgrass's definition is that the slice of result of implementing a temporal query on a historical database is equal to the result of implementing a regular query on the slice of historical database at the same time.

We give the definition of snapshot reducibility as follows.

Definition 3.19 (Snapshot reducibility) Let db^v be a temporal relation. Apply a temporal operator (or temporal query sentence) q^v to db^v and obtain the result r^v. q^v is said to be snapshot reducible to q (q is a non-temporal operator or non-temporal query sentence) when it satisfies: for any valid time t_1 and transaction time t_2, the two-dimensional slice of r^v is equivalent to the result of applying q to the two-dimensional slice of db.

Figure 3.3 illustrates this reduction as follows.

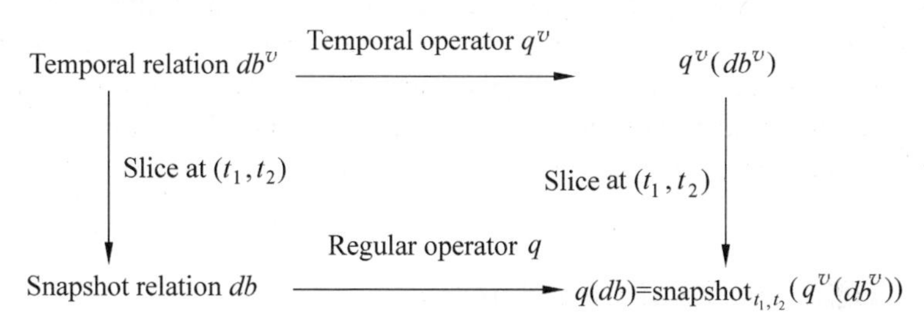

Figure 3.3 Snapshot reducibility

For example, Table 3.22 is a temporal relation about employees' information in a historical database. Table 3.22 shows the slice of this relation at time 2008-11-3, which is a regular relation.

Table 3.22 Employee's information

ID	Name	Birthday	Salary	Department	VTS	VTE
100	John	1978-01-01	$6000	Sales	2003-01	2004-06
100	John	1978-01-01	$6500	Marketing	2004-07	2006-06
100	John	1978-01-01	$8000	Marketing	2006-07	2008-12

Table 3.23 Slice of employee's information

ID	Name	Birthday	Salary	Department
100	John	1978-01-01	$8000	Marketing

Given a temporal query sentence "validtime select ID, Name of employees whose salary>6000", the result of implementing it on temporal relation of Table 3.22 is shown in Table 3.24.

Given a regular query sentence "select ID, Name of employees whose salary >6000", the result of implementing it on slice relation of Table 3.23 is shown in Table 3.25.

Table 3.24 Result of implementing query on temporal relation

ID	Name	Birthday	Salary	Department	VTS	VTE
100	John	1978-01-01	$6500	Marketing	2004-07	2006-06
100	John	1978-01-01	$8000	Marketing	2006-07	2008-12

Table 3.25 Result of implementing query on regular relation

ID	Name	Birthday	Salary	Department
100	John	1978-01-01	$8000	Marketing

It is clear that the slice of temporal relation of Table 3.23 at time 2008-11-3 is equal to regular relation in Table 3.25. On the other hand, if we implement "slice" operation at every time point on historical relation in Table 3.22 and implement regular query at every slice relation, the union of these results is equal to the temporal relation in Table 3.24.

3.4.2 Temporal Semi-Completeness

Michael Bohlen and Robert Marti (1994) introduced temporal semi-completeness and completeness for temporal query languages in 1994. The theory of temporal completeness is mainly used to evaluate the design of temporal query languages.

Definition 3.20 (Temporal semi-completeness) Let $M = (DS, QL)$ be a non-temporal data model. Here, DS denotes the data model and QL denotes a query language on it. Let $M^v = (DS^v, QL^v)$ be a temporal (valid time) data model. Here, DS^v denotes the temporal data model and QL^v denotes a temporal query language on it. M^v is temporally semi-complete to M, if and only if the following conditions are satisfied:

(1) For any relation r on DS, there exists a temporal (valid time) relation rv on DSv and a time point t, that is, $r = \tau_t(rv)$, where $\tau_t(rv)$ means the slice of rv on t.

(2) For every query q in QL, there exists a snapshot reducible q^v in QL^v.

(3) There exists two strings S_1 and S_2 (may be empty) such that for every query pair (q, q^v), q^v has a format of S_1qS_2, where q^v is snapshot reducible to q.

The first condition makes sure that every relation on non-temporal database can be obtained by making slice on a relation of temporal database at a special time point. This means that if non-temporal database supports duplicate tuples, temporal database should also support duplicate tuples. If union operator on temporal database model M^v eliminates duplicate tuples, M^v does not satisfy this condition.

The second condition demands that every non-temporal query q on M can be implied by a temporal query q^v on M^v. This means that the result of q on M is equal to the slice of result of q^v on M^v at a time point. If M supports an operation, M^v should support the temporal extension of q. Thus, it can satisfy this condition.

The last condition points out the semantic forms that the temporal query language should follow. Every temporal query sentence has a form that is obtained by adding prefix and suffix to the non-temporal query sentence it implies. This gives an easy transition method from non-temporal language to temporal language.

3.4.3 Temporal Completeness

Temporal completeness has much more functional and syntax restrictions than temporal semi-completeness.

Definition 3.21 (Temporal completeness) Let $M = (DS, QL)$ be a non-temporal data model, $M^v = (DS^v, QL^v)$ be a temporal (valid time) data model. M^v is temporal complete to M, if and only if the following conditions are satisfied:

(1) M^v is temporally semi-complete to M.

(2) For any snapshot reducible query q^v in QL^v, there are two methods to remove its snapshot reducibility. This means we should omit valid time when applying a query. The two methods are:

① Delete syntax extension that is used to support snapshot reducibility.

② Revise q^v to format of S_1qS_2, where S_1 and S_2 may be empty and rely on QL^v but not on q^v.

Temporal completeness is mainly used to evaluate temporal query languages. If a temporal language fulfills some requirements, it is said to be temporal complete.

References

[1] Allen JF (**1983**) Maintaining knowledge about temporal intervals. Communication of the ACM 26(11): 832 – 843

[2] Bohlen M, Marti R (**1994**) On the completeness of temporal database query languages. In: Proceedings of the First International Conference on Temporal Logic: 283 – 300

[3] Clifford J, Croker A (**1987**) The Historical Relational Data Model (HRDM) and algebra based on lifespans. In: Proceedings of the International Conference on Data Engineering: 528 – 537

[4] Codd EF (**1970**) A relational model of data for large shared data banks. Communications of the ACM 13(6): 377 – 387

[5] Codd EF (**1972**) Relational completeness of database sublanguage. Computer Sciences, Data Base Systems: 65 – 98

[6] Frank DA, Robert AM, Rita VR (**1988**) A temporal relational calculus. Lecture Notes in Computer Science 406, AI'88: 177 – 186

[7] Lacroix M, Pirotte A (**1977**) Domain-oriented relational languages. In: Proceedings of the Third International Conference on Very Large Data Bases: 370 – 378

[8] Lee JY, Elmasri RA (**2001**) A temporal algebra for an ER-based temporal data model. In: Proceedings of the 17th International Conference on Data Engineering, pp 33 – 40

[9] Morris RA, Al-Khatib L (**1991**) An interval-based temporal relational calculus for events with gaps. Journal of Experimental & Theoretical Artificial Intelligence 3(2): 87 – 107

[10] Orgun MA (**1999**) A temporal relational algebra based on multiple time-lines. Temporal representation and reasoning. In: Proceedings of the 6th International Workshop on TIME-99, pp 100 – 105

[11] Rose E, Segev A (**1993**) TOOA: a temporal object-oriented algebra. ECOOP'93, LNCS 707: 297 – 325

[12] Snodgrass RT (**1991**) Evaluation of relational algebras incorporating the time dimension in databases. ACM Computing Surveys 23(4): 501 – 543

[13] Soo MD, Jensen CS, Snodgrass RT (**1995**) An algebra for TSQL2. http://www.cs.auc.dk/~csj/Thesis/pdf/chapter21.pdf

[14] Tansel AU, Atay CE (**2006**) Nested bitemporal relational algebra. ISCIS 2006, LNCS 4263: 622 – 633

[15] Wang LB, Tang Y, Yu Y (**2004**) A bitemporal relational algebra based on BCDM. Journal of Computer Research and Development 41(11): 1949 – 1953

[16] Zhang Shi-chao (**1993**) The equivalence between temporal relational algebra and temporal tuple calculus. Chinese J Computers 16(12): 936 – 939

[17] Zhang Shi-chao (**1994**) Interval-gap-based representation of temporal knowledge. Journal of Software 5(6): 13 – 18

[18] Zhang Shi-chao, Zhang Cheng-qi (**1997**) Query languages for 1NF temporal databases. TENCON'97, IEEE Region 10 Annual Conference. Speech and Image Technologies for Computing and Telecommunications. Proceedings of IEEE, Volume 1, 211 – 214

Part Ⅱ Database Based on Temporal Information

- Temporal Data Model and Temporal Database Systems
- Spatio-Temporal Data Model and Spatio-Temporal Databases
- Temporal Extension of XML Data Model
- Data Operations Based on Temporal Variables

4 Temporal Data Model and Temporal Database Systems

Chengjie Mao[1], Hui Ma[2], Yong Tang[1,2+], and Liangchao Yao[2]

[1] Database Lab, South China Normal University, Guangzhou 510631, P.R. China
[2] Department of Computer Science, Sun Yat-sen University, Guangzhou 510275, P.R. China

Abstract This chapter introduces fundamental concepts in temporal database. We first discuss the time dimensions. We mainly explain the user-defined time, valid time and transaction time. Subsequently, we present four temporal database types based on the support of valid time and transaction time. Then, we introduce three temporal data models. Finally, we give a short comparison of temporal database and real-time database.

Keywords *valid time, transaction time, bitemporal database, real-time database*

4.1 Time-Dimensions

Time is a core concept as we study temporal database. To understand "time" in a better way, we first examine the time dimensions. We mainly introduce three time dimensions that are widely studied in temporal database: user-defined time, valid time and transaction time (Soo 1991; Jensen et al. 1992; Jensen 2000).

4.1.1 User-Defined Time

Just as its name implies, user-defined time is the "time" where the meaning is interpreted not by the DBMS (Database Management System), but by the users according to their application requirements. Hence, different user-defined times may have different meanings.

Example 4.1 A freshman enrollment relation may have *Birth date* as its attribute. A freshman born at October 1' 1980 may have "1980-10-01" as its *Birth*

---

issty@mail.sysu.edu.cn

date value. In addition, the relation may have another attribute *Enrollment date*. Here, both *Birth date* and *Enrollment date* are user-defined times, and they have different meanings.

The DBMS takes the user-defined time as a usual attribute. For transactions like insertion, deletion, update and query, the DBMS does not make any discrimination or give special treatment to the user-defined time as compared to other usual attributes. It is similar to the relation of age and integer. The value of user-defined time is managed by the users (applications). In most cases, the value of user-defined time is a time stamp. Conventional DBMS already supports the DataTime type. Thus, it has no special mechanism support for user-defined time.

4.1.2 Valid Time

Before we proceed to consider valid time, we will examine some related terms. In reality, database only records a part of information, which is termed either *modeled reality* or the *mini-world*. The fact is a logical statement that can be assigned with a truth value by the information from the database records. The valid time is associated to the fact, which means that when the fact is true in the mini-world, valid time is of significance. It is widely used in applications, for example, the guarantee period of food.

Valid time can span the past, present and the future. The value of valid time is a set of collected times. It can be a single time point, a period (time intervals), a set of time points, a set of periods or even the entire time domain.

Example 4.2 Ross's credit card is issued at 2009-01-01 and expires at 2012-01-01. Therefore, the valid time of Ross's credit card is [2009-01-01, 2012-01-01]. At 2008-12-01, Ross got a salary increase and from then on he earns forty thousand dollars a year. Hence, the valid time of his salary that equals to 40,000 is [2008-12-01, *Now*].

As opposed to the user-defined time, valid time is interpreted by the database system. With reference to Example 4.2, when query statements like "how much does Ross earn at 2009-01-01" or "how much does Ross earn at present" arises, the system should return the answer 40,000. The value of valid time is usually provided by the users and it may be system generated as well.

It is not necessary that all valid time of facts be recorded in temporal database. For example, some valid time may not be known. Some facts are constants, for example, the birth date of a person. Recording the valid time of such facts is irrelevant to the application.

Other alternative names of valid time are real-world time, intrinsic time, logical time, and data time (Jensen et al. 1992).

4.1.3 Transaction Time

Transaction time is the time when the change of fact has occurred in the database. Here change means the insertion, deletion, or update transaction that changes the state of the database. The transaction time is a time dimension that corresponds to the historical state changes of a database. The transaction time is a time stamp. It is automatically determined by the system time when the transaction occurs and is entirely application independent. In other words, the user cannot change its value. Since the transaction time captures the real transaction commitment time, it cannot exceed the current time. The transaction time starts when the database is created. The time granularity is determined by the system.

From the valid time point of view, the valid time of a fact stored in a database spans from the time when the fact was inserted or updated in the database, until the next time when the fact was changed. During this period, the fact remains unchanged.

Other alternative names of transaction time are registration time, extrinsic time, and physical time (Jensen et al. 1992).

4.1.4 Two Temporal Variables: *Now* and *UC*

Now and *UC* are two special temporal variables. In the following, we give a short description of the semantics of these two variables.

1. ***Now***

Now is a valid time marker that indicates the associated fact is valid until present. *Now* is a variable because its value changes as time progresses.

Now is very useful and widely applied in temporal database. For example, a faculty relation records a faculty and his/her title information. Suppose Ross got a promotion as an associate professor at 2009-02-02. From that day onward, Ross is an associate professor, and this fact will remain true if Ross' title has not changed. Suppose the time granularity of that database is day. At 2009-03-01, the valid time of this fact should be [2009-02-02, 2009-03-01]. One day later (on 2009-02-03), the valid time of this fact should be changed to [2009-02-02, 2009-03-02]. It is quite a time consuming job to update all this information every time granularity. Introducing *Now* into temporal database can solve this problem. With reference to the instance mentioned above, we can record [2009-02-02, *Now*] as valid time.

However, *Now* also brings some problems. One problem is the semantics distortion of *Now*. With reference to the instance mentioned above, if the current time is 2009-03-01 and the query is "Who will be the associate professor on 2009-03-02", then the result will not include "Ross" because the valid time of "Ross is an associate professor" is [2009-02-02, 2009-03-01] (as 2009-03-01 is the current time). Actually Ross should be included since he will be an associate professor on 2009-03-02 but the temporal database has not recorded any changes of this fact. In this situation, the semantics of *Now* is distorted.

Another problem is that the start valid time point is later than the current time. For instance, Ross will be an associate professor at 2009-03-01 and the current time is 2009-02-18. This fact is to be inserted in the database. If the valid time of this fact is recorded as [2009-03-01, *Now*], then it will bring a contradiction during the period from 2009-02-18 to 2009-02-28 because the ending point of valid time exceeds the starting point of valid time if the variable *Now* is explained as the current time.

Besides, there are also some considerations for the query, insertion, update, and deletion transactions on facts with temporal variable *Now*. Readers may further refer to (Clifford et al. 1997; Ye and Tang 2005).

2. *UC*

UC is short for "until changed". A transaction time marker indicates that the associated fact remains part of the current database state. When a fact is inserted or updated in temporal database, the ending time point of transaction time is *UC*, which indicates that this fact is part of the database until changed. For example, "Ross is an associate professor" is inserted into the database at 2008-02-18 and this fact is not changed. Therefore, the ending point of the transaction time is *UC*. There are also some semantic distortion and transaction problems with *UC*.

4.1.5 An Illustration

In the sections above, we introduced user-defined time, valid time, and transaction time. Other times have also been proposed, such as the decision time (Kim and Chakravarthy 1994). However, latest researches of temporal database mainly consider the three time dimensions presented above. Out of these three dimensions, valid time and transaction time are studied most of the times.

The valid time dimension and the transaction time dimension record the temporal information of facts. We give an additional illustration to get a better understanding. Note that there are many data models like the relational data model, object data model. The subject of associating valid time and transaction time to data models has been widely studied. In this chapter, we concentrate on the semantics of valid time and transaction time. Thus, we take the widely used relational database as an example in the rest of this chapter.

A relational database supports two dimensions: the attribute dimension that refers to the attributes of the relation and the tuple dimension that refers to the records. Usually the attribute dimension is presented horizontally as shown in Fig. 4.1 and the tuple dimension vertically. Upon the two dimensions, the valid time dimension and transaction time dimension are added to construct a four -dimensional database. As for the concrete physical implementation, the valid time may be assigned to attribute(s) or the entire tuple. However, we focus on the conceptual layer. Therefore, in the rest of this chapter, we take an easy to understand way to assign valid time to the entire tuple.

Example 4.3 Figure 4.1 shows a faculty salary relation. The transaction time dimension goes from top to bottom. There are four tables. Each table is ordered by the transaction time and this shows the change in the state of database as time moves on. In the tables, the rows show the tuple dimension. The columns show the attribute dimension that consists of four attributes (Name, Birth date, Title, Salary). The following *Valid time* dimension indicates the logical valid duration of the associated fact (the corresponding state of a tuple).

Transaction time

2003-12-18

Name	Birth date	Title	Salary	Valid time
Alice	1960-09-25	Lecturer	1000	[2003-09, *Now*]
Bob	1962-11-11	Associate prof.	1600	[2003-03, *Now*]

(a) A snapshot at 2003 - 12 -18

2004-05-20

Name	Birth date	Title	Salary	Valid time
Alice	1960-09-25	Lecturer	1000	[2003-09, 2004-03]
			1300	[2004-04, *Now*]
Bob	1962-11-11	Associate prof.	1600	[2003-03, *Now*]

(b) A snapshot at 2004-05-20

2005-07-10

Name	Birth date	Title	Salary	Valid time
Alice	1960-09-25	Lecturer	1000	[2003-09, 2004-03]
			1300	[2004-04, *Now*]
Bob	1962-11-11	Associate prof.	1600	[2003-03, 2005-03]
		Professor	2200	[2005-04, *Now*]

(c) A snapshot at 2005-07-10

2006-02-20

Name	Birth date	Title	Salary	Valid time
Alice	1960-09-25	Lecturer	1000	[2003-09, 2004-03]
			1300	[2004-04, 2006-03]
		Associate prof.	1800	[2006-04, *Now*]
Bob	1962-11-11	Associate prof.	1600	[2003-03, 2005-03]
		Professor	2200	[2005-04, *Now*]

(d) A snapshot at 2006-02-20

Figure 4.1 An additional illustration of a faculty salary temporal relation

Birth date is a user-defined time. It is a temporal invariable since the birth date of a person may not change with time. Therefore, the valid time of *Birth date*

covers the time domain. Opposite to this, the *Title* and *Salary* attribute may vary as time moves on.

Figure 4.1(a) shows that at the time 2003-12-18, the fact that "Alice, a lecturer, was born at 1960-09-25 and earned 1000 dollars per month ever since September 2003 (we use 'Alice, 1960-09-25, Lecturer, 1000', for short)" was stored in the database. At the time this record was inserted (or updated), the user had no idea how long would the fact "Alice, 1960-09-25, Lecturer, 1000" last. Therefore, the user used a temporal variable *Now* to indicate that this fact happens until the present time and may remain true in the future. It seems unreasonable since Alice's title or salary may change. Hence, to capture the real state of the fact, the tuple needs frequent updates once Alice's title or salary is changed.

Figure 4.1(b) shows an **update** transaction. Alice got a salary increase at April 2004. The previous record's valid time is set to [2003-09, 2004-03], because "Alice, 1960-09-25, Lecturer, 1000" was no longer true after March 2004. Note that, by the logical meaning, the fact changed at April 2004. However, such "change" was not recorded into the database until 2004-05-20 when the update transaction occurred. 2004-05-20 is the transaction time. The fact "Alice, 1960-09-25, Lecturer, 1000" was recorded in database from 2003-12-18 to 2004-05-20. In other words, the valid time of "Alice, 1960-09-25, Lecturer, 1000" stored in database is [2003-12-18, 2004-05-20]. The tuple was updated. Therefore, the current state of database ever since 2004-05-20 tells the fact that "Alice, 1960-09-25, Lecturer, 1300 since 2004-04 to now".

Figure 4.1(c) shows that at 2005-07-10, an update occurred on Bob's title and salary: "Bob was an associate professor and earned 1600 dollars from 2003-03 to 2005-03", and "Bob is now a professor and earns 2200 dollars since 2005-04". Similarly, Fig. 4.1(d) shows that "Alice would be an associate professor and would earn 1800 dollars from 2006-04". The valid time of the database state shown in Fig. 4.1(d) starts from 2006-02-20, and lasts until changed, namely [2006-02-20, UC].

From the illustration we can see that both valid time and transaction time capture the temporal aspect of database states and they are important.

The valid time is widely used in applications. It tells us when the fact is valid. The transaction time captures the time varying states of a database. When updating a tuple, the old version(s) of that tuple is stored in database and is accessible. These old version(s) also provides important temporal information. For example, the user would like to know Alice's salary at 2005-05 according to the 2004-01-01 version of the database.

At first glance, it seems that the valid time and the transaction time may have intersections. For example, when we submit an update transaction to database, at that transaction time, the updated facts should be true, namely, the valid time of that fact contains the transaction time. In fact, we will see that the proposition above is incorrect. The valid time and the transaction time are two orthogonal dimensions. They cannot substitute each another. Take Example 4.3 for instance. As shown in Fig. 4.1(a), the fact "Alice, 1960-09-25, Lecturer, 1000" became valid at 2003-09. However, "Alice, 1960-09-25, Lecturer, 1000" is not necessarily stored in the database at 2003-09. Besides, as shown in Fig. 4.1(b), the fact "Alice, 1960-09-25, Lecturer, 1000" is not deleted when it becomes false, namely, at 2004-04. Even in the mini-world, the fact changes. It may still be stored in the database as long as it is not deleted. The valid time has no constraints on the transaction time.

Valid time needs transactions to change, and therefore the transaction generates transaction time. If there is no transaction performed, the change of valid time cannot be recorded in a database, hence the database cannot tell the fact. For example, the valid time of the fact "Alice, 1960-09-25, Lecturer, 1000" stops at 2004-03. However, in Fig. 4.1(a), the first tuple does not contain such information. The change of the valid time of "Alice, 1960-09-25, Lecturer, 1000" occurs in the transaction at 2004-05-20. Without committing such a transaction, the valid time of "Alice, 1960-09-25, Lecturer, 1000" is not changed in the database, even though "Alice, 1960-09-25, Lecturer, 1000" is already false in the mini-world.

4.2 Temporal Database Types

The combination of valid time and transaction time derives from the following temporal database types (Soo 1991; Jensen et al. 1992; Jensen 2000):

- **Snapshot database** that supports neither valid time nor transaction time;
- **Historical database** that supports only valid time;
- **Roll-back database** that supports only transaction time;
- **Bitemporal database** that supports both valid time and transaction time.

In the following, we introduce these four types of databases.

4.2.1 Snapshot Database

A snapshot database is the conventional database that supports neither valid time nor transaction time. Table 4.1 represents a conventional faculty salary relation. The snapshot relation only contains the attribute dimension and tuple dimension. It captures a certain state of a fact, although the fact is time-varying. The transition of database states is done by transactions. Once a transaction is done, the database state is changed and the previous state is deleted. Therefore, we cannot know the historical changes of the facts. The query, insert, update, and delete transactions can only be performed on the current state of the database.

Table 4.1 A faculty salary relation

Name	Birth date	Title	Salary
Alice	1960-09-25	Associate Prof.	1600
Bob	1962-11-11	Associate Prof.	1600

The snapshot database cannot record and handle temporal information. It cannot answer questions such as (see Table 4.1):

Question 1: Historical query that questions the facts' time-varying history from the mini-world, for example, "Was Alice a lecturer five years ago?" (Since in Table 4.1, Alice is now an associate professor)

Question 2: Rollback query that questions the state of database in some past time point, for example, "In 2005-9-20, what is Bob's title?" This is not stored in the database.

Generally, the facts recorded in the snapshot database are true in the mini-world. The relations concerning valid time, transaction time and the states of snapshot database are shown in Fig. 4.2. The cylindrical object represents a database state.

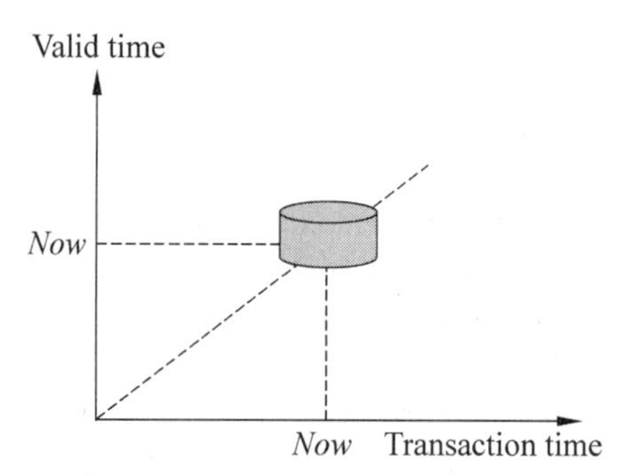

Figure 4.2 The state of snapshot relation concerning valid time and transaction time

4.2.2 Historical Database

The historical database supports only valid time. The historical relation supports three dimensions: the attribute dimension, the tuple dimension and the valid time dimension. The historical database records the facts' valid time varying information. An example is shown in Table 4.2(a). The historical database can tell the valid time information such as "Was Alice a lecturer five years ago".

Since the historical database does not support transaction time, all the transactions are performed on the current state. Once the state is changed, the old version of the state is replaced by the updated version, and the old version is no longer stored in the database.

Example 4.4 Alice would get a salary increase to 1200 since 2004-04. This information was recorded in the historical database. The new record was inserted as shown in Table 4.2(b). However, some days later, the financial department found that it was a mistake since Alice should get a salary increase to 1300. Hence, the new record was directly fixed on the second tuple and the result is as shown in Table 4.2(c). From Table 4.2(c), we can see the history of changes of Alice's salary. However, we have no idea of the history of states of the database, for example, which tuple has been updated or which tuple has been deleted.

Table 4.2 A faculty salary historical relation

(a) A faculty salary historical relation

Name	Birth date	Title	Salary	Valid time
Alice	1960-09-25	Lecturer	1000	[2003-09, *Now*]
Bob	1962-11-11	Associate prof.	1600	[2003-03, *Now*]

(b) Alice gets a salary increase

Name	Birth date	Title	Salary	Valid time
Alice	1960-09-25	Lecturer	1000	[2003-09, 2004-03]
Alice	1960-09-25	Lecturer	1200	[2004-04, *Now*]
Bob	1962-11-11	Associate prof.	1600	[2003-03, *Now*]

(c) Alice's salary is fixed

Name	Birth date	Title	Salary	Valid time
Alice	1960-09-25	Lecturer	1000	[2003-09, 2004-03]
Alice	1960-09-25	Lecturer	1300	[2004-04, *Now*]
Bob	1962-11-11	Associate prof.	1600	[2003-03, *Now*]

The historical database states concerning valid time and transaction time are shown in Fig. 4.3. The valid time may vary, but the transaction time stays at the *Now* time point.

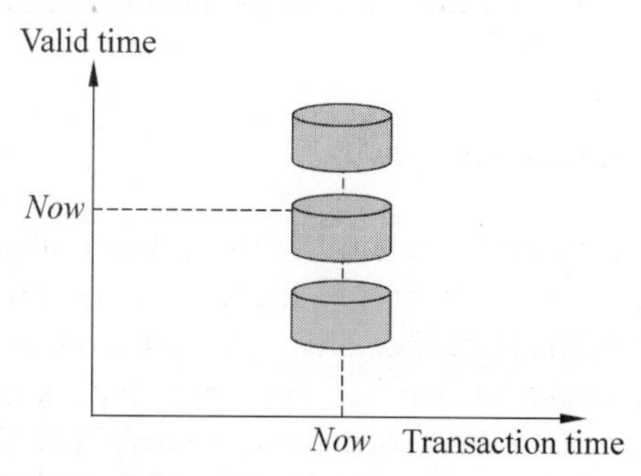

Figure 4.3 The states of the historical database concerning valid time and transaction time

Figure 4.4 shows a detailed view in a coordinate system. The x-axis, y-axis and z-axis represent the attribute dimension, tuple dimension, and valid time dimension, respectively. The z-axis starts from a certain time that is acceptable by the database system and application, and extends to *Now*. T_1, T_2, $\cdots$, T_6 may be time points or periods. If we perpendicularly slice the z-axis, the cross-section that is parallel to the x-y-plane shows the ⟨attribute, tuple value⟩ pair, which are true at that time point or during that period. For example, if T_2 represents the period [2004-01, 2004-02], as for the relation in Table 4.2(b), ⟨Name, Alice⟩, ⟨Birth Date, 1960-9-25⟩, ⟨Title, Lecturer⟩, ⟨Salary, 1000⟩, ⟨Name, Bob⟩, ⋯, ⟨Salary, 1600⟩ are true. If we pick one point in the x-y-plane (which means a concrete value for a concrete attribute, for example, Alice's salary is 1000), and watch the line cross this point along the z-axis, we can see the valid times for this value.

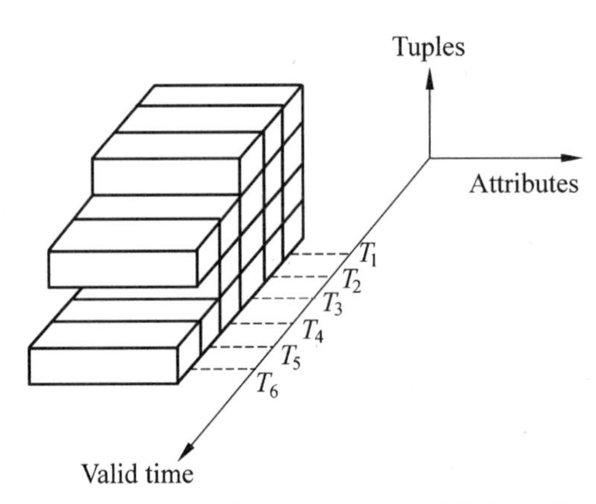

Figure 4.4 Relation of attribute, tuple and valid time dimension in a coordinate system

Historical database associates valid time to attribute(s). It can reflect the time-varying changes of a fact in the mini-world. Note that there is data redundancy. For example, in Table 4.2(b), the attribute value "Alice, 1960-9-25, Lecturer" appears in multiple tuples. Besides, it cannot record the historical state of the database.

4.2.3 Rollback Database

The rollback database supports transaction time. The rollback relation supports three dimensions: the attribute dimension, the tuple dimension, and the transaction time dimension. The rollback database records the historical states of a database. Any historical state could be queried. However, the modification transaction could only be performed on the current state, namely: Let S be the current state of database, and T be a transaction that may modify S, for example, an update transaction. T should only be performed on S. Let S' be the derived state from S

by T. S' is newly added to the database, and S remains unchanged in the database. After committing T, the current state turns to S'. In other words, once the state is changed, the updated version of the state is newly added to the database, while the old version of the state remains unchanged. Besides, not all the old states can be changed, even if there is an error in the old state. The remedy is to create a new correct state. Similarly, the data cannot be deleted. If we want to delete a record, the only way is to commit a transaction to set the ending point of the transaction time to the transaction committing time.

Example 4.5 Table 4.3(a) is a faculty salary rollback relation. The transaction time item represents the duration when this fact was stored in the database. [2001-12-01, *UC*] means that the fact "Alice, 1960-09-25, lecturer, 1000" was stored in database since 2001-12-01, and has not been modified since then. Alice got a salary increase and the new fact was stored in the database at 2004-01-01. Different from the snapshot database, the new fact cannot be directly modified on the current state. Opposite to this, it was inserted in the database as shown in Table 4.3(b). It indicates that the fact "Alice, 1960-09-25, lecturer, 1000" was

Table 4.3 A faculty salary rollback relation

(a) A faculty salary rollback relation

Name	**Birth date**	**Title**	**Salary**	**Transaction time**
Alice	1960-09-25	Lecturer	1000	[2001-12-01, *UC*]
Bob	1962-11-11	Associate prof.	1600	[2001-08-30, *UC*]

(b) Alice gets a salary increase

Name	**Birth date**	**Title**	**Salary**	**Transaction time**
Alice	1960-09-25	Lecturer	1000	[2001-12-01, 2004-01-01]
Alice	1960-09-25	Lecturer	1200	[2004-01-01, *UC*]
Bob	1962-11-11	Associate prof.	1600	[2001-08-30, *UC*]

(c) Alice's salary is fixed

Name	**Birth date**	**Title**	**Salary**	**Transaction time**
Alice	1960-09-25	Lecturer	1000	[2001-12-01, 2004-01-01]
Alice	1960-09-25	Lecturer	1200	[2004-01-01, 2004-01-02]
Alice	1960-09-25	Lecturer	1300	[2004-01-02, *UC*]
Bob	1962-11-11	Associate prof.	1600	[2001-08-30, *UC*]

(d) Bob's record is deleted

Name	**Birth date**	**Title**	**Salary**	**Transaction time**
Alice	1960-09-25	Lecturer	1000	[2001-12-01, 2004-01-01]
Alice	1960-09-25	Lecturer	1200	[2004-01-01, 2004-01-02]
Alice	1960-09-25	Lecturer	1300	[2004-01-02, *UC*]
Bob	1962-11-11	Associate prof.	1600	[2001-08-30, 2005-02-05]

stored in the database from 2001-12-01 to 2004-01-01. Logically, the fact is no longer in the database. However, it is physically stored in the database. The next day, the financial department found that it was a mistake and corrected it, as shown in Table 4.3(c). At 2005-02-05, a transaction was committed to delete Bob's record. Actually, the record is not deleted from the database. Instead, its transaction ending time is set to the transaction occurring time, as shown in Table 4.3(d), which indicates that this record is logically deleted. With transaction time, the rollback database can tell the historical state of the database like "At 2005-08-04, what was Alice's salary as stored in the database?"

The rollback database states concerning valid time and transaction time are shown in Fig. 4.5. The database contains a series of states in chronological order.

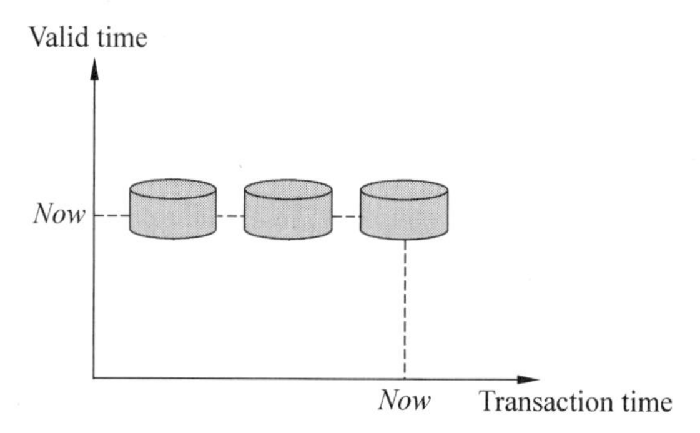

Figure 4.5 The states of the transaction database concerning valid time and transaction time

Figure 4.6 shows a more detailed view in a coordinate system. The x-axis, y-axis and z-axis represent the attribute dimension, tuple dimension and transaction time dimension, respectively. T_1, T_2, T_3, and T_4 are time points on z-axis. $t_1, t_2, \cdots$ are tuples. If we perpendicularly slice the z-axis, the cross-section, which is parallel to the x-y-plane, shows the tuples that are stored in the database at that time point. For example, at T_2, the tuples t_1, t_2 and t_3 are stored at the database, and at T_4 there are t_1, t_2, t_3 and t_5. If we pick one point in the x-y-plane (which means a concrete value for a concrete attribute, for example, Alice's salary is 1000), and watch the line cross this point along the z-axis, we can see the transaction information of this fact, i.e., when this fact is stored in the database.

The rollback database cannot tell the valid time information in the mini-world. The old states can only be queried. Any modification transaction should be performed on the current state. A derived state should be added into the database and the old states should not be modified or deleted, in spite of a tiny modification. Thus, the transaction database may contain a huge redundancy. Besides, since the transaction database provides only "logical delete" and not the "physical delete" transaction, its space would expand and never shrink.

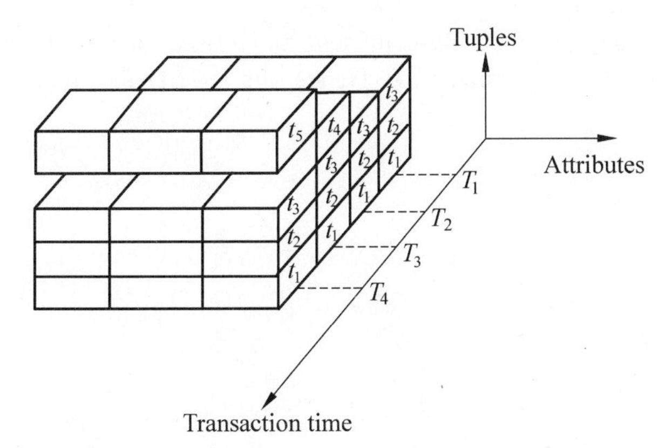

Figure 4.6 The relation of attribute, tuple and transaction time dimension in a coordinate system

4.2.4 Bitemporal Database

The Bitemporal database inherits both historical database and transaction database. The bitemporal relation supports four dimensions: attribute dimension, tuple dimension, valid time dimension, and transaction time dimension. A bitemporal relation can be viewed as a series of historical relations. Actually, Example 4.3 in Section 4.1.5 shows a bitemporal relation illustration. *Now*, we give another illustration from the view of combining valid time and transaction time.

Example 4.6 An illustration of bitemporal relation concerning the valid time and the transaction time dimensions is shown in Fig. 4.7. The *x*-axis represents transaction time and *y*-axis represents valid time. $T_1, T_2, \cdots$ are transaction times and $t_1, t_2, \cdots$ are time points or periods. The coordinate shows the historical states of the time-varying information of the modeled fact "Alice gets a salary of 1000". At transaction time T_1, the fact is true at t_1, t_2, t_3. At T_2, another transaction is committed but it does not modify the fact's valid time, hence its valid time remains at t_1, t_2, t_3. At T_3, the valid time is modified to t_1, t_3.

Valid time			
	t_3	t_3	t_3
	t_2	t_2	
	t_1	t_1	t_1
	T_1	T_2	T_3 Transaction time

Figure 4.7 An illustration of the valid time and the transaction time of a fact

With valid time and transaction time, the database records the facts' time-varying information in the mini-world and its historical changes as well. The bitemporal states concerning valid time and transaction time are shown in Fig. 4.8. Note that the valid time may span the past, present and future, but the transaction time cannot exceed the current time *Now*. Hence, the state cannot transcend the time point *Now*.

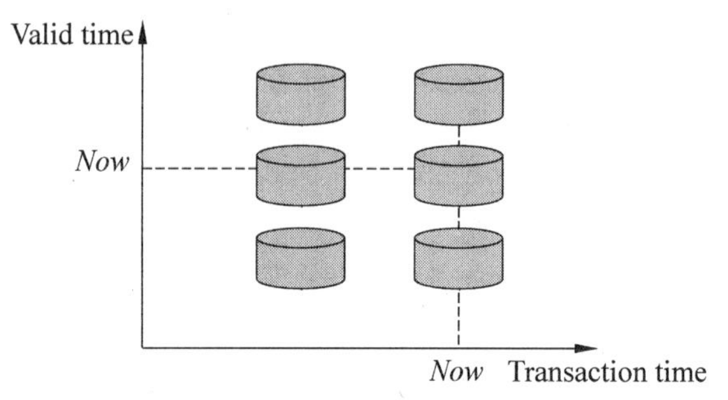

Figure 4.8 The bitemporal states concerning the valid time and the transaction time

Bitemporal database can answer both the historical queries and the rollback queries, and the combination of the two, such as "What is Alice's salary at May 2005 recorded in database at 2005-03-05".

Since the bitemporal database inherits the transaction database, all the new states are inserted, and the old states are preserved in the database. Hence, the huge redundancy is also a fatal disadvantage to the bitemporal database.

4.3 Temporal Data Models

The temporal data models describe how to associate temporal aspects to data entities. From different conventional data models, several temporal data models have been derived. Roughly, the temporal data models can be further divided into subcategories: temporal data model based on relational data model, temporal data model based on entity-relationship model, temporal data model based on semi-structure data model, and temporal data model based object-oriented data model. In the following, we mainly introduce three temporal data models: **bitemporal time stamps** based on relational data model, **BCDM** (bitemporal conceptual data model) as a conceptual data model and **TERM** based on ER model.

4.3.1 Bitemporal Time Stamps

As discussed above, the bitemporal relation contains large redundancy. More specifically, let $A_1, A_2, \cdots, A_n$ be the usual attributes, V be the valid time and T be the transaction time. Then, the bitemporal relation can be $A_1 \times A_2 \times \cdots \times A_n \times V \times T$. Obviously, many records only differ at $T \times V$ and they have the same values projected on $A_1 \times A_2 \times \cdots \times A_n$. As in Example 4.3, "Alice, 1960-09-25, Lecturer,

1000" appears four times (at each snapshot). Hence, we should devise a compact and efficient storage mechanism.

One possible optimization is to group the discrete transaction times by the records with the same attribute values and valid time value. Table 4.4 shows a storage method for Example 4.3. Transaction time is the time when a transaction is done to the tuple at the current state, although the transaction has not modified its value. For example, at 2005-07-10 Bob's title and salary are modified. Since the tuple "Alice, 1960-09-25, Lecturer, 1000, [2003-09, 2004-03]" is at the current state when the transaction is done, the transaction time is recorded, though that tuple is not modified. In Fig. 4.1 there are totally 14 records in the four snapshots, while in Table 4.4 there are 8 records, which is less, especially if the usual attribute value is large, for example, an image type. However, there are still some records that are the same when projected on the usual attributes.

Table 4.4 A way to store bitemporal data

Name	Birth date	Title	Salary	Valid time	Transaction time
Alice	1960-09-25	Lecturer	1000	[2003-09, *Now*]	2003-12-18
Alice	1960-09-25	Lecturer	1000	[2003-09, 2004-03]	2004-05-20, 2005-07-10, 2006-02-20
Alice	1960-09-25	Lecturer	1300	[2004-04, *Now*]	2004-05-20, 2005-07-10
Alice	1960-09-25	Lecturer	1300	[2004-04, 2006-03]	2006-02-20
Alice	1960-09-25	Associate prof.	1800	[2006-04, *Now*]	2006-02-20
Bob	1962-11-11	Associate prof.	1600	[2003-03, *Now*]	2003-12-18, 2004-05-20
Bob	1962-11-11	Associate prof.	1600	[2003-03, 2005-03]	2003-07-10, 2006-02-20
Bob	1962-11-11	Professor	2200	[2005-04, *Now*]	2005-07-10, 2006-02-20

Many works have been done on the storage organization of bitemporal data (Segev and Shoshani 1988; Rose and Segev 1991; Jensen et al. 1994; Worboys 1994; Jensen and Snodgrass 1996; Tryfona and Jensen 1999). In the following sections, we present a bitemporal time stamping way for relational temporal data model (Jensen and Snodgrass 1996). The idea is to attach bitemporal time stamp(s) to each tuple in a relation consisting of usual attributes.

Definition 4.1 (Bitemporal time stamp) Bitemporal time stamp is pair $\langle TT, VT \rangle$, where *TT* is the transaction time and *VT* is the valid time. Usually *TT* is a time point. *VT* can be time point and period. The granularities of *TT* and *VT* depend on the application.

With bitemporal time stamp, the bitemporal relation in Example 4.3 can be represented in Table 4.5.

Table 4.5 Bitemporal time stamp representation of a faculty salary bitemporal relation

Name	Birth date	Title	Salary	Bitemporal label
Alice	1960-09-25	Lecturer	1000	⟨2003-12-18, [2003-09, *Now*]⟩ ⟨2004-05-20, [2003-09, 2004-03]⟩ ⟨2005-07-10, [2003-09, 2004-03]⟩ ⟨2006-02-20, [2003-09, 2004-03]⟩
Alice	1960-09-25	Lecturer	1300	⟨2004-05-20, [2004-04, *Now*]⟩ ⟨2005-07-10, [2004-04, *Now*]⟩ ⟨2006-02-20, [2004-04, 2006-03]⟩
Alice	1960-09-25	Associate prof.	1800	⟨2006-02-20, [2006-04, *Now*]⟩
Bob	1962-11-11	Associate prof.	1600	⟨2003-12-18, [2006-04, *Now*]⟩ ⟨2004-05-20, [2006-04, *Now*]⟩ ⟨2005-07-10, [2003-03, 2005-03]⟩ ⟨2006-02-20, [2003-03, 2005-03]⟩
Bob	1962-11-11	Professor	2200	⟨2005-07-10, [2005-04, *Now*]⟩ ⟨2006-02-20, [2005-04, *Now*]⟩

Table 4.6 is an improved bitemporal time stamp representation. The tuple at the current state is associated with a bitemporal time stamp ⟨*UC*, *VT*⟩ indicating that the tuple is at the current state of database. If at 2003-01-01, we delete the third tuple "Alice, 1960-09-25, Associate Prof., 1800", then the bitemporal time stamp should be "⟨2006-02-20, [2006-04, *Now*]⟩, ⟨2003-01-01, [2006-04, *Now*]⟩", which indicates that the tuple is logically deleted.

Table 4.6 Improved bitemporal time stamp representation of a faculty salary bitemporal relation

Name	Birth date	Title	Salary	Bitemporal label
Alice	1960-09-25	Lecturer	1000	⟨2003-12-18, [2003-09, *Now*]⟩ ⟨2004-05-20, [2003-09, 2004-03]⟩ ⟨2005-07-10, [2003-09, 2004-03]⟩ ⟨2006-02-20, [2003-09, 2004-03]⟩ ⟨*UC*, [2003-09, 2004-03]⟩
Alice	1960-09-25	Lecturer	1300	⟨2004-05-20, [2004-04, *Now*]⟩ ⟨2005-07-10, [2004-04, *Now*]⟩ ⟨2006-02-20, [2004-04, 2006-03]⟩ ⟨*UC*, [2004-04, 2006-03]⟩
Alice	1960-09-25	Associate prof.	1800	⟨2006-02-20, [2006-04, *Now*]⟩ ⟨*UC*, [2006-04, *Now*]⟩
Bob	1962-11-11	Associate prof.	1600	⟨2003-12-18, [2006-04, *Now*]⟩ ⟨2004-05-20, [2006-04, *Now*]⟩ ⟨2005-07-10, [2003-03, 2005-03]⟩ ⟨2006-02-20, [2003-03, 2005-03]⟩ ⟨*UC*, [2003-03, 2005-03]⟩
Bob	1962-11-11	Professor	2200	⟨2005-07-10, [2005-04, *Now*]⟩ ⟨2006-02-20, [2005-04, *Now*]⟩ ⟨*UC*, [2005-04, *Now*]⟩

Figure 4.9 shows the bitemporal information change of Alice's title and salary with reference to Table 4.6. Note that the birth date attribute is a temporal invariant. It is not presented in the figure.

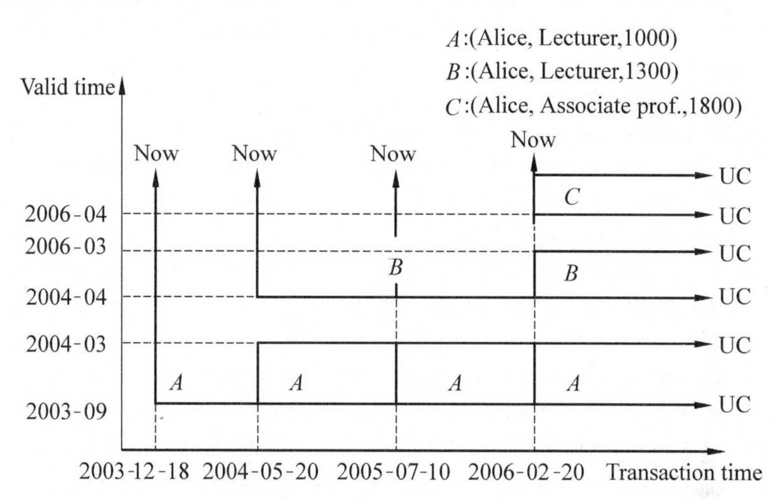

Figure 4.9 The bitemporal information of Alice's title and salary

With bitemporal time stamp representation, the bitemporal information can be preserved. Each tuple's usual attribute value is recorded once. The redundancy can be reduced.

4.3.2 BCDM

BCDM was proposed by Jensen and the others (Jensen et al. 1994). After studying five proposed bitemporal time stamp data models, they proposed that there are many requirements for temporal data models, and each designed temporal data model satisfies a subset of these requirements. However, none of these models is ideal to satisfy all the requirements. Hence, they tried not to propose a brand new temporal data model. Instead, they tried to consider the temporal semantics-related information only. BCDM is a conceptual layer of bitemporal data model. The idea of BCDM is to capture the temporal semantics of facts in database, while making this model as simple as possible. They propose using the snapshot equivalent relation to map the temporal operator and temporal relation entities from one model into equivalent ones from another model. They studied five bitemporal data models and their relation mappings with BCDM. BCDM eliminates the differences of presentations and physical storage mechanisms from these five bitemporal time stamp data models. By mappings, the presentation and physical storage of temporal data can be implemented by different temporal data models. Hence, suitable temporal data models can be applied for different requirements.

4.3.3 Temporal Entity-Relationship Data Model

The entity-relationship data model (ER model) has become increasingly popular in many aspects of computer applications. Hence, many temporal entity-relationship data models have been proposed for the past few decades. The first temporal ER model, TERM, is proposed by M. R. Klopprogge (Klopprogge 1981). After that, many other temporal ER models have emerged, such as Enhanced Entity Relationship model, Semantic Temporal EER model (Elmasri and Wuu 1990), Entity-Relation-Time model (Theodoulidis et al. 1991), Temporal EER model (Batra et al. 1990). In this section, we briefly introduce the original temporal entity-relationship data model TERM.

TERM, the Temporal Entity-Relationship Model, is the first model to temporally extend the ER model. Its main motivation, by the TERM's author, is "to provide database designers with a model for data definition and data manipulation that allows a general and rigorous treatment of time" (Klopprogge 1981).

TERM introduces the notion of a history, which is a function from a time domain to a value domain. With this concept, histories are used for modeling time-varying aspects. Then, the values of attributes are not just simple values any longer. They become history.

In TERM, the representation of time is distinguished from the real time. Each real time concept can have many different representations.

Domains are termed *structures*. For example, in Fig. 4.10, we first define two structures, str_name and int_ID, and then define the time concept "month". Finally, we define a relation "before_month".

```
Structure
   str_name = packed array [1..30] of char
Structure
   int_ID = integer
Structure
   month =
        record y,m:integer end
             where
                  this.y > 1600 and
                  this.m >= 1 and this.m <=12;
Relations
   function before_month (m1,m2:month): boolean;
   begin
        before_month := m1.y < m2.y or
                          (m1.y = m2.y and m1.m < m2.m);
   end
```

Figure 4.10 Simple values and time structure in TERM

Attributes of entities and roles of relationships in TERM have "atomic" histories while the entire entities have "composite" histories. Composite histories are sets of histories. An entity history is composed of an existence history of the entity itself and the set of all its attributes' histories.

Having the time definition and entity history, we could further consider the association of histories with time-varying items. The construction elements of TERM are there in ER model. Entities model the objects from the world and values model the properties of the objects. The values are associated with the entities by attributes. If an attribute never changes its value once it is assigned, it is said to be a constant attribute, otherwise it is a time variable. Constant attributes are represented by tuple (attribute, value) and variable attributes are represented by tuple (attribute, history).

Schema in TERM consists of a set of entity type definitions and a set of relationship type definitions. Figure 4.11 shows an entity example, in which "standard_existence" is a structure representing a continuous period of time. To keep the book brief, its definition is not given here.

```
Entity Type
   Student
   existence
        variable
             standard_existence
   attributes
        Stuednt_Num constant int_ID
        Height constant real;
```

Figure 4.11 Entity definition in TERM

There is a four-step bottom-up procedure for designing TERM schemas. Firstly, define all the needed component values, as shown in Fig. 4.10. The second step is to define history. The third step is to define patterns. A pattern is a value structure together with at least one assertion, at most one derivation function and zero or more approximation functions. The final step is to define entity and relationship types as shown in Fig. 4.11.

TERM has a Pascal-like syntax. Designers could use TERM to devise a model with temporal aspects by using the history structures as values of attributes and relationship-type roles.

4.4 Difference from Real-Time Database

Temporal database manages temporal information. Meanwhile, real-time database (Bihari et al. 1992; Kao and Garcia-Molina 1994; Purimetla et al. 1995) is also highly concerned with time. In real-time database, many transactions have deadlines or time constraints, for example, "This transaction must be done within two minutes". Real-time database is usually employed in the application with strong transaction temporal constraints, such as the stock market, banking and monitoring devices. A real-time system's temporal characteristics can be mainly categorized into two types: ① state and behavior required of the system (e.g. constraints on process deadlines), ② state and behavior exhibited by the system (e.g. process execution time) (Bihari et al. 1992). Though temporal database and real-time database both closely consider temporal information, they have differences and emphasize on different areas (Ozsoyoglu and Snodgrass 1995).

Valid time is mainly considered in temporal database. Though real-time database does not explicitly distinguish among various time dimensions, it also concerns valid time. Valid time is used for data items that are closely monitored, and the transactions record the changes of value caused by external events, since the real-time system uses the most recent data item values. In that sense, the transaction time can almost stand for the valid time.

Transaction time in temporal database provides information related to historical database state. In real-time database, transaction time is a tight constraint for transactions. Transactions must be done before their deadlines and should satisfy transaction constraints, especially for hard transactions.

So far, various temporal data models have been proposed for temporal database. Besides, there are many temporal languages and some of them are well developed, such as TSQL2 (Snodgrass et al. 1994). Some real-time data models (Prichard et al. 1994; Selic et al. 1994) and languages (Jahanian and Mok 1994) are proposed for real-time databases also. Although some work on real-time database discusses temporal data and temporal consistency constraints, a few of them utilize temporal data models and temporal query languages. Some researchers argue that the temporal research achievements can be well applied on real-time database research, since the real-time database also considers temporal data semantics (Ozsoyoglu and Snodgrass 1995). Some temporal research achievements have been applied to the area of real-time database (Koymans 1990; Ramamritham et al. 1996).

References

[1] Batra D, Hoffler J, et al. (**1990**) Comparing representations with relational and EER models. Communications of the ACM **33**(2): 126 – 139

[2] Bihari T, Gopinath P, et al. (**1992**) Object-oriented real-time systems: Concepts and examples. Computer **25**(12): 25 – 32

[3] Clifford J, Dyreson C, et al. (**1997**) On the semantics of "now" in databases. ACM Transactions on Database Systems (TODS) **22**(2): 171 – 214

[4] Elmasri R, Wuu G (**1990**) A temporal model and query language for ER databases. In: Proceedings of the 6th International Conference on Data Engineering, pp 76 – 85

[5] Jahanian F, Mok A (**1994**) Modechart: a specification language for real-time systems. IEEE Transactions on Software Engineering **20**(12): 933 – 947

[6] Jensen C (**2000**) Temporal database management. Department of Computer Science, Aalborg University

[7] Jensen C, Clifford J, et al. (**1992**) A glossary of temporal database concepts. ACM sigmod Record **21**(3): 35 – 43
[8] Jensen C, Snodgrass R (**1996**) Semantics of time-varying information. Information Systems **21**(4): 311 – 352
[9] Jensen C, Soo M, et al. (**1994**) Unifying temporal data models via a conceptual model. Information Systems-Oxford-Pergamon Press **19**: 513 – 513
[10] Kao B, Garcia-Molina H (**1994**) An overview of real-time database systems. Nato Asi Series F Computer and Systems Sciences **127**: 261 – 261
[11] Kim S, Chakravarthy S (**1994**) Modeling time: adequacy of three distinct time concepts for temporal databases. Lecture Notes in Computer Science 823: 475 – 475
[12] Klopprogge M (**1981**) TERM: an approach to include time dimension in the Entity-Relationship Model. North-Holland Publishing Co. Amsterdam, The Netherlands
[13] Koymans R (**1990**) Specifying real-time properties with metric temporal logic. Real-Time Systems **2**(4): 255 – 299
[14] Ozsoyoglu G, Snodgrass R (**1995**) Temporal and real-time databases: A survey. IEEE Transactions on Knowledge and Data Engineering **7**(4): 513 – 532
[15] Prichard J, DiPippo L, et al. (**1994**) RTSORAC: a real-time object-oriented database model. Lecture Notes in Computer Science **856** (601 – 610): 326
[16] Purimetla B, Sivasankaran R, et al. (**1995**) Real-time databases: issues and applications. Advances in Real-Time Systems: 487 – 507
[17] Ramamritham K, Sivasankaran R, et al. (**1996**) Integrating temporal, real-time, an active databases. ACM SIGMOD Record **25**(1): 8 – 12
[18] Rose E, Segev A (**1991**) TOODM: a temporal object-oriented data model with temporal constraints. In: Proceedings of the 10th International Conference on Entity Relationship Approach, pp 205 – 229
[19] Segev A, Shoshani A (**1988**) The representation of a temporal data model in the relational environment. Lecture Notes in Computer Science 339, Springer, pp 39 – 61
[20] Selic B, Gullekson G, et al. (**1994**) Real-time object-oriented modeling, John Wiley & Sons, Inc. New York, NY, USA
[21] Snodgrass R, Ahn I, et al. (**1994**) TSQL2 language specification. ACM SIGMOD Record **23**(1): 65 – 86
[22] Soo M (**1991**) Bibliography on temporal databases. ACM SIGMOD Record **20**(1): 14 – 23
[23] Theodoulidis C, Loucopoulos P, et al. (**1991**) A conceptual modelling formalism for temporal database applications. Information Systems **16**(4): 401 – 416
[24] Tryfona N, Jensen C (**1999**) Conceptual data modeling for spatiotemporal applications. Geoinformatica **3**(3): 245 – 268
[25] Worboys M (**1994**) A unified model for spatial and temporal information. The Computer Journal **37**(1): 26 – 34
[26] Ye X, Tang Y (**2005**) Semantics on "*Now*" and calculus on temporal relations (in Chinese). Journal of Software **16**(5): 838 – 845

5 Spatio-Temporal Data Model and Spatio-Temporal Databases

Xiaoping Ye[1,2], Zewu Peng[2], and Huan Guo[2]

[1] Computer School, South China Normal University, Guangzhou 510631, P.R. China
[2] Department of Computer Science, Sun Yat-sen University, Guangzhou 510275, P.R. China

Abstract Spatio-temporal database can deal with the objects' temporal and spatial properties uniformly, and it can manage data that is evolving with time, i.e., the information of spatial objects whose shape and position evolve with time. In spatio-temporal database, spatial data has one more time dimension, which increases the complexity of data management. In this chapter, we first introduce the concepts of spatio-temporal database, and then introduce the spatio-temporal data model, query types of spatio-temporal data and the architecture of spatio-temporal database system.

Keywords *spatio-temporal database, spatio-temporal data model, query types of spatio-temporal data, architecture of spatio-temporal database system*

5.1 Introduction

Spatio-temporal database is the combination of temporal database and spatial database. Its basic idea is to create a three-dimensional or four-dimensional database by adding time constraints for spatial database, which is mainly used to store and manage various types of spatial objects with temporal information. The main purpose of spatio-temporal database is to deal with spatio-temporal information, which usually involves the expression of spatio-temporal objects, its modeling, indexing and query of spatio-temporal data, the system structure, prototype systems and application of spatio-temporal database, and so on.

Spatio-temporal database can support complex applications with **spatio-temporal objects** and their relationships by the database core, so it has a wide range of applications. According to the types of spatio-temporal objects, the application of spatio-temporal database can be divided into the following categories:

- The application of spatio-temporal objects with continuous movement: In such

mcsyxp@mail.sysu.edu.cn

applications, the location of spatio-temporal objects continuously changes with time, but their shapes remain unchanged. The traffic-related spatio-temporal database applications, such as vehicle traffic management, vessel traffic management and aircraft flight management, can be classified as such applications.

- The application of spatio-temporal objects with discrete change: The spatio-temporal objects involved in this application have a spatial location, and their spatial attributes, such as shape and location, may change discretely over time. Such applications include cadastral management, city zoning management and surface vegetation changes, as well as virus, disease region detection and so on.
- The application of spatio-temporal objects with continuous movement and shape changing: Such applications usually refer to environment-related applications, such as storm monitoring and prediction, forest-fire monitoring, offshore oil pollution monitoring, as well as migration of species and so on. In addition, the biological information processing also belongs to such applications, for example, a cell's shape may change during its moving process.

5.2 Spatio-Temporal Data Model

In order to devise a spatio-temporal data model, we first discuss the basics of spatio-temporal objects.

5.2.1 Spatio-Temporal Object

The spatial objects whose spatial locations or extents change with time are called spatio-temporal objects, for example, flying aircraft, highway vehicles, the forest whose boundaries change (broadened or narrowed) over time and so on. Many practical applications need effective storing and management of spatio-temporal objects to query the objects' locations and their information in the past, at present, and to speculate their behavior in the future.

Spatio-temporal databases store time-varying spatial objects, and there are three kinds of basic spatial objects: points, lines and regions. Points represent the spatial objects whose spatial extension are zero or the ones for which we only concern on their spatial locations rather than their sizes, for example, a city in a large-scale map. Lines, including the space curves, are used to describe the facilities' traces in space or the space connection, such as roads, rivers, telephone lines, wires. Objects with certain space range are called regions, for example, a country's administration regional divisions, a mountain or lake and so on. Currently, the study of spatio-temporal data objects is mainly focused on two basic cases: moving points and moving regions and the study of line (curve) is usually converted to the study of moving point's trajectory.

A spatio-temporal object O_i can be expressed by tuple $\langle oid, p_i(t), e_i(t), t \rangle$, where *oid* is object O_i's unique identifier, $p_i(t)$ and $e_i(t)$ correspond to the location and extent of object O_i in time t, respectively. The application of spatio-temporal database is usually to insert and delete objects at any time. The time period from an object's insert time to its delete time is called the object's survival time. During the survival time, the object is alive. Spatio-temporal database's state $S(t)$ in time stamp t includes all activity objects in time t, and $S(t)$ can be seen as the snapshot of spatio-temporal object's spatial location and extent in time t. Spatio-temporal databases store time-varying spatial objects, so it needs to preserve all of the spatio-temporal statuses $S(t)$, and can query on any time's state $S(t)$. This means that in spatio-temporal database, the deletion of spatio-temporal data object is only a logical deletion. The records of deleted objects are still preserved in the database.

Spatio-temporal database can be seen as spatial database's extension in the time dimension, but the time dimension itself has the following characteristics:

- Monotony increment: Time information's change is always monotone increasing.
- Two time dimensions: Temporal information of data object usually has two time dimensions—valid time and transaction time. Spatio-temporal database needs to support valid time or transaction time or both.
- Discrete and continuous: Spatial objects always change with time evolution. Its representation method in database is related to its changing frequency and can be generally divided into two cases: discrete and continuous. In the discrete case, the changing frequency of spatial objects is slow. We can use the snapshot of spatio-temporal state to express spatio-temporal data and update data when the spatio-temporal object is changed. In the continuous case, the changing frequency of spatial objects is rapid. The corresponding spatial information change can be expressed as a location function $p_i(t)$ and scope function $e_i(t)$, respectively. t is variable time and an update only happens when corresponding time function changes.

5.2.2 Basic Considerations of Spatio-Temporal Modeling

At present, there are no uniform modeling standards for spatio-temporal data. Most of them are driven by specific applications and are devised to solve a specific problem. Judging from the current work, there are three major phases for spatio-temporal data modeling. First, determine the smallest-scale GIS unit in certain scope. Second, devise a reasonable, accurate and tight spatio-temporal relationship expression. Third, determine the model's layered architecture.

1. Spatio-temporal data information unit

Spatio-temporal information belongs to multi-dimensional data, which generally can be divided into space dimension, time dimension and attribute dimension.

- The basic element of space dimension is the spatial coordinates and it determines the spatial relationship between spatio-temporal entities (adjacent, intersection, separation and contains, etc.). For example, arbitrary point on the earth is uniquely determined by the spatial coordinates (X, Y, Z), where X, Y and Z are the longitude, latitude and altitude coordinates, respectively. In order to facilitate the query, the entire earth's surface can be cut into a number of relatively independent parts by the projection. The corresponding relationship can also be transformed from three-dimensional space to two-dimensional plane.
- Time dimension, as important data information, describes the creating, developing and deleting of spatio-temporal entities, and it can be seen as the life cycle of spatio-temporal entity or as a valid time or version information.
- Properties in attribute dimension can be divided into two types: core property and non-core property. The core property is the only identifier of the spatio-temporal entities or spatio-temporal process. The changes of the core property imply the deleting of the spatio-temporal objects (entity or process).

Only when studied with a certain scope, spatio-temporal information can be meaningful. In spatio-temporal modeling process, it is necessary to determine the basic unit in the spatio-temporal system research. There is a "constant" relationship between space and process, which is the smallest unit of spatio-temporal information: spatial and temporal information unit. As the unit of the smallest "space" with the shortest "time", spatio-temporal information unit is the appropriate "granularity" to study this problem, and it can further guarantee the "heterogeneity" of space and the "indivisibility" of time. The spatio-temporal information unit can be seen as same unit in the spatio-temporal system analysis.

The key to characterize the spatio-temporal information unit is through a suitable spatio-temporal data model. With rich, accurate expressive power, it can describe the complex, dynamic and interrelated spatio-temporal information unit. Theoretically, spatial and temporal information unit can be abstracted as the following function: $\mathrm{STIC} = F(x, y, z, A, T)$. STIC (spatial temporal Information cell) is the temporal and spatial information unit, where, x, y, z are the coordinates of spatial location, A is the set of properties and T is time.

2. Description of spatio-temporal relations

Study on spatio-temporal data entities is in fact the study on spatio-temporal relations between the information units, whose spatial relationship is the topological relationship, and time relationship is the temporal relationship. With reference to spatial databases, Egenhofer describes topological relations of two-dimensional space as {adjacency, contained, including, covering, covered, overlap, separation} (Egenhofer 1991). With reference to temporal databases, Allen divided the time relationship into {equal, before, meet, overlap, start, period, end} (Allen 2005). In spatio-temporal database, it is necessary to consider spatio-temporal relationship unitedly. Therefore, the spatio-temporal relationships that may exist between spatio-temporal data elements are shown in Fig. 5.1.

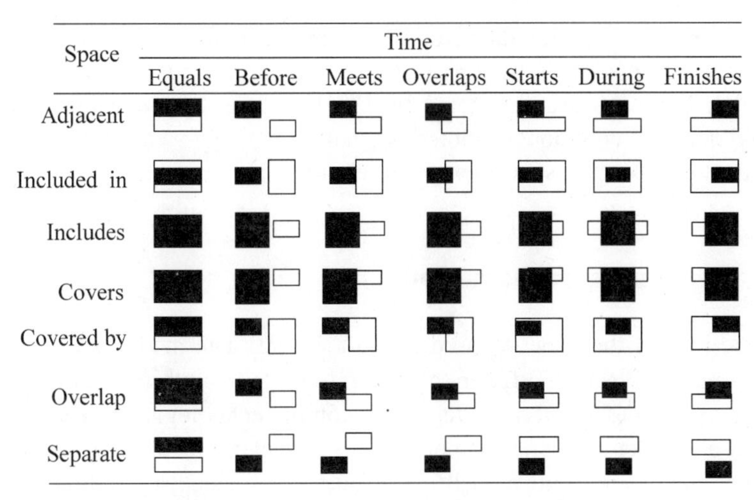

Space
Time
Equals
Before
Meets
Overlaps
Starts
During
Finishes
Adjacent
Included in
Includes
Covers
Covered by
Overlap
Separate

Figure 5.1 Spatio-temporal relationship

3. Hierarchical structure of spatio-temporal data model

After abstraction, classification, calculation and association, spatio-temporal information units can be implemented in the computer system. Similar to general data modeling, spatio-temporal data modeling also has a stepwise layer process, which is shown in Fig. 5.2.

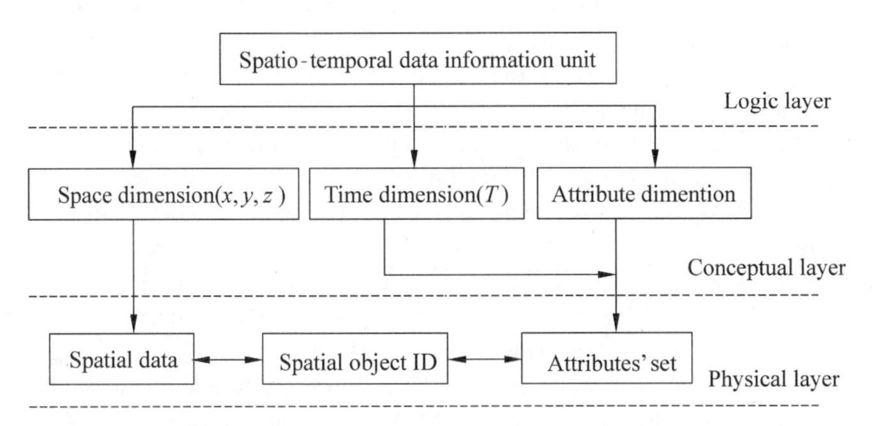

Figure 5.2 Modeling layers of spatio-temporal data

- Logic layer: The main purpose of this layer is to abstract the spatio-temporal information units from different dimensions, i.e., give their representations in space, time or property dimensions.
- Conceptual layer: The main purpose of this layer is to express the concept of spatio-temporal information units and to establish the relationships between them.
- Physical layer: The main purpose of this layer is to devise a physical model that is applicable in computer hardware. Most of the existing spatio-temporal

systems manage spatial data and common attributes separately, while the temporal information is seen as part of common attributes. Spatial data and common attributes are connected by ID. However, the connection between spatial data and common attributes is usually relatively weak and it is difficult to reflect the complex spatio-temporal relations.

5.2.3 Version Based Data Model

The basic idea of the version based spatio-temporal data model is to record the states of spatial objects in different events in order to track the spatial information changing over time. Based on different versioning technologies, researchers have proposed a number of spatio-temporal data models: sequential snapshots model, base state with amendments model, space-time cubic model, space-time composite model, object-oriented spatio-temporal model.

1. Sequential snapshots model

Armstrong (1988) discussed **sequential snapshots model.** The sequential snapshots model adopts the database versioning technology, using a series of database snapshots to represent the evolving process of spatial object with time changes. Each snapshot records the database state at that time. The sequential snapshots are a series of snapshots. Each of these snapshots corresponds to a layer in state diagram at some time (as shown in Fig. 5.3). Spatio-temporal data is organized according to sequential snapshot to store all data at some time or after an interval as a new layer in the database.

In a snapshots model, each layer stores all the information at a specific time point. It shows a changing spatio-temporal distribution over time. Therefore, there is no definite temporal relation between layers. The time interval between any two layers may be different. Moreover, it is uncertain whether there are any changes in objects in any two layers. Generally, sequential snapshot has two categories: vector snapshots model and grid snapshots model. Temporal Map Sets (TMS) may be regarded as an expansion of the snapshots model. The advantage of sequential snapshots model is that it can be achieved in current spatio-temporal information system directly. The disadvantage is that massive changeless spatio-temporal data in different layers cause data redundancies and this may lead to data inconsistency. In addition, the effect is not direct for snapshot when representing changes. In order to obtain different states at two different times, complete comparison between corresponding snapshots is indispensable.

2. Base state with amendments model

To avoid storing changeless data in consecutive snapshots, Langran and Chrisman (1988) proposed the **base state with amendments model**. Base state with

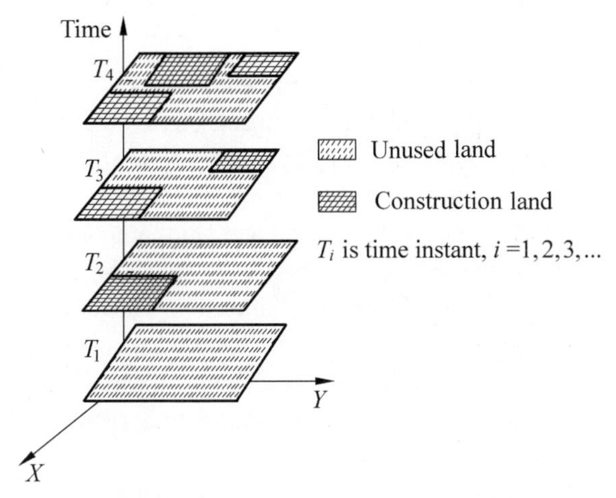

Figure 5.3 Sequential snapshots model

amendments model is based on original database state (snapshots). It just stores a basic data state (base state) at a certain time, and then records all the changes relative to previous state after every predefined time interval. This model updates data only when the data is changed, and it only records the differences between current state and previous state. Compared to the spatio-temporal snapshot model, the base state with amendments model decreases the data redundancy. However, every time, in order to calculate the current state of database, it needs to retrieve the records of base state and all the changes, so the query efficiency is still low. The base state model is shown in Fig. 5.4.

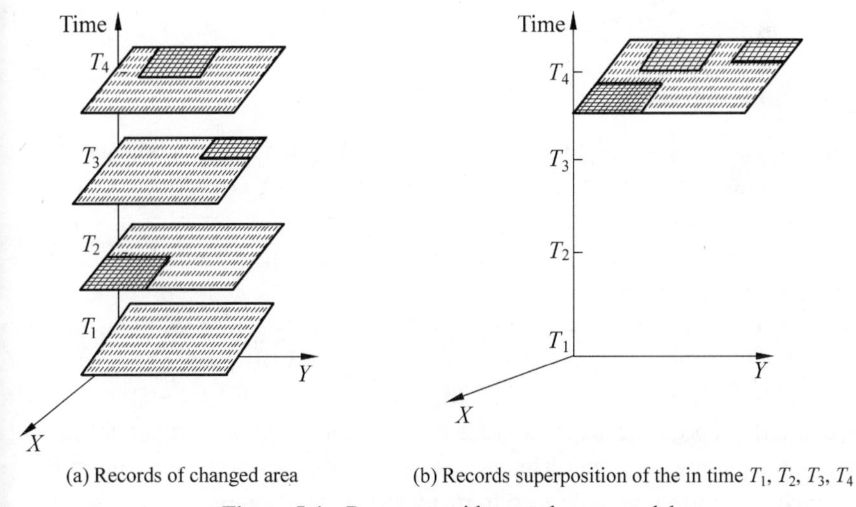

(a) Records of changed area (b) Records superposition of the in time T_1, T_2, T_3, T_4

Figure 5.4 Base state with amendments model

3. Space-time cube model

Hagerstrand first proposed the **space-time cube model** (Hägerstrand 1970). Usually the space-time cube model is composed of two-dimensional space data and one-dimensional time data. It describes the evolving procedure of the two-dimensional objects along the time dimension. The evolutional process of any spatial object is an entity in the space-time cube. Spatio-temporal query is to retrieve the corresponding sub-cube. The characteristic of space-time cube model is that it uses the geometric features of the time-dimension and represents the changes of the spatio-temporal state in a simple and clear way. However, it is difficult to implement the three-dimensional cube. Moreover, with the increase of data, the operations in the cube will become more and more complicated. Space-time cube model is shown in Fig. 5.5.

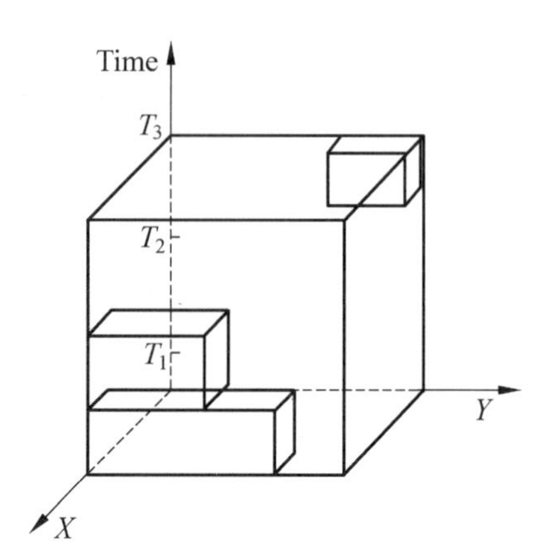

Figure 5.5 Space-time cube model

4. Space-time composite model

Langran proposed the **space-time composite model** (Langran and Chrisman 1988), which combines the characteristics of database versioning and object versioning. It marks up the versioning information on a combinational complex of time and space. The complex of time and space is the combination of a number of objects that are isomorphic in space and consistent in time. Spatio-temporal database, in this case, is a collection of these space-time complexes. Each space-time complex can individually store its states. This model can effectively answer the queries of the spatial and temporal changes over time. However, each space-time complex independently represents its own information. Hence, it is difficult to answer queries for relations between different space-time complexes. In addition, if there is an update operation, it could lead to many space-time complex reconstructions. Space-time composite model is shown in Fig. 5.6.

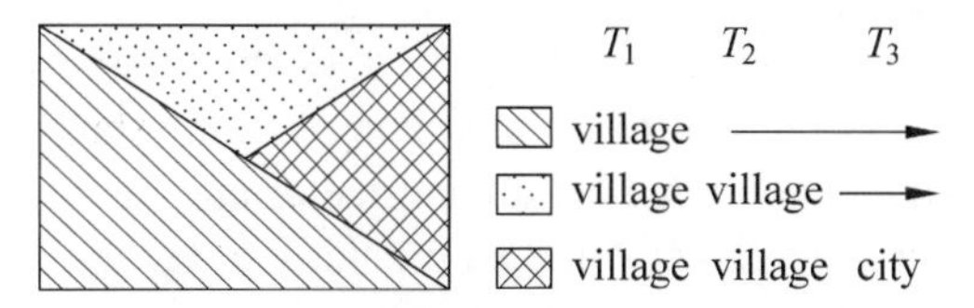

Figure 5.6 Space-time composite model

5. Object-oriented spatio-temporal model

Object-oriented spatio-temporal model (object oriented spatial temporal data model) organizes spatio-temporal data based on the object-oriented idea. Each spatio-temporal data object is packaged independently and each entity has a unique mark. It packages the object's temporal and spatial characteristics, the field properties, related operations and relations with other objects. Object-oriented spatio-temporal model presents the spatio-temporal database model as a collection of spatio-temporal objects and each object contains a number of space-time spatio-temporal atoms. A spatio-temporal atom is a part of some special properties of the spatio-temporal object. These properties remain unchanged for a long time. Although the spatio-temporal atoms themselves do not express any temporal and spatial variation, but projecting the spatio-temporal atom of a spatio-temporal object onto the time dimension or the space dimension, we can get the states of the spatio-temporal object for different times. Therefore, it can express the states and changes of spatio-temporal objects.

Worboys in 1992 presented object-oriented spatio-temporal model based on three-dimensional spatio-temporal characteristics (Worboys et al. 2006). Its basic idea was to use generalization, inheritance, aggregation, composition and an orderly combination of the object-oriented technology to expand data modeling methods based on the entity. The idea was also to use (two-dimensional) spatial objects together with the information of its time axis to compose a complete three-dimensional spatio-temporal object. Since then, people present or improve the object-oriented spatio-temporal data model. These spatio-temporal data models use or partly use object-oriented methods of collection, organization and ordered organization, describe the level hierarchy relationship (such as a large object by a number of small objects) and the mutual reference relationship (such as a piece of land was owned by an individual, a person living in a city). This is done by defining the appropriate object's characteristic structures. The model, however, ignores the data expression patterns of the interaction course between the objects. Therefore, it can be seen that this model was founded partly based on the object-oriented idea.

Object Data Management Group has put forward a series of standards about storing objects, including specification of the object-definition language and query-objects language (Group ODM 2005). In the standards, we use series of changes in the expression of spatio-temporal data to describe the object state. Object

Behavior is a set of predefined operations. It uses the program logic to achieve some necessary and special features to help complete the standardization of data management, as well as to obtain a certain degree of intelligent data management. In fact, with the use of the object-oriented methods and techniques, the standard becomes an expanded program for the features of object-oriented database. It builds an object-oriented data management model. At present, some commercial DBMS offers some basic functions of object-oriented data management. Spatio-temporal object model is shown in Fig. 5.7.

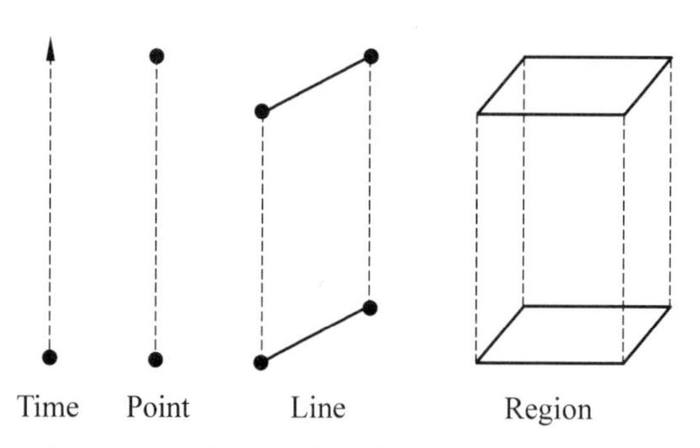

Figure 5.7 Object-oriented spatio-temporal model

Object-oriented spatio-temporal model data structure is simple. If we take full advantage of object-oriented software technology, it is helpful in expansion and the temporal operation of the spatio-temporal data model. However, the spatial and temporal object model is similar to the space-time snapshot model and the space-time composite model. It can only express some discrete changes, but cannot express the continuous changes. At present, the object-oriented space system is rare, because many questions about the model are unresolved.

5.2.4 Event-Based Data Model

Because spatio-temporal data model based on versioning has many disadvantages, researchers began to model the spatio-temporal objects through event or process. The **ESTDM** and **Three Domains Model** are models of this type. Event is a fact in a specific time. The basic idea of models based on event is to track the spatio-temporal changes in different events and every event records a change of a spatio-temporal object. In the event-based model, every event records a change and its state information before and after the change. The characteristic of event-based spatio-temporal data model is to explicitly keep all the spatio-temporal information changes, but it also imposes great overhead on the system. An example of spatio-temporal object is depicted as in Fig. 5.8.

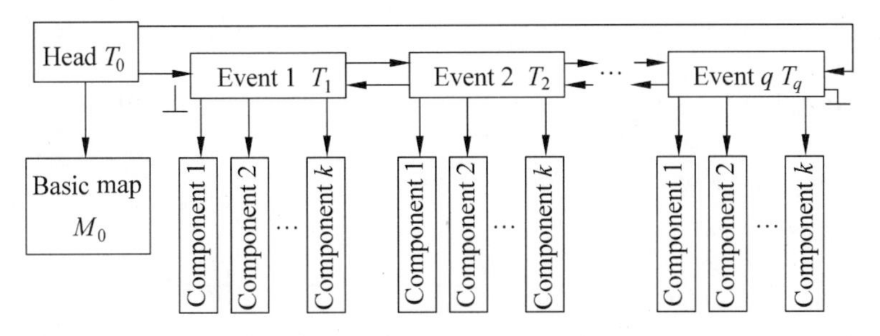

Figure 5.8 Event-based spatio-temporal data model

1. ESTDM and TRIAD

Peuquet and Duan (1995) proposed an event-based spatio-temporal data model. They believe that the spatio-temporal model based on location or trait is not fit for analyzing the spatio-temporal relation. Therefore, they proposed a new data model based on raster named ESTDM (event-based spatio-temporal data) to organize the spatio-temporal information about location changes. ESTDM depicts the spatio-temporal observation of a single event in the spatio-temporal list by organizing the levels based on time stamp. Compared to time list snapshot, the advantages of ESTDM are efficiency in dealing with data and analyzing the spatio-temporal model and relation. The snapshot only stores one state and the related changes from the former state. One head file in ESTDM stores the information of state and pointers in ESTDM point to basic map and start-end event list. The basic map appears on the initial snapshot of the content needed in a spatio-temporal area. An event-based list includes the spatio-temporal information in this spatio-temporal area. Every event has a time stamp and is associated with the time component list. The event component list stores the change of a unit in a predefined location and a given time.

ESTDM effectively supports the temporal query as follows:

- At ESTDM's specified time, change location to the specified value.
- After a specified time interval, change location to a specific value.
- After a specified time interval, change region to the union of some regions with specified values.

If we are using ESTDM in a system based on vector, then it needs to redesign event component. When spatial object or topological relation is changed, the historical information of entity may be sliced as transition information. This mechanism needs event component to track the spatial information and change in entity properties for a predefined entity.

As the extension of ESTDM, the model named TRIAD uses an integrated view (including trait, location and time) to represent spatio-temporal data. It can be implemented through object-oriented method. TRIAD implements a universal and general representation by integrating attribute dimension, space dimension and time dimension. The characteristics of TRIAD are as follows:

- It can support query according to any combination of time, location and properties.
- The specific accessing sequence of view can be determined easily according to query.

The pointer structure of ESTDM is shown in Fig. 5.9.

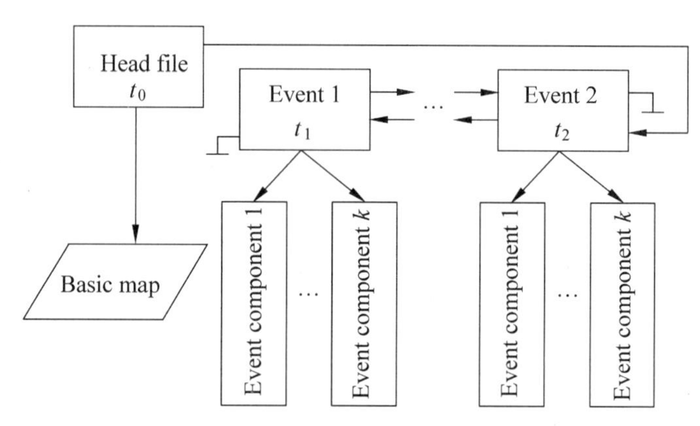

Figure 5.9 ESTDM's pointer structure

2. Three Domains Model

Three Domains Model is proposed to meet the needs of analyzing spatio-temporal information during the process of research on wildfire (Yuan 1994). The information ring of wildfire includes many kinds of data representation, such as snapshot state, fire entity process, snapshot change of entity, representation history of fire block. They can be utilized for research on fire-alarm prediction, fire behavior, and fire impact and fire history of wildfire. Because the application data are separated on concrete semantic, relevant spatio-temporal information needs to support this representation dynamically. Three Domains Model defines semantic and spatio-temporal object in each independent area. In the snapshot model of time list, time is an attribute of location. In space-time composite model and space-time object model, time is a part of space entity. In Three Domains Model, time models as an independent concept. It represents the real world from three angles: making location, entity or time as center. Hence, it can depict various basic changes of spatio-temporal information, such as attribute change, static space distribution, static space change, dynamic space change, transit of process and movement of entity. Three Domains Model is shown in Fig. 5.10.

Three Domains Model uses its dynamic linking relative object to represent spatio-temporal entity. These links between objects can be numerical type or fuzzy member function. Member function is especially very effective in representing dynamic boundary, such as the change of oil distribution, seasonal lake boundary and coastline.

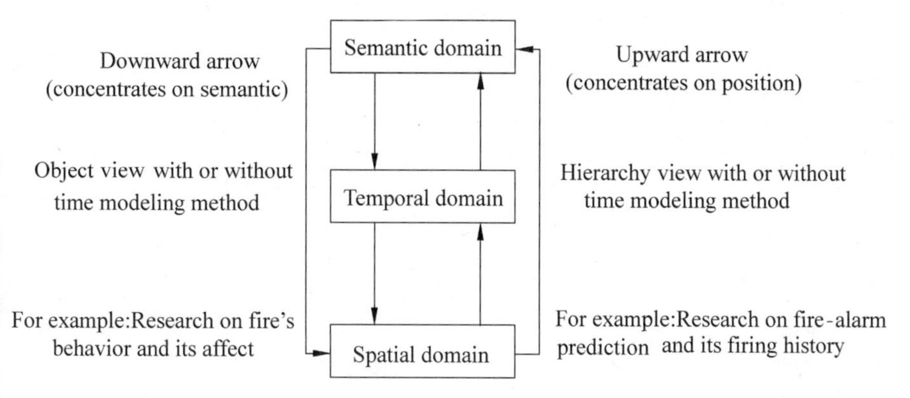

Figure 5.10 Three domains model

5.2.5 Constraint-Based Data Model

As a new data model, **constraint-based data model** extends traditional relational data model through "generalized tuple". A constraint tuple of constraint relation R is defined as a constraint conjunct on a variable set. For example, "$((1<x<3) \wedge (2<y<5))$" can be a constraint tuple defined on variable set $\{x, y\}$ and can correspond to rectangular region of two-dimensional plane. Every constraint tuple can depict a point set, which may be infinite. Therefore, constraint data model can represent multi-dimensional information, such as spatio-temporal data in the form of constraint. Constraint-based spatio-temporal model can adopt relational algebra or relational calculus as its data operation. Compared to other models, constraint-based spatio-temporal data model has better presentation ability. It can not only represent continuous changes, but also depict spatio-temporal semantics based on version.

The performance of spatio-temporal data model is shown in Fig. 5.11.

5.2.6 Moving Objects Data Model

At present, the technique of MOD (Moving Objects Databases) has already become hot spots of spatio-temporal database. Erwig and others believe the essence of spatio-temporal database is to store the moving objects (Erwig et al. 1998). MOD is a database that manages the moving objects and location. MOD can be used in many domains such as civil air regulation, traffic management, military command and location-based information service. *Now*, many researchers have studied the key points of this domain, such as representation, modeling, index and query of moving objects.

1. Continuous and discrete model

Moving object is the geometry object whose figure, location and domain attribute

	Snapshot (Langran and Chrisman 1988)	Base state with amendments (Langran and Chrisman 1992)	Space-time composite (Langran and Chrisman 1988)	Space-time object (Worboys et al. 1992)	Three domains (Yuan 1994)	Event-based (ESTDM)
Data redundancy	big	small	small	small	small	small
Data query	weak	weak	weak	relative strong	strong	weak
Data obtaining	easy	difficult	easy	easy	easy	difficult
Change obtaining	difficult	easy	difficult	easy	easy	easy
Space dependent	yes	yes	yes	yes	no	yes
Attributes change	partly	partly	partly	no	yes	partly
Discrete change	yes	yes	yes	yes	yes	yes
Continuous change	no	no	no	yes	yes	no
Motion support	no	no	no	no	yes	no
Event consideration	no	no	no	no	no	yes
Time participation	edition map layer	edition map layer	edition space object	edition space object	independent time object	edision event
Time dimensionality	VT	VT	VT	BT	BT	VT
Space expression	vector/grid	vector/grid	vector	vector	vector	grid

Figure 5.11 Summarization of spatio-temporal database

is changing along with the time. These geometry objects can be divided into moving points and moving regions. The change of moving objects is either continuous or discrete. Hence, the continuous model and discrete model can be established considering only the change of moving objects.

- Continuous model depicts the moving object as a set of infinite moving points and takes moving points as a continuous curve in a three dimensional space. This model can depict the moving information of moving object precisely. Actually, computer cannot store and operate on infinite points, so this model will have problems in implementation.
- Discrete model depicts the moving object as a set of finite moving points and takes moving points as a continuous broken line in a three dimensional space. Discrete model can depict the moving information of moving object by approximate value. Although the change of actual moving objects' location is a continuous concept, modeling using discrete concept is still needed and is feasible for the limitation of computer system resources.

2. Moving data model

For time and space, discrete model is a high-level criterion of DBMS of data structure. In the process of implementing a spatio-temporal subsystem, discrete model needs to be mapped to actual data structure as the attribute data type of object. Spatio-temporal data are of four types: basic data, spatial data, basic spatio-temporal data and moving spatial object data.

- **Basic Data Type**: Basic data type is the standard data type in regular DBMS, such as *int*, *real*, *boolean* and *string*. It can be supported, making full use of current database platform.

- **Spatial Data Type**: Spatial data type is the data type in spatial database, such as *points*, *lines* and *regions*. Every spatial object has a corresponding *SID*. Spatial object deals in the Realms Criterion. A complex moving spatial object, such as *area*, is constituted of hundreds or even thousands of points. Its data structure should support the basic operations on *point*, *thread* and *area* of moving spatial objects, such as increase and delete. At the same time, it should also support most of the operations in spatial analyze. In spatial database, balanced binary tree can be chosen as the basic data structure of moving spatial object. The time interval of every unit is regularly defined as right open and the time factor is the start time of this interval. Following the regulation of discrete model, unit function x (for continuous change) or unit object value (for discrete change) should be defined in the time interval of this unit to decide the object's value in this unit.
- **Basic spatial-temporal data type**: This type is represented as an ordered tuple of time factor and basic data type, such as *mint* (instant, int), *mreal* (instant, real) and *mstring* (instant, string). In practice, unit function of linear simulation is usually used for *mreal* class.
- **Moving spatial object data type**: This type is represented as an ordered tuple of time factor and the spatial data type. For example, many moving objects can be expressed as *mpoint* (instant, point), *mline* (instant, lines) and *mregions* (instant). In implementation of the algorithm, unit function of linear simulation can be used for *mreal* type. However, for the moving spatial object, such as *mpoint* and *mregion*, the selection of time factor must follow the discretization criterion for the constraint of implementation condition of computer.

3. Framework of abstract data type

Currently, people have done a lot of work on modeling of moving objects. Erwig and Güting (1998) define a frame of abstract data type for moving objects. This frame includes standard data type (*int* and *Boolean*), spatial data type (*point* and *area*) and spatio-temporal type (*time* and *interval*). Through type construction, a new type is created from basic types.

In fact, the modeling method by Erwig defines two models on different levels, abstract and discrete models. Abstract model considers the independence of a system, generality and consistency of operation, closure and consistency of the relation between non-temporal and temporal data structure and operation. Simple query operation can be designed using abstract model. Discrete model is a higher-level standard of data structure. It defines the possibility of implementation of the corresponding data operation in the computer system when the representation is limited. This data type of moving object can be used as the extended type in traditional relation model and could also be integrated into object-oriented or object-relational model. In the meantime, it can support the analyzing operation of spatio-temporal data, based on DBMS, and it can apply the defined spatio-temporal analyzing function in the query.

5.3 Query on Spatio-Temporal Data

Since a long time, the expression and modeling of spatio-temporal data are the most popular problems in the area of spatio-temporal database. In recent years, the research of queries on spatio-temporal data has become a hot spot. The types of current spatio-temporal queries are as follows:

- **Window query**: Given query area *QR* and time interval *QT*, to find all the objects that have intersection with *QT* and *QE*.
- **Recent adjacent query**: Given query point *qP* and time interval *qT*, to find all the adjacent objects with the smallest time distance to *QT*.
- **Spatio-temporal join query**: Given two sets S_1, S_2 of moving objects, a future time stamp t_q and a distance threshold *d*, a spatio-temporal join retrieves all pairs of objects that are within distance *d* at t_q.
- ***k*-CPQ (*k* Closest Pair Query)**: It discovers the *k* pairs of spatial objects formed from two data sets that have the *k* smallest distances between them, where $k \geqslant 1$.
- **Navigational WQ**: Based on historical databases, Pfoser et al. (2000) proposed navigational WQ, that is, given two query regions QR_1, QR_2, and two time stamps QT_1 and QT_2, retrieve all the objects that intersect with region QR_1 at time QT_1 and intersect with region QR_2 at time QT_2.
- **TP (time-parameterized) query**: For predictive spatio-temporal database, Tao and Papadias (2002) point out that the results of traditional query (such as WQ, *k*NN, WDJ, *k*CP) may change because of the movement of projects. Therefore, results of traditional query are not enough for spatio-temporal database. According to this situation, TP query is proposed. This kind of query can be applied to any traditional query method, and the query result could return the result *R* by normal traditional query, the expiry time *TC* of the result *R* and the change result *C* after *T*. Then Tao and Papadias extended the TP concept to continuous query, intending to track the change of query result until some conditions are satisfied.
- **LB (location-based) query**: Zhang et al. (2004) puts LB query forward. This query could be applied to WQ and *k*NN queries to get query results and valid range of query.

In addition to the above normal spatio-temporal queries, some researchers of spatio-temporal database have done some researches on the model of spatio-temporal data query cost. Choi proposes a model of spatio-temporal cost on TPR-tree. Tao and Papadias (2001) put forward a cost model according to overlapping B-tree and multi-version B-tree. This model can evaluate the query tree, accessing point and selection size of query. Lately, Zhang et al. (2005) proposed a cost model of NN query.

5.3.1 Spatio-Temporal Data Query

There are several kinds of spatio-temporal data queries, such as window query, nearest neighbor query and TP and LB queries.

1. Window query

Window query can be represented as $\{O \in DB \mid O(t) \in queryR \wedge t \in [t_1, t_2]\}$. $O(t)$ represents the spatial location of object O at time t. *queryR* represents a rectangular window. The semantic of window query is: find all the moving objects whose locations are in *queryR* during t_1 to t_2. If the window query does not indicate the query time directly, it means that the query time is current time.

2. Nearest neighbor query

The mathematical definition of nearest neighbor query is: for any given object q and object set $P = \{p_1, p_2, \cdots, p_m\}(m \geqslant 1)$, find the p_i, which makes $|q, p_i|$ the smallest. $|q, p_i|$ represents the spatial distance between object q and object p_i. In traditional NN query, all the objects are stationary. Spatio-temporal database uses the concept of nearest neighbor query in spatial database and extends the concept that all objects can be either stationary or moving. Nearest neighbor query on moving objects is a critical technique in spatio-temporal database. It is widely used in many domains, such as intelligent navigation, modern communication, traffic control and weather forecast. It is the key point of spatio-temporal database. The nearest neighbor query of moving objects can be abstracted as: Given query object q and its movement status (velocity and direction), find a series of nearest neighbors of q from start time s to end time e. If the shape of the object is not considered, the object can be treated as a point. Then the nearest neighbor query is the nearest neighbor query of moving points.

3. Approximate query

Approximate query can be represented as $\{O \in DB \mid dist(Q(t), O(t)) \leqslant \varepsilon\}$, where $Q(t)$ is the object to be queried, *dist* the "distance" between object $Q(t)$ and $O(t)$, ε the given range of distance, and $t \in [t_1, t_2]$. The semantic of approximate query is: retrieve all the objects $O(t)$ whose distance from $Q(t)$ is in the given range ε during t_1 to t_2.

5.3.2 Moving Data Query

Moving data query can be divided into several basic types, and those are queries on historical location, current location and future location. The future location query is to predict the future location of moving objects based on the current location, its velocity and direction. Historical and current location query can be further divided into two subcategories: coordinate-based query and trajectory-based query. Coordinate-based query includes time slice query and time interval query. Trajectory-based query includes topological query, navigation query, and combination query.

1. Coordinate-based query

- **Time slice query**: Time slice query can be represented as $Q=(R, t)$, where R is a hypercube at time t. For example, finding the traffic cars in area R at time t is a simple and regular spatio-temporal query of traffic monitoring. Time slice query could be seen as the product of the select operation in spatial database and select operation in temporal database.
- **Time interval query**: Time interval query can be represented as $Q=(R, t_1, t_2)$, where R is a hypercube of k-dimensional space, and Q a $(k+1)$-dimensional hypercube constructed by R and time interval (t_1, t_2). In the application of traffic monitoring, this kind of query returns all the vehicles passing through the area R during t_1 to t_2.

2. Trajectory-based query

- **Topology query:** Topology query retrieves part of or the whole trajectory of moving objects. The basic predicates of topology query are "center", "cross", "pass by", and "leave". In order to judge whether a moving object satisfies those predicates, multiple line segments need to be examined.
- **Navigation query:** The information of moving objects that is not directly kept in database but can be inferred from its trajectory is called navigation information. For example, the average and maximum speed of the object can be obtained by the distance it made in some time interval. The moving vector can be calculated by two vectors at different locations. The queries on speed and moving vector are very popular in applications. For example, "What is the current speed of this car? What is its maximum speed?" The first question is related to current time and the second question is related to the data in some time interval.
- **Combination query:** The trajectory-based query turns out to be more complicated in practical applications. It could be a combination of several query types. For example, if every car has an exclusive *ID* number, there may be queries like: "Where will the vehicles on the 5th Zhongshan Road from 9 to 11 o'clock this morning be in the following one hour?" In this case, it needs to find all the vehicles that are on the 5th Zhongshan Road from 9 to 11 o'clock and then it needs to predict where those vehicles would be in the following one hour.

5.3.3 Spatio-Temporal Database Language

Spatio-temporal database language and spatio-temporal data model are closely related. Every spatio-temporal data model needs its query language. There are mainly two directions in current spatio-temporal database language: spatio-temporal database based on SQL and spatio-temporal database based on OQL. As the popular language in relational database, SQL has been supported by most of commercial databases such as Oracle and SQLServer. OQL is an object-oriented database query language proposed by ODMG and has become the standard language of object-oriented database. It has been supported by OODBMS such as O2 and Versant. STSQL is a query language of spatio-temporal database based on SQL. Because SQL3 has scalability in its abstract data types, the spatio-temporal data model based on data type can be combined with SQL3.

The characteristic of STSQL is that it is compatible with SQL. It does not extend clauses of SQL language, but just extends data types and operations. Therefore, STSQL is applicable in many applications.

STQL is another spatio-temporal database query language based on SQL. The data model of STQL is based on relational model and it represents the spatio-temporal change in the form of tuple versioning. Since STQL extends the clauses of SQL, it is incompatible with SQL3. The query clause of STQL extends the traditional SQL in two ways. The first one is using WHEN clause to represent the temporal query condition and the second one is to express the spatial query condition by adopting spatial operator and spatial predicate. Therefore, STQL uses the combination of two kinds of conditions (space and time) to represent a spatio-temporal query. Compared to STSQL, STQL introduces transaction time-dimension. It can deal with bitemporal data. However, it splits the spatial and temporal query criteria artificially and its spatio-temporal data model is based on versioning. It cannot deal with continuous spatio-temporal query.

For the spatio-temporal data model and spatio-temporal database language based on ODMG, object-oriented data model is more expressive in semantics than relation model (such as polymorphism, inherit and aggregation). Therefore, it has some advantages in the representation of spatio-temporal data.

5.4 Structure of Spatio-Temporal Database System

Because of the special requirements in spatio-temporal applications, it is critical to design an effective system structure that supports STDBMS (spatio-temporal database management systems). The system structure of spatio-temporal database system utilizes the achievements of the former spatial database and temporal database. There are three basic types of system structures: complete type, layered type, and extended type.

5.4.1 Structure of Complete Type

The complete type implements an STDBMS from bottom to top. This type needs to implement all the modules of DBMS including query compilation and execution, transaction management, storage management, and even the drivers for the spatio-temporal database. Its workload is too heavy for many spatio-temporal applications.

5.4.2 Structure of Layered Type

Implementation of layered structure (such as in Fig. 5.12) adds a spatio-temporal layer on the traditional RDBMS. Dealing with spatio-temporal data in spatio-temporal layer, it does not need to change the core of DBMS. Spatio-temporal layer does all the work related with spatial temporal information, such as translation of spatio-temporal database language and SQL, spatio-temporal query optimization. However, the SQL translated from spatio-temporal query is very complicated. It is very difficult for the RDBMS to do the query optimization. Because all the queries need to be transformed to standard SQL by the spatio-temporal layer first, this layer probably will become the bottleneck of applications.

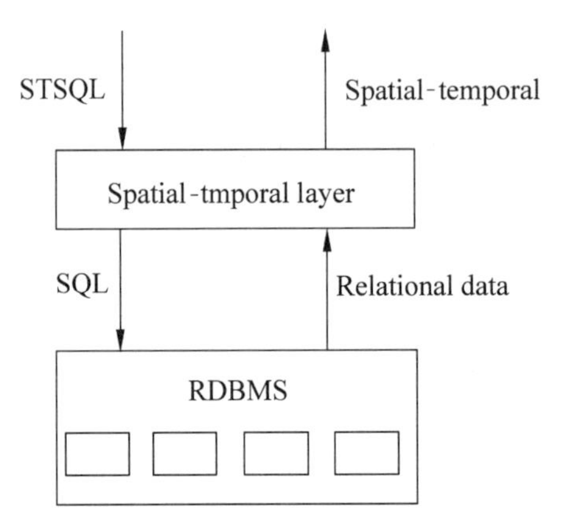

Figure 5.12 Layered architecture of STDBMS

5.4.3 Structure of Extended Type

The extended structure is to execute a spatio-temporal extension on the ORDBMS. Figure 5.13 shows such a case. Since ORDBMS provides the function of UDT (user-defined data type) and UDR (user-defined routine), it can be used to extend new spatio-temporal types and operations. This structure is the most popular one right now. The extended system structure makes the implementation and application of spatio-temporal DBMS possible. Its main problem is: though original DBMS can accelerate the query by extended spatio-temporal index, the strategies of query optimization is still the old ones of relation database, and it is not appropriate for spatio-temporal query.

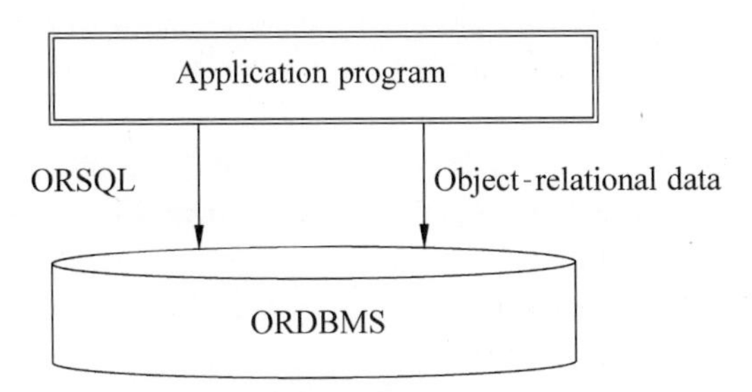

Figure 5.13 Extending architecture of STDBMS

Reference

[1] Allen J (**2005**) Towards a general theory of action and time. The Language of Time: A Reader: 251

[2] Armstrong M (**1988**) Temporality in spatial databases. In: Proceedings of GIS/LIS'88, pp 880 – 889

[3] Cho H J, Chung C W (**2005**). An efficient and scalable approach to CNN queries in a road network. In: Proceedings of the 31st International Conference on Very Large Data Bases (VLDB), Trondheim, Norway, 865 – 876

[4] Egenhofer M (**1991**) Reasoning about binary topological relations. Springer-Verlag London, UK

[5] Erwig M, Güting RH, et al. (**1998**) Abstract and discrete modeling of spatio-temporal data types. In: Proceedings of the 6th ACM International Symposium on Advances in Geographic Information Systems. Washington, D.C., USA

[6] Group ODM (**2005**) The standard for storing objects. Retrieved 1-18, 2005. http://www.odmg.org

[7] Hägerstrand T (**1970**) What about people in regional science? Regional Science **24**(1): 6 – 21

[8] Langran G, Chrisman N (**1988**) A framework for temporal geographic information. Cartographica: The International Journal for Geographic Information and Geovisualization **25**(3): 1 – 14

[9] Peuquet D, Duan N (**1995**) An event-based spatiotemporal data model (ESTDM) for temporal analysis of geographical data. International Journal of Geographical Information Science **9**(1): 7 – 24

[10] Pfoser D, Jensen CS, Theodoridis Y (**2000**) Novel approaches in query processing for moving object trajectories. In: Proceeding of the International Conference on Very Large Databases (VLDB), 395 – 406

[11] Tao Y, PaPadias D (**2001**) MV3R-tree: a spatio-temporal access method of timestamp and interval queries. In: Proceedings of the 27th International Conference on Very Large Databases (VLDB), 431 – 440
[12] Tao Y, Papadias D (**2002**) Time parameterized queries in spatio-temporal databases. In: Proceedings of the ACM International Conference on Management of Data (SIGMOD), 334 – 345
[13] Worboys M, Hearnshaw H, et al. (**2006**) Object-oriented data modeling for spatial databases. Classics from IJGIS: Twenty Years of the International Journal of Geographical Information Science and Systems **4**(4): 119
[14] Yuan M (**1994**) Representation of wildfire in geographic information systems. State University of New York at Buffalo
[15] Zhang J, Mamoulis N, PapadiasD, Tao Y (**2004**) All-nearest-neighbors queries in spatial databases. In: Proceedings of the 15th IEEE International Conference on Scientific and Statistical Database Management (SSDBM), 297 – 306
[16] Zhang J, Papadias D, Mouratidis K, Zhu M (**2005**) Query proceeding in spatial databases containing obstacles. International Journal of Geogrophic Information Science(IJGIS), **19**(10): 1091 – 1111

6 Temporal Extension of XML Data Model

Na Tang[1,2], Hanjiang Lai[1], Yong Tang[1,2+], and Xiaoping Ye[1,2]

[1] Computer School, South China Normal University, Guangzhou 510631, P.R. China
[2] Department of Computer Science, Sun Yat-sen University, Guangzhou 510275, P.R. China

Abstract The supporting technology of web applications are mostly limited to static snapshots, but the structure and the content of the XML documents are constantly changing over time. After the introduction of the "temporal expression", the temporal XML itself can record a series of traces of the XML document's modifications. It provides a feasible and efficient method to the XML version management. On the other hand, because of the poor flexibility of "traditional relational model" and low scalability of "SQL Language", it has not yet solved how to fully support temporal on the existing commercial database system. Compared to the traditional relational model, XML and XQuery can support the expression of temporal information and temporal queries in a better way. Whether XML data documents are needed to deal with a series of problems that are caused by the time-varying changes, or traditional temporal database is required to look for new opportunities in the commercialization process, the version management of XML documents and the temporal query of the content are becoming the next generation of web information system applications.

Keywords *temporal XML, model expansion, temporal information expression, temporal query*

With the development of networking and the Internet, the ability of data exchange has become an important requirement of new application systems. With the scope of information sharing and data exchange expanding, the traditional relational database is also facing challenges. The applications of database technology are set up in the database management system. Databases belong to high-end applications, commanding high prices and major runtime environment. The heterogeneity of various database management systems and operating systems, which database management systems rely on, severely limit the scope of information sharing and data exchange. Database technology has poor capability for describing semantics,

[+] Corresponding author: isstn@mail.sysu.edu.cn

mostly through technical documentation. It is very difficult to achieve the durability and transitivity of the data semantics. The data exchange and information sharing are based on semantics. When data exchange is heterogeneous, it is not conducive for computation to automatically retrieve and apply the correct data based on semantics.

6.1 Motivation

6.1.1 XML Temporal Driven

Extensible Markup Language (XML), which is similar to HTML, is a tag-based markup language, which inherits most of SGML's functionality, but uses a less complex technology. HTML only provides the general method to show the information on the page (no context-sensitive dynamic features), but XML provides the context-sensitive functions to data, which has more advantages than HTML. The greatest advantage is the management of various data. Any system can use XML parsers to read XML data, so data can be accessed everywhere, and we do not have to worry about the compatible problem of the system. XML's advantages are mainly extendibility, flexibility, self-description, conciseness, internationalization, and efficient retrieval. Because of these advantages, XML is a simple and standard method to define text data. The benefit of XML is that it is data-exchangeable (portable), while XML also has the following advantages in the data applications: ① as a plain text file, an XML document is not limited by the operating system or software platform. ② XML has self-describing semantic function based on schema, which makes it easy to describe the semantics of data. This description can be understood and automatically handled by computers. ③ XML can not only describe structured data, but also effectively describes semi-structured or even unstructured data. It has been called "the ASCII code on Web". You can use one of your favorite programming languages to create any kind of data structure and then share it with other people who use different languages in their computation platforms.

As the next generation of markup language of data expression and data exchange for the Internet, XML plays an increasingly important role in new web applications, such as the new generations of e-government, and e-commerce. As the latest technology in the field of data access, XML data management has become the research hot spot in information systems (Gray 2004; Meng et al. 2004).

Now, the supporting technology of web applications is mostly limited to static snapshots, but XML documents are rarely static. The structure and the content of documents are changing over time. For example, many documents are usually modified over time. Although many methods to store XML documents have been proposed (Tian et al. 2002), the traces of a series of amendments that are done to the documents by the user cannot be retrieved. Therefore, the old version must be saved in order to avoid the loss of important data. This causes a series of questions such as, "how to keep a series of amended traces on the document", "how to query a version in a certain time period". Therefore, many applications also need to query historical information of XML documents, retrieve legal data information in a certain time, query which parts of the document are changed and modified, and so on.

XML documents have two research fields related to temporal attributes.

1. Version management

For some documents, content changes over time, such as software documentation, product catalogs. However, the usual methods dealing with such documents are versions management, so these documents are also known as versioned document.

There are two methods to study version management.

Method Ⅰ: All versions of the changes are retained. All versions of a document are based on the commonalities and differences of two successive versions. Most of these methods will save the modified XML document as a new document, for example, XML Diff algorithm is used to compare the differences between the versions of XML documents and edited scripts are generated, which show changes between them. Then, we can use such edited scripts to express the new version of XML documents and use a renewed algorithm to reconstruct the version of XML documents (Chien et al. 2000; Wong and Lam 2003). However, this type of algorithm cannot break through direct access to any version of XML documents.

Another kind of object is the reference-based model. This version is expressed as a series of records of the actual object and it marks the greatest common reference record of the old and new versions (Chien et al. 2001). This means that the original version would be completely preserved. When it is modified, only the inserted parts are stored and other parts with no change, point to the old version. As the cost of storage is very little, some scholars (Nørvåg 2004) proposed to retain all versions, and through usage of compression, merge temporal granularity and some other kinds of technology to reduce the storage space. This method can be direct access to any version of XML file. The version management of XML had also attracted scholars' attention and a number of scholars participated in it.

Method Ⅱ: All changes will be retained in one version, using transaction time of the temporal database to record the history of the change.

Method Ⅱ is more suitable for frequent modifications and inquiries, while Method Ⅰ is suited to less frequent modifications, such as the data warehouse of web page, when the version of the page changes.

2. Temporal information management

Method Ⅱ contains the temporal information (the information of temporal database uses XML to release historical document) or the content of documents related to the temporal information. Such information is often managed by the temporal database. If this type of data uses the XML document for storage, the biggest feature is the data model object (or node or edge) tagged by time stamp. The realization is to add temporal information into XML document and record the entire history of data changes in a document. Since "XML documents" are similar to "traditional relational database", both can be used to store data. We can introduce "temporal" into XML, through the introduction of "temporal expression" to record a series of history traces in "Temporal XML" document. Literature (Wuwongse and Yoshikawa 2004) studied how to use multiple versions to retain historical data.

In this section, we discuss the second method of version management and how to release and manage temporal information, that is, through temporal XML documents. The document contains temporal information, the objects of that document tagged by time stamp.

6.1.2 Commercial-Driven Temporal Database

Time is the ubiquitous attribute in nature. All information has the corresponding temporal attributes, so many of the real-world data applications are related to time-varying data, and many information systems have to deal with the new demand of temporal information. It plays an increasingly important role in e-government, e-commerce, medical systems, data warehouses, data mining, decision support systems, CAD and other information systems. In some systems, the temporal information also plays a key role. For example, in the intelligent decision support system of salary, not only does the data have temporal attributes, the knowledge also has the same attributes. One staff's salary is determined by related information, such as his qualifications, position and when he accesses that position (temporal data) and the salary policies for the corresponding period (temporal knowledge). For example, the application based on RFID electronic tags will have large historical data about the state of physical equipment. With "temporal information process" becoming increasingly popular, there has been a much discussion on how to research "temporal" in the support of the "traditional relational database" over the past few years, that is, "Temporal Relational Database". Literature (Grandi 2004) was also formulated on the temporal web applications.

After decades of development, the trends of the temporal database are:

First, the scope of the application of the temporal information is very wide, infiltrating through all the aspects of scientific experiment information systems, multimedia information systems, geographic information systems, municipal information systems, telecommunications information systems, e-government, e-commerce, data warehousing and so on.

Second, the temporal data models develop towards unification and standardization of temporal database query language in the direction of "Products". There is also a lot of work to be done in the production of temporal information.

Third, the requirements of the temporal information become increasingly complex. It is difficult to achieve unified temporal data model, but there would still be unified temporal data models in a variety of fields. Temporal information products will be moving toward "Embedded", based on the mainstream of the database technology and platform, domain-orientation, middleware of application-orientation, software components and so on.

The progress of research on temporal data query shows that if the traditional DBMS can be enhanced to support the processing of the temporal information, these applications can gain many advantages. For example, the application code will be greatly simplified and easily maintained, and thus the output code will be improved. At the same time, we can get better performance if we let DBMS deal with the temporal information. In the past twenty years, although many researches are related with temporal database, it has not yet resolved how to support effective and feasible temporal attributes in the existing commercial database systems, because of the poor flexibility of the "traditional model" and the low scalability of the "SQL Language". At present, support of the temporal attributes by the sophisticated database products is limited to the definition of the temporal data type. For example, Informix TimeSeries DataBlade, Oracle8 TimeSeries cartridge and some new system temporal data types can be defined by users. However, the requirements of temporal information applications are not only to provide relevant temporal data types, but also to expand the query language itself.

Clifford et al. divided temporal data model into "temporally ungrouped" (Table 6.1) and "temporally grouped" (Table 6.2) models. The literature also proved that the second type of temporal data model was more expressive. Since it was historically-oriented, it was more natural to use it to express temporal information. From Table 6.1 and Table 6.2, it is clear that the first data model will cause database redundancy, and sometimes the inquired results will be carried out using extra "temporal merging". Traditional relational data model can only support the "structured" data, so it cannot support the "temporally grouped" data. However, different from the traditional relational data model, XML supports "unstructured" data. It can be very good at supporting "temporally grouped" data model. Another difference is that XML query language—XQuery has the characteristic of being "Turing complete" and natively extensible, that is, XQuery can use itself to define many complex temporal queries, without having to rely on the extension of "Standards Association" (relational databases and XML comparison, see Table 6.3 and Table 6.4).

Now, in fact the majority of commercial database systems support to adopt the XML view to release and view the underlying relational database table. This function is used to translate the queries of the XML view into the queries that are supported by the underlying relational database. At present, the database manufacturers and standards organizations will add such capacity into SQL. Therefore, some database researchers, developers and SQL standard-setting associations will start studying how to use XML to release and query the history of database (Wang et al. 2009), how to extend the SQL to support XML and so on.

Table 6.1 Temporally ungrouped employee table

ID	Name	Salary	Title	DeptNo	Start	End
1001	Bob	60000	Engineer	d01	1995-01-01	1995-05-31
1001	Bob	70000	Engineer	d01	1995-06-01	1995-09-30
1001	Bob	70000	Sr Engineer	d02	1995-10-01	1996-01-31
1001	Bob	70000	TechLeader	d02	1996-02-01	1996-12-31

Table 6.2 Temporally grouped employee table

<table>
<tr><th>ID</th><th>Name</th><th>Salary</th><th>Title</th><th>DeptNo</th></tr>
<tr><td rowspan="3">1995-01-01
1001
1996-12-31</td><td rowspan="3">1995-01-01
Bob
1996-12-31</td><td>1995-01-01
60000
1995-05-31</td><td>1995-01-01
Engineer
1995-09-30</td><td>1995-01-01
d01
1995-09-30</td></tr>
<tr><td rowspan="2">1995-06-01
70000
1996-12-31</td><td>1995-10-01
Sr Engineer
1996-01-31</td><td rowspan="2">1995-10-01
d02
1996-12-31</td></tr>
<tr><td>1996-02-01
TechLeader
1996-12-31</td></tr>
</table>

Table 6.3 Comparison of data structure

XML	Relational databases
unstructured	Structured
well supports "temporally grouped"	only supports "temporally ungrouped"

Table 6.4 Comparison of query language

XQuery	SQL
"Turing complete"	Difficult to describe the complex temporal queries
natively extensible	Difficult to expand
Does not rely on the changes of standards	Expansion depends on the change of standards

Whether it is necessary for XML to deal with the time-varying data document, or the traditional temporal database required to look for new opportunities in the business, the version management of XML document and the temporal query of content are becoming important applications of the next generation of web information systems. It plays an increasingly important role in information systems especially in e-government, e-commerce, data warehousing, digital libraries, collaborative editing and so on.

6.2 Temporal Research of the Semi-Structured Data

For the temporal information, there are three common data types:

- **Valid time**: Valid time refers to the time period during which an object (event) happens and it is maintained in the real world, or the object is true in the real world.
- **Transaction time**: Transaction time refers to the time during which a database object is operated. It denotes the time period during which a database fact is stored in the database.
- **User-defined time**: User-defined time denotes the time that is defined according to the user's own needs or understanding.

Valid time and transaction time are common data types that are used in temporal information. There are some models that support valid time, some other models support transaction time, and some support both.

The data of each web site is separate and there is no specific model. There are some self-descriptive, dynamic variability and a certain degree of hierarchy in web data itself. Such characteristics of web data are called semi-structured. Semi-structured is opposed to the fully-structured traditional structured data. The semi-structured are the most prominent feature of web data. XML is a way to express semi-structured data.

In order to express the time-varying semi-structured data, most of the works use the graph-based data model to extend temporal.

Chawathe et al. (1999) proposed a general model DOEM to express the time-varying semi-structured data. The difference from his previous work (Chawathe et al. 1996) is that the changes are not retained for the comparison of two versions, but all changes will be retained in the graph-based model in order to let users directly query the changes. He also proposed a language called Chorel, which can query the data and the change of the table in DOEM. The main contributions are focused on how to query and express the change of content.

Literature (Chawathe et al. 1999; Oliboni et al. 2001) proposed a graph-based semi-structured temporal data model TGM. This model is used to add the time interval of the existence of object into the object to extend the valid time. It can use a query language similar to SQL.

Literature (Dyreson et al. 1999) considered that the edge of the DBJ model have the user-defined property. Edge is a more complex structure. Edge can bind with valid time, transaction time and a variety of meta-data. The biggest difference from other models was that edge structure could express metadata. (Combi et al. 2004) further discussed the temporal constraints based on GSMM model and used GSMM element model to express the above DOEM, TGM, DBJ models and compared GSMM element model with these three models.

6.3 Temporal XML Model and Query Mechanism

How to express and query temporal information triggers many temporal scholars' attention. With reference to research on the expansion of the XML data model, Ali and Pokorny (2006) compared a variety of temporal XML data models. The literature (Currim et al. 2004) had made the expansion of the schema to express temporal data. Some work is done on the expansion of the query language. This paper, in accordance with the work of the literatures, has divided the research work into four categories, depending on whether it extends the model and whether it extends the query language.

The first category extends the temporal into both the XML data model and XML query language (such as XPath). In Vaisman and others' view (2004), temporal XML document was expressed as a directed acyclic graph, through the introduction of the concept of "version node" and each edge was added a "temporal element tag" to support "valid time." In that paper, they achieved the temporal query through the expansion of the XPath syntax, increasing some "temporal functions" (built-in functions). In addition, they also discussed and achieved four approaches to map the data model into a temporal XML. After the mapping, any property or element of the XML document will have a valid time interval string. Finally, they also accord the size of the mapped XML documents to compare the four mapping methods, and increase the property of ID, IDREF, as well as the thought of version node.

In contrast to the "increase in temporal tags to the edge", Zhang and Dyreson (2002) increased the valid time stamp for each node. In additon, they increased a "valid timeline" for Xpath to achieve temporal query.

Different from the above-mentioned papers, Dyreson et al. (2001) first conducted a research on the method to support "transaction time" in XML. Different from all the above models, they proposed a model that added the transaction time stamp to each node and edge. In addition, they added some "transaction timeline", "the test node of transaction time" and "the context of transaction time" for XPath to support the temporal queries.

These methods, studied through increased temporal tag for the node or edge of the XML data model to introduce the temporal information, discussed on the method to map these new tags to the XML document and expressed the temporal of these new tags. Finally, these methods increased the number of axes, node test and the extension of XPath to support temporal queries.

The second category is not to extend any of the XML data model and Xpath/XQuery.

This type of research is represented by Wang Fusheng and Zaniolo (2009). This type of work did not extend any data model, but used the self-defined and scalability of XML to describe the property of temporal information element and then used XQuery to support the user-defined function. The research defined some function of time to simplify and facilitate the temporal query.

Compared to the traditional relationship table and SQL, Wang Fusheng considered XML and XQuery to better support the temporal information. Therefore, their research of temporal XML focused on how to use temporal XML to achieve temporal database (how to use XML to express the traditional relationship table and the historical changes of XML database). It also focused on the document version management of the web (Wang et al. 2005b). It increased the *tstart* and *tend* property of each element of the XML document to describe the temporal information (the XML document is called the H-document).

In the literature, they introduced the concept of "H-table" between the temporal XML and the traditional "Rollback Database" (supporting transaction time). The "H-table" can release the traditional "Rollback Database" for the temporal XML and these temporal XML documents are called "H-documents". All the "H-tables" are in line with "temporally grouped" of the temporal data model. "H-document" and the traditional "rollback database" can be mutual converted by the "H-table". Finally, they also introduced the method to use XQuery to query temporal information in the "H-document".

In addition, Wang Fusheng and others (Wang and Zaniolo 2003, 2004; Wang et al. 2005) applied the thought of the literature to the traditional "historical database" (supporting the valid time) and "bitemporal database" (supporting the valid time and transaction time), respectively. They introduced the method to release the traditional "historical database" and "bitemporal database" for the "V-document" and "BH-document".

In the "H-document", "V-document" and "BH-document", Wang Fusheng also increased *tstart*, *tend* and *vstart*, *vend* of the temporal properties to each element of XML document to describe the transaction and valid time.

In addition, Wang and Zaniolo (2009) proposed how to express the transaction changes of the relational database in XML, and based on this thought, used the temporal XML to implement the rollback database system ArchIS, which supports transaction time. (Wang and Zaniolo 2009) was the system's technical report. In that system, users can use XQuery to query the released temporal XML (that is "H-document"). These XQuery statements are translated into the corresponding SQL statements of the H-table. Finally, the system will release the query results of the H-table for the temporal XML and then put temporal XML back to the user. In addition, ArchIS also used the "clustering" technology of traditional relational database to improve the speed of temporal query.

Grandi and Mnadreoli (2000) introduced the valid time for the web documents, without changing any existing web technologies, but by only directly using the existing XML standards. It used XML schema to define a ⟨valid⟩ element tag. Then, they added ⟨valid⟩ sub-elements to all elements of the temporal XML document to describe the valid time of the elements. Then, in accordance with the user-defined valid time text, used XSL technology to filter temporal information in order to achieve some of the temporal inquiries.

The third category is not to extend the XML data model, but extend the Xpath/XQuery. The ideas of problem-solving are: to develop a sophisticated DBMS that supports the temporal information processing is a very difficult job. If we use the traditional XQuery engine to achieve temporal query, not only will the code be very prolix, but also will bring a great deal of inconvenience to programmers. Therefore, the feasible solution to extend the query language is to use the sophisticated existing DBMS, and add a middle layer between the user applications and the traditional DBMS. The middle layer will translate the extended temporal XML query command and pass it to DBMS, without changing the underlying DBMS. The representatives of this type of work are by C. Gao and Kjetil Nørvåg.

Gao and Snodgrass's work (Gao and Snodgrass 2003a; 2003b) focused on extending the XQuery grammar to form τXQuery language. Therefore, the new temporal query language can maintain compatibility with the non-temporal data query, and the original XQuery grammar was almost unchanged and was only extended.

Nørvåg and others (2003) added the temporal operator into SQL and created a TEXOR system based on Oracle. The system extended the transaction time.

Whether τXQuery or TXSQL language, the essence is to translate the new temporal XML Query language into XQuery by adding a middle layer. The benefits of this method can make use of the existing XML data management system. It is more economical in this way for the organization who has already invested in the database system. However, this method is difficult to optimize the temporal XML document query and design the specialized index.

The fourth category is the XML data model with temporal expansion but with no expansion of the query language.

This method learned from the XML documents the method to store in the relational database and the method to query using SQL. It stored the XML document to express the valid time in temporal relational databases.

Amagasa et al. (2001) expanded the Xpath data model of temporal for the valid time, to form the TXPath model and based on the temporal relational database, TXPath model implemented the temporal XML database. Their main work focused on how to use the temporal relational database to store the temporal XML information and used the method of (Shimura et al. 1999) to store the XML documents in relational database. In order to reduce the cost of updating, TXPath pattern was designed. It used SQL to achieve the query of document data.

In addition, De Capitani (2002) designed a model that can access the temporal XML documents. However, his research focused on authorizing, rather than releasing the information of the temporal XML documents.

References

[1] Ali K, Pokorny J (**2006**) Acomparison of XML-based temporal models. In: Advanced Internet Based Systems and Applications: Second International Conference on Signal-Image Technology and Internet-Based Systems, SITIS 2006, Hammamet, Tunisia

[2] Amagasa T, Yoshikawa M, Uemura S (**2001**) Realizing temporal XML repositories using temporal relational databases. CODAS: 63 – 68

[3] Chawathe SS, Abiteboul S, Widom J (**1999**) Managing historical semistructured data. Theory and Practice of Object Systems 5(3): 143 – 162

[4] Chawathe SS, Rajaraman A, Garcia-Molina H, Widom J (**1996**) Change detection in hierarchically structured information. SIGMOD: 493 – 504

[5] Chien SY, Tsotras VJ, Zaniolo C (**2000**) Version management of XML documents. In: WebDB Workshop

[6] Chien SY, Tsotras VJ, Zaniolo C (**2001**) Copy-based versus edit-base version management schemes for structured documents. In: RIDE

[7] Combi C, Oliboni B, Quintarelli E (**2004**) Specifying temporal data models for semistructured data by a constraint-based approach. In: Proceedings of the 19th ACM Symposium on Applied Computing (SAC), Track on Internet Data Management, ACM Press, pp 1103 – 1108

[8] Currim F, Currim S, Dyreson C, Snodgrass RT (**2004**) A tale of two schemas: creating a temporal schema from a snapshot schema withτXSchema. In: EDBT

[9] De Capitani S (**2002**) An authorization model for temporal XML documents. In: Proceedings of SAC'02, pp 1088 – 1093

[10] Dyreson C, Böhlen MH, Jensen CS (**1999**) Capturing and querying multiple aspects of semistructured data. In: Proceedings of the Conference on Very Large Databases (VLDB), Edinburgh, Scotland, September , pp 290 – 301

[11] Dyreson CE, et al (**2001**) Observing transaction-time semantics with TTXPath. In: Proceedings of the 2nd International Conference on Web Information Systems Engineering (WISE'01), Volume 1, 193

[12] Gao C, Snodgrass R (**2003a**) Syntax, semantics and query evaluation in the τXQuery temporal XML query language. Time Center Technical Report TR-72

[13] Gao C, Snodgrass R (**2003b**) Temporal slicing in the evaluation of XML queries. In: VLDB, pp 632 – 643

[14] Grandi F (**2004**) Introducing an annotated bibliography on temporal and evolution aspects in the world wide web. SIGMOD Record 33(2): 84 – 86

[15] Grandi F, Mandreoli F (**2000**) The valid Web: an XML/XSL infrastructure for temporal managementof web documents. In: ADVIS

[16] Gray J (Microsoft) (**2004**) The next database revolution. In: Proceedings of the ACM SIGMOD International Conference on Management of Data. Paris, France. SIGMOD, Jun 13 – 18, pp 1 – 4

[17] Meng XF, Zhou LX, Wang S (**2004**) State of the art and trends in database research. Journal of Software 15(12): 1822 – 1836

[18] Nørvåg K (**2004**) The design, implementation, and performance of the V2 temporal document database system. Journal of Information and Software Technology 46(9): 557 – 574

[19] Nørvåg K, Limstrand M, Myklebust L (**2003**) TeXOR: temporal XML database on an object-relational database system. Ershov Memorial Conference: 520 – 530

[20] Oliboni B, Quintarelli E, Tanca L (**2001**) Temporal aspects of semistructured data. In: Proceedings of the 8th International Symposium on Temporal Representation and Reasoning (TIME-01), IEEE Computer Society Press, pp 119 – 127

[21] Shimura T, Yoshikawa M, Uemura S (**1999**) Storage and retrieval of XML documents using object-relational databases. In: Proceedings of the 10th International Conference on Database and Expert Sysrems Applications (DEXA'99), LNCS 1677, pp 206 – 217

[22] Tian F, DeWitt DJ, Chen J, Zhang C (**2002**) The design and performance evaluation of various XML storage strategies. In: SIGMOD Record

[23] Vaisman A, Mendelzon AO, Molinari E, Tome P (**2004**) Temporal XML: Data model, query language and implementation. http://www.cs.toronto.edu/~avaisman/ papers.html

[24] Wang F, Zaniolo C (**2003**) Temporal queries in XML document archives and web warehouses. In: TIME-ICTL

[25] Wang F, Zaniolo C (**2004**) XBiT: an XML-based bitemporal data model. In: ER

[26] Wang F, Zaniolo C (**2009**) Temporal queries and version management for XML document archives. Journal of Data and Knowledge Engineering (DKE)

[27] Wang F, Zaniolo C, Zhou X (**2009**). ArchIS: an XML-based approach to transaction-time temporal database systems. The International Journal on Very Large Data Bases (VLDBJ)

[28] Wang F, Zhou X, Zaniolo C (**2005a**) Efficient XML-based techniques for archiving, querying and publishing the histories of relational databases. Time Center TeEchnical Report

[29] Wang F, Zhou X, Zaniolo C (**2005b**) Temporal XML? SQL strikes back! In: Proceedings of the 12th International Symposium on Temporal Representation and Reasoning (TIME'05), pp 47 – 55

[30] Wong K, Lam N (**2003**) Efficient re-construction of document versions based on adaptive forward and backward change deltas. Lecture Notes in Computer Science 2736, Database and Expert Systems Applications (DEXA), Germany, pp 266 – 275

[31] Wuwongse V, Yoshikawa M (**2004**) Temporal versioning of XML documents. ICADL: 419 – 428

[32] Zhang S, Dyreson C (**2002**) Adding valid time to XPath. In: DNIS

7 Data Operations Based on Temporal Variables

Xiaoping Ye[1,2], Yong Tang[1,2+], and Bin Xiang[2]

[1] Computer School, South China Normal University, Guangzhou 510631, P.R. China
[2] Department of Computer Science, Sun Yat-sen University, Guangzhou 510275, P.R. China

Abstract This chapter discusses the processing of temporal variables in bitemporal database. The use of temporal variables *Now* and *UC* can be greatly useful in temporal database, but at the same time, also brings some problems. This chapter first extends the RDM data model proposed by Snodgrass, and builds a temporal variables based bitemporal data model VTRDM. Subsequently, this chapter proposes the bitemporal data update algorithm in VTRDM. Besides, it presents a data-querying algorithm (selecting, joining and projecting) by way of studying the complicated semantics of the temporal variables, which may result in the indeterminacy of the query. The bitemporal query and update method are closed in VTRDM.

Keywords *variables-based bitemporal relation data model, temporal data updating, complicated semantics, closing theorem*

7.1 Introduction

The bitemporal database supports both valid time and transaction time. Usually bitemporal database uses two symbols *Now* and *UC* as temporal variables. As constrained variables, *Now* and *UC* are different from the widely used default symbol *Null*. The introduction of temporal variables is a great help to temporal database, but at the same time it also brings some problems. One of these problems is to design a bitemporal algebra data model with temporal variable semantics and its implementation with TSQL2.

Valid time (VT) and transaction time (TT) are two elemental concepts. Valid time reflects the life circle of the fact in modeled reality, while the transaction time reflects the life cycle of the fact stored in database system. Both the temporal variables can enhance the efficiency and reliability of database systems.

Usually the transaction time is determined by system. Hence, the semantics of *UC* is relatively simple. However, the valid time is usually provided by users (or

---

mcsyxp@mail.sysu.edu.cn

applications). In practical applications the temporal variable *Now* may derive several meanings. Theoretically, the valid time and transaction time are orthogonal, but in practical processing, it should consider their relationships.

Historical database only supports valid time. Although the research on historical database provides valuable hints to the research on bitemporal database, the bitemporal database is not a simple coupling with historical database and transaction time. If a temporal database contains temporal variables, then it is called a (temporal) variable database, else it is called a (temporal) ground database.

The research on variable database mainly states two aspects. Firstly, there is some work on the semantics of temporal variable and its temporal data operations. The importance of *Now* is outlined and the semantics of temporal variable *Now* on different applications is studied (Clifford et al. 1997, 2000; Torp et al. 2004). Because of the difficulty of dealing with the variables, the concepts of now-relative time and now-relative indeterminate time are introduced as the basis for the binding of the variables, which may carry out the corresponding temporal data operations by using the commercial database language such as SQL or temporal database language such as TSQL (Clifford et al. 1997; Ye and Tang 2005). In this way, the query (Clifford et al. 1997) and the update (Ye and Tang 2005) are studied separately. In addition, the expression of temporal tuples with *Now* is presented to process the current time semantic of this variable, which uses the point querying language (Stantic et al. 2004). Secondly, there are some issues on practical processing technologies related to the variable databases. The temporal grouping and coalescing of now-relative time are studied, which play an important role in operations of temporal relational projects and the temporal XML data processing (Dyreson 2003). The technology of temporal indexing with the temporal variables is proposed by consulting the special indexing, such as GR tree and 4R tree (Bliujute et al. 1998, 2000; Stantic et al. 2004). The method of transforming the variables databases into temporal XML documents is studied, which may change the querying of the temporal data into the querying of the temporal XML data (Wang and Zaniolo 2003a, 2003b).

Some basic issues need to be studied further.

- It is necessary to propose a perfect model based on the temporal variables, in order to deal with the data with variables, and complete the temporal operations efficiently in the framework of variables.
- It is needed to study the variable *Now* in the light of the combination of the valid time and the transaction time because it is quite different between the combination and separation discussion of these two kinds of time.
- The semantics of the temporal variables should be fully studied. However, the current work focuses on the now-relative time situation, which is only one of the various semantics of variable *Now*.
- It is necessary to simplify the arithmetic of binding the temporal variables in order to implement conveniently in the commercial databases.

Aimed to resolve the issues mentioned above, the main works in this chapter are as follows:

- It presents a relatively more refined bitemporal data model, which is suitable to the temporal variables, by proposing the data schema and introducing the temporal order relation. The model is closed to the temporal operations (updating and querying).
- It studies the variables of the valid time and the transaction time uniformly based on the combination of these two kinds of time. It then devises the bitemporal updating operations (inserting, deleting and modifying).
- It studies the binding operations of the temporal variables within the temporal query (connecting, projecting and selecting) by analyzing several semantics on the variables. The temporal coalescing in the temporal project may provide the positive references to other related researches such as temporal object oriented databases and temporal XML databases.

As variable database is an important issue of the general databases, the temporal database with variables is also an important issue of the temporal databases. The practical application of temporal databases with the indeterminate time stamps may result in the use of the temporal variables. The relationship between the valid time and transaction time in the databases results in various semantics of the variables. Because the temporal data processing needs bind specific values to temporal variables, we should study different temporal binding methods in different semantic situations. Therefore, the study on the temporal variables not only concerns the essential relationship of valid time and transaction time, but is also necessary to study the integrality and systematization of the application of temporal databases. It is noticeable that the work in the references (Bliujute et al. 1998, 2000; Stantic et al. 2003, 2004; Wang and Zaniolo 2003a, 2003b) discussed the practical issues in terms of the single semantic of the variable *Now*. The work of this chapter may be significant to the study and application of the temporal indexing and temporal XML related to the references mentioned above.

This chapter is organized as follows. The next section proposes a bitemporal data model, BRDM, extended from the data model BDM, which is suitable to temporal variables. In section 7.3, we discuss the update method of temporal variable data in bitemporal database. In section 7.4, we analyze the temporal variable semantics, and discuss the determinate and indeterminate query results. Subsequently, we study the temporal querying technology and propose an algorithm.

7.2 Data Model Based on Temporal Variables

7.2.1 Order and Temporal Variables

7.2.1.1 Periods and temporal order

The minimal time unit in a system is called time granularity. The set of the instants between the maximum system supported time *maxS* and minimum system supported time *minS* is a time domain of the given system and is denoted as $\Gamma = [minS, maxS]$.

$t_1, t_2 \in \Gamma(t_1 \leqslant t_2)$, the set of the instants between t_1 and t_2 is defined as the time period whose starting point is t_1 and ending point is t_2, and is denoted as $T = [t_1, t_2] = [T_s, T_e]$. A single instant t_0 may be regarded as the period whose starting point equals to the ending point. The periods are the basic time elements in Γ, and the set of all the periods in Γ is denoted as *TP*.

An element in the power set 2^Γ of Γ is called a time element, which consists of the set of instants and the set of periods. The periods are only considered in this paper and related discussions may be extended to the general time elements, even though the extension may be quite complex.

Definition 7.1 (Orders of periods) Let $T_1 = [T_{1s}, T_{1e}]$, $T_2 = [T_{2s}, T_{2e}] \in TP$, where the T_{is} and T_{ie} are the starting point and the ending point of T_i $(i = 1,2)$, respectively.

① **Relation met behind "$\precsim$":** $T_1 \precsim T_2 \Leftrightarrow T_{1e} = T_{2s}$, where T_2 is called the period met behind T_1, and is denoted as $T_2 = (\precsim T_1)$.

② **Relation succeeded behind "$\prec$":** $T_1 \prec T_2 \Leftrightarrow T_{1e} + 1 = T_{2s}$, where T_2 is called the period success behind T_1, and is denoted as $T_2 = (\prec T_1)$.

③ **Relation adjacent behind"$\backsim$":** $T_1 \backsim T_2 \Leftrightarrow T_{1e} < T_{2s}$, where T_2 is called the period adjacent behind, and is denoted as $T_2 = (\backsim T_1)$.

④ **Relation adjacent behind directly"$\underline{\backsim}$":** Given two periods T_i and T_j in *TP*, if $T_i \backsim T_j$, and there is no $T_k \in TP$ so that $T_i \backsim T_k$ and $T_k \backsim T_i$, then T_i is called directly adjacent behind T_j in *TP*. We also call T_j the directly adjacent behind period of T_i, and denote it as $T_j = (\underline{\backsim} T_i)$.

⑤ **Temporal connection relation "~":** Let Ω be the set of all periods of valid time and u, v be two tuples. If $VT(u) \cap VT(v) \neq \varnothing$, we call u and v temporal join with each other and denote it as $\langle u, v \rangle$. If there are some tuples $u_1, u_2, \cdots, u_m$ with $\langle u, u_1 \rangle$, $\langle u_1, u_2 \rangle \cdots$, $\langle u_{n-1}, u_n \rangle$, $\langle u_n, v \rangle$, we call u and v as temporally connected with each other, and denote it as $u \sim v$. It can be proved that the temporal connection "~" is an equivalent relation, which has the properties of reflexivity, symmetry and transition in the set of all temporal tuples.

7.2.1.2 Temporal variables

As mentioned above, *VT* describes the validity period of a fact in modeled reality. *TT* is the time of systems, which denotes the time during which the fact is stored in the database. These two periods are denoted as $VT = [VTs, VTe]$ and $TT = [TTs, TTe]$, respectively.

The ending points of the period may be divided into two types. If the ending points of the periods are all determinate, we call the corresponding temporal databases the ground databases, otherwise, we call them the variable database. The symbols *Now* and *UC* (until changed) are used to mark the indeterminant ending points of the valid and transaction time, respectively. *Now* states that the life cycle of a given data is true until the current time (*CT*), and *UC* states that the data is not changed until the current time. Obviously, the ground database is one special case of the variable database.

It is noticeable that the variable *UC* may be bound to *CT* by the system automatically in practical operations because the transaction time is generated by the systems and has a simple semantic. However, since the variable *Now* is the time provided by users, it has other semantics resulting in different contexts besides the original semantic ("at the present time"). Therefore, it is more complex than the variable *UC*.

7.2.2 Main Body Instances

7.2.2.1 Temporal independent attributes

In a given temporal relation *TR*, there may be some attributes that are unrelated with time. Let *A* be an attribute in *TR*. The set of the valid time periods of the values of *A* is called the set of valid time of *A*, and is denoted as Ω_A. Let *f* be a mapping from the attribute field V_A to Ω_A: f: $V_A \rightarrow \Omega_A$. If, for $\forall x \in V_A$, $f(x) = \cup \Omega_A$, then *A* is called a **temporal independent attribute**, else, the temporal dependent attribute. For example, in temporal relation (name, ID, title, department, salary, *VTs*, *VTe*, *TTs*, *TTe*), name and ID are the temporal independent attributes, and title, department and salary are the temporal dependent attributes. In a given temporal relation, some temporal attributes do not vary as the time evolves. That is to say, they are unchanged relatedly in a certain circumstance. The attributes that do not contain the time stamps are called non-temporal attributes and the temporal independent attributes are a kind of the non-temporal attributes, which are more strongly constrained.

7.2.2.2 Subject attributes and main body instances

Definition 7.2 (Subject attributes and main body instances) Let *A* be the temporal independent attribute of a given temporal relation $\mathcal{R}$, and T_1, $T_2 \in TR$. We define an equivalent relation among the set of *TR* in the light of *A*: $T_1 \mathcal{R} T_2 \Leftrightarrow \Pi_A(T_1) = \Pi_A(T_2)$,where $\Pi_A(T)$ is the projection of the tuple *T* on attribute *A*. Tuples in *TR* are grouped by the value of *A*, such that a large temporal relation on *TR* is further divided into small ones. *A* is called the main body attribute of *TR*, and the equivalent class obtained by $\mathcal{R}$ is called the **main body instance** (*MI*) of *TR* on *A*.

Example 7.1 A temporal relation Employee is shown in Table 7.1.

Here, *Name* is a temporal independent attribute. The first three tuples describe the state of the same main body and they constitute an *MI*. The 4th to 6th tuples also consist of an *MI* instance, and the last tuple an *MI*.

Table 7.1 Bitemporal relation "Employees"

Name	Dept	Title	Salary	Vs: Ve	Ts: Te
Bob	Math	Vice-leader	6000	2004-01-01: *Now*	2004-01-10: 2005-01-09
Bob	Math	Vice-leader	6000	2004-01-01: 2004-12-31	2005-01-10: *UC*
Bob	Math	leader	7500	2005-01-01: *Now*	2005-01-10: *UC*
White	IS	Leader	7500	2003-01-01: *Now*	2003-01-10: 2006-01-10
White	IS	Leader	7500	2004-02-01: 2004-10-01	2006-06-10: *UC*
White	CS	Leader	8000	2006-01-01: *Now*	2006-01-10: *UC*
Raul	CS	Leader	7500	2003-10-01: *Now*	2003-01-10: *UC*

7.2.2.3 Snapshot equivalent groups

For a given MI, the tuples in it may also be grouped according to their non-temporal parts being the same or not. The tuples in the same group only differ at their time stamps. Such groups are called snapshot equivalent groups.

Definition 7.3 (Snapshot equivalent groups, *SG*) If the set of tuples $\{T_1, T_2, \cdots, T_k\}$ in a given *MI* satisfies the following conditions, we call the set a **snapshot equivalent group** of *MI*, and denote it as *SG*.

(1) The non-temporal attribute values of the tuples are all the same.

(2) The transaction time of the tuples satisfy the succeeded behind relation:

$$TT(T_1) \prec TT(T_2) \prec \cdots \prec TT(T_k).$$

(3) The valid time of the tuples satisfy:

$$VT(T_1) \cap VT(T_2) \cap \cdots \cap VT(T_k) \neq \varnothing.$$

(4) There is at most only one tuple whose ending point of transaction time is *UC*.

If we order the tuples in *SG* by(2), *SG* may be denoted as $SG = List\{T_1, T_2, \cdots, T_k\}$.

From condition (2), $TTe(T_1)$ is the least and $TTe(T_k)$ is the largest in $\{TTe(T_1), TTe(T_2), \cdots, TTe(T_k)\}$. We denote $T_1 = \min(SG)$ and $T_k = \max(SG)$. $T_k = \max(SG)$ is called the latest tuple in *SG*. The tuple *T* with ending point *UC* is the latest tuple. However, on the reverse, $TTe(\max(SG))$ may not be necessarily *UC*. The latest tuple is unique. In *MI*, the tuple *T*, which satisfies $VTe(T) = Now$ and $TTe(T) = UC$, is called the current tuple of *MI*, and is denoted as $Tnu(MI)$. If there is $Tnu(MI)$, then $Tnu(MI)$must be unique.

Definition 7.4 (Succeeded behind element of *SG* and succeeded behind *SG*) Let T_i, T_j be two elements in the same *SG*. If $TT(T_i) \prec TT(T_j)$, then T_j is called the succeeded behind tuple of T_i, and is denoted as $T_i \prec T_j$. Let SG_1 and SG_2 be two snapshot groups in *MI*. If $VT(\max(SG_1)) \backsimeq VT(\max(SG_2))$ then SG_2 is called the succeeded behind *SG* of SG_1, and is denoted as $\max(SG_1) \backsimeq \max(SG_2)$.

In Example 7.1, insert a tuple into *MI* 'White':

(white, Is, leader, 7500, 2003-01-01: 2004-01-01, 2006-06-10: *UC*)

The new obtained *MI* is shown in Table 7.2.

Table 7.2 MI White of inserting new tuple

Name	Dept	Title	Salary	Vs: Ve	Ts: Te
White	IS	Leader	7500	2003-01-01: 2004-01-01	2006-01-10: *UC*
White	IS	Leader	7500	2003-01-01: *Now*	2003-01-10: 2006-01-10
White	IS	Leader	7500	2004-02-01: 2004-10-01	2006-06-10: *UC*
White	CS	Leader	8000	2006-01-01: *Now*	2006-01-10: *UC*

The *MI* may be divided into three *SG*. The first tuple constitutes SG_1, the second and third SG_2, and the fourth SG_3, according to Definition 7.3. The latest tuple of SG_2 is the third tuple, and the current tuple is the fourth tuple in this *MI*, according to Definition 7.4. SG_2 is the succeeded behind tuple of SG_1, and SG_3 is the succeeded behind tuple of SG_2.

7.2.3 Bitemporal Relation Model Based on Variables

The schema of Snodgrass's Representational Data Model (RDM) is $\{TR(A_1, A_2, \cdots, A_n; VTs, VTe, TTs, TTe)\}$ (Snodgrass 1987), where $A_1, A_2, \cdots, A_n$ are the non-temporal attributes in *TR*, *VTs*, *VTe* and *TTs*, *TTe* represent the starting points and end points of the valid and transaction time, respectively, and these time stamps satisfy the requirement of 1NF. In addition, the *VTe* and *TTe* may be the temporal variables *Now* and *UC*. For physical implementation, the values of *Now* and *UC* evolve along with time increase in Γ, and the maximum (Γ) may be enacted, for instance, maximum (Γ) = 9999-12-31. The idea in RDM is that the time stamps are regarded as one of the relational attributes.

Definition 7.5 (Bitemporal data relational model with variables, BRDM) If a *TR* in RDM satisfies the following conditions, we call *TR* a **bitemporal relation model** with variables:

① There is a main body attribute *A*, and the set of main body instances MI divided by the equivalent relation "$\mathcal{R}$" result from *A*.

② Every *MI* is divided into the set of *SG*, that is, the *MI* consists of $SG_1 = \{T_{11}, \cdots, T_{1k1}\}$, $SG_2 = \{T_{21}, \cdots, T_{2k2}\}, \cdots, SG_n = \{T_{n1}, \cdots, T_{nkn}\}$, where, $T_{1k1}, T_{2k2}, \cdots, T_{nkn}$ are the latest tuples, respectively.

③ The latest tuples $T_{1k1}, T_{2k2}, \cdots, T_{nkn}$ of $SG_1, \cdots, SG_n$ hold: $VT(T_{1k1}) \backsim \cdots \backsim VT(T_{nkn})$ and $VT(T_{1k1}) \cap \cdots \cap VT(T_{nkn}) = \varnothing$, that is, the periods of the valid time do not intersect each other.

④ There is at most one current tuple in an *MI*.

The *SG* in *MI* can be ordered as $MI = \text{List}\{SG_1, \cdots, SG_n\}$ by the relation "$\backsim$".

With reference to the conditions mentioned above, the first point guarantees that the relations are relatively simple and regular, the second further construct parts from every *MI*, the third guarantees the separability among the elements in the set of *SG* and the elements are ordered, the fourth guarantees that the model is consistent with the conventional intution. The TR in BDM in Example 7.1 is a TR in BRDM.

Let $MI = \text{List}\{SG_1, SG_2, \cdots, SG_n\}$. If $SG_1 = \text{List}\{T_{11}, T_{12}, \cdots, T_{1q}\}$, T_{11} is called a minimum tuple in MI, and is denoted as $T_{\min}(MI)$.

Definition 7.6 (Versions of temporal data) Let T_i, $T_j \in MI$. If $TT(T_i) \subseteq TT(T_j) \vee TT(T_j) \subseteq TT(T_i)$, then T_i and T_j are in the same version. If $TTe(T_0) = UC$, T_0 is called the **current version data**, else, the **non-current version data**.

7.3 Data Updating

The operations (updating or querying) on temporal variables *Now* and *UC* are to bind specific values to them according to specific applications, so that the variables are transformed into constants. Three main updating operations in relation data are inserting, deleting and modifying. The inserted tuple is the latest tuple in the corresponding SG, and the deleting and modifying operations must be performed on the latest tuples.

7.3.1 Data Inserting

Let T be the to-be-inserted tuple. The key of inserting T is to determine the inserting position, so that the model is closed on the operation of inserting. We may take the following steps to complete the insertion.

Step 1: Check the integrity. If it violates the integrity, then stop.

Step 2: Search corresponding MI. If the result returns *null*, then go to Step 5, else, continue.

Step 3: If $VT(T) \backsim VT(T_{\min}(MI))$, then the inserting position is before the first tuple in MI. In this case, go to Step 5, else, search the tuple T_0 in all the latest tuples of MI, which satisfies $VT(T_0) \backsim VT(T) \backsim VT(\backsim(T_0))$. If there is a tuple T_0, then go to Step 4, else, $\{T\}$ consists of a new SG, and its inserting position is after the last SG. In this case, go to Step 5.

Step 4: If T_0 is non-current version, the inserting position of T is after T_0. In this case, go to Step 5.

If T_0 is current version,

(1) Bind UC to the current time CT, when the operation has occurred.

(2) Insert T_0' with $v(T_0') = v(T_0)$; $VTs(T_0') = VTs(T_0)$; $VTe(T_0') = VTs(T) - 1$; $TTs(T_0')$ is bound to the operating time in system $TTe(T_0') = UC$.

(3) The inserting position of new tuple T is after T_0'. *Now*, go to Step 5.

Step 5: Insert new tuple T.

Example 7.2 Let TR = (Name, Dept, Title, Salary; VTs, VTe, TTs, TTe) be an employee relation. Black works at department EN as an engineer with salary 7500 from $VTs = 2003\text{-}01\text{-}01$. The corresponding tuple is stored in the database at $TTs =$ 2003-01-10, that is, the lagging transaction time to the valid time is $\Delta = 9$(let the time granularity be 'day'). *TR* is shown in Table 7.3.

Table 7.3 Black's record(1)

Name	Dept	Title	Salary	Vs: Ve	Ts: Te
Black	EN	Engineer	7500	2003-01-01: *Now*	2003-01-10: *UC*

Black is promoted at 2004-01-01 and the relation is updated at 2004-01-10. Insert a new tuple *T*:

(Black, En, Sr Engineer, 8500, 2004-01-01, *Now*, 2004-01-10, *UC*).

Since the original record is a single tuple, we only explain Step 4 and Step 5.

Step 4: Keep the original tuple as historical version. At the same time, the ending point of the original record *UC* is bound to $CT = 2004\text{-}01\text{-}10$ automatically by the system, which indicates that the version of data ends at this moment. In addition, *Now* in the original tuple is bound to $VTs - 1 = 2003\text{-}12\text{-}31$. In this way, we obtain the tuple (Black, EN, Engineer, 7500, 2003-01-01, 2003-12-31, 2004-01-10, *UC*) and store it as a current version data. The result is shown in Table 7.4.

Table 7.4 Black's record (2)

Name	Dept	Title	Salary	Vs: Ve	Ts: Te
Black	EN	Engineer	7500	2003-10-01: *Now*	2003-01-10: 2004-01-09
Black	EN	Engineer	7500	2003-01-01: 2004-12-31	2004-01-10: *UC*

Step 5: The latest record *T* is inserted, which is kept as the current version data.

Table 7.5 Black's record (3)

Name	Dept	Title	Salary	Vs: Ve	Ts: Te
Black	EN	Engineer	7500	2003-01-01: *Now*	2003-01-10: 2004-01-09
Black	EN	Engineer	7500	2003-01-01: 2003-12-31	2004-01-10: *UC*
Black	EN	Sr. Engineer	8500	2004-01-01: *Now*	2004-01-10: *UC*

The first two records describe two different versions of the same state. The first record is the non-current version of the state and the second is the current version of the same state. The third record is the new inserted data and the symbols *Now* and *UC* indicate that the data belongs to the latest version.

If Black's salary is changed to 9000, the fact is updated at $CT = 2005\text{-}01\text{-}10$. We need to insert another new tuple T_1:

(Black, En, Sr Engineer, 9000, 2005-01-01, *Now*, 2005-01-10, *UC*)

We only explain Steps 3~5.

Step 3: The third tuple is the target T_0. In this case, the inserting position is after T_0.

Step 4: Bind $TT(T_0) = UC$ to $CT - 1 = $ 2005-01-09 as a "new" T_0 in the original position. Bind $VTe(T_0) = Now$ to $CT - \varDelta = $ 2005-01-01 as another T_0' stored after "new" T_0.

Step 5: The newly inserted tuple T_1 is inserted behind T_0'.

The result is shown in Table 7.6.

Table 7.6 Black's recording (4)

Name	Dept	Title	Salary	Vs: Ve	Ts: Te
Black	EN	Engineer	7500	2003-01-01: *Now*	2003-01-10: 2004-01-09
Black	EN	Engineer	7500	2003-01-01: 2003-12-31	2004-01-10: *UC*
Black	EN	Sr. Engineer	8500	2004-01-01: *Now*	2004-01-10: 2005-01-09
Black	EN	Sr. Engineer	8500	2004-01-01: 2004-12-31	2005-01-10: *UC*
Black	EN	Sr. Engineer	9000	2005-01-01: *Now*	2005-01-10: *UC*

The examples above show that *Now* is in fact bound with the past time, that is, *Now* has the semantics of the past time. In addition, as required by the definition of bitemporal database, we usually need to "insert" two records, which are the non-current version and current version of the original data, respectively. Therefore, compared to the historical databases, the operations in bitemporal database with variables are more subtle and more complex.

In the following section, we give a formal presentation of the inserting operation.

Let T_0 be the to-be-inserted tuple of $TR = \{MI_1, MI_2, \cdots, MI_m\}$. The data operations, in fact, transform one relation into another. In this way, we define a transformation operator Ω of the inserting operation as follows:

$$\Omega_{T_0}(TR) = \text{list}\{MI_1, \cdots, MI_j, \cdots, MI_{jp}\}$$

(1) If any $T_{jk} \in MI_j$, $v_A(T_j) \neq v_A(T)$, that is, T does not belong to any MI, then

$$\Omega_{T_0}(TR) = TR \cup \{T_0\}$$

(2) If there is $MI_j \in TR$ so that $v_A(T_0) = v_A(MI_j)$, that is, T should be inserted into MI_j, then, after inserting operation, MI_j will change:

$$\Omega_{T_0}(TR) = \{MI_1, \cdots, MI_{j-1}, MI_j', MI_{j+1}, \cdots, MI_m\}, \text{ and}$$

Case 1. if $VT(T) \backsim T_{\min}(MI_j)$, then $MI_j' = \text{List}\{\{T\}, MI_j\}$.

Case 2. if $VT(T) \backsim T_{\min}(MI_j)$ does not hold, but $\exists SG_{j0} \in MI_j$, where $TTe(T_{\max}(SG_{j0})) \neq UC$:

- if $VT(T_{\max}(MI_j)) \backsim VT(T) \backsim VT(T_{\max}(SG_k))(j_{0+1} \leqslant k)$,

$$MI_j' = \text{List}\{SG_{j1}, \cdots, SG_{j0}, \{T\}, SGj_{0+1}, \cdots, SG_{jp}\}$$

- if $VT(T_{max}(MI_j)) \backsim VT(T)$ but $VT(T) \backsim VT(T_{max}(SG_k))(j_{0+1} \leqslant k)$ does not hold,

$$MI_j' = \text{List}\{SG_{j1}, \cdots, SG_{jp}, \{T\}\}$$

Case 3. If $\exists SG_{j0}$, $TTe(T_{max}(SG_{j0})) = UC$, $[VTs(T_{max}(SG_{j0})), VTe(T) - 1] \backsim VT(T)$. Let $MI_j = \text{List}\{SG_{j1}, SG_{j2}, \cdots, SG_{jp-1}, SG_{jp}\}$, obviously, $T_{max}(SG_{j0}) \in SG_{jp}$. In addition, let SG_{jp} has k tuples. $MI_j' = \text{List}\{SG_{j1}, SG_{j2}, \cdots, SG_{jp-1}, SG_{jp}'\}$. In this case, $SG_{jp}' = \text{List}\{SG_{jp} \backslash \{T_{max}(SG_{j0}), \{T_k\}, \{T_{k+1}\}, \{T_{k+2}\}\}$, where:

$v(T_k) = v(T_{max}(SG_{j0}))$, $VT(T_k') = VT(T_{max}(SG_{j0}))$, $TTs(T_k') = TTs(T_{max}(SG_{j0}))$, $TTe(T_k') = CT$;

$VTs(T_{k+1}) = VTs(T_{max}(SG_{j0}))$, $VTe(T_{k+1}) = VTs(T) - 1$, $TTs(T_{k+1}) = CT$, $TTe(T_{k+1}) = UC$;

$T_{k+2} = T$.

7.3.2 Data Deleting

According to the definition of bitemporal data, the deleting operation of bitemporal relation only concerns the current version data. Besides, the "deleting" operation of bitemporal data is actually "logical" deletion instead of "physical" deletion. Hence, the deleting operation, in fact, turns the current version data into non-current version data. The operation can be done in the following steps:

Step 1: Search the *MI* and *SG* to which the to-be-deleted tuple belongs.

Step 2: Search the current tuple of the *SG*, bind *UC* to the *CT* when the operation occurred.

Step 3: Make the bound tuple the non-current version.

Example 7.3 Delete the last record in Table 7.5, where $CT = 2004\text{-}06\text{-}10$. The result is shown in Table 7.7.

Table 7.7 New relation after deleting

Name	Dept	Title	Salary	Vs: Ve	Ts: Te
Black	EN	Engineer	7500	2003-01-01: *Now*	2003-01-10: 2004-01-09
Black	EN	Engineer	7500	2003-01-01: 2003-12-31	2004-01-10: *UC*
Black	EN	Sr. Engineer	8500	2004-01-01: *Now*	2004-01-10: 2004-06-10

In the following section, we give a formal description of deleting:

Let $T_{j0} = \{v(T_{j0}), VT(T_{j0}), TT(T_{j0})\}$ be the to-be-deleted tuple in $SG_j = \text{List}\{T_1, T_2, \cdots, T_{j0-1}, T_{j0}, T_{j0+1}, \cdots, T_{jp}\}$. SG_j is transformed into a new snapshot equivalent group $SG_j' = \Delta_{tj0}(SG_j)$, where the transformation $\Delta_{tj0}(SG_j)$ is defined as follows:

$$\Delta_{tj0}(SG_j) = \text{List}\{T_1, T_2, \cdots, T_{j0-1}, T_{j0}', T_{j0+1}, \cdots, T_{jp}\}$$

where, $v(T_{j0}') = v(T_{j0})$, $VT(T_{j0}') = VT(T_{j0})$, $TTs(T_{j0}') = TTs(T_{j0})$, $TTe(T_{j0}') = CT$.

7.3.3 Data Modifying

In bitemporal database, the modification of current version data is, in fact, first "logically" delete the original record, and insert the updated new record to the current version. The main steps are as follows.

Let T_0 be the to-be-modified tuple and T_1 the new tuple.

Step 1: Search the to-be-modified tuple T_0 and logically delete it. The logically deleted tuple is T_0'.

Step 2: Insert T_1 behind T_0'.

Step 3: If there is a succeeded behind *SG*, in the *SG* to which T_0 belongs, then the newly inserted tuple T_1 may not satisfy the condition "③" in Definition 7.5. In this case, the modification needs to be propagated backward, that is, Step 1 and Step 2 have to be repeated.

Example 7.4 Let the to-be-modified tuple in Table 7.6 be T= (EN, Engineer, 7500, 2003-01-01:2003-12-31, 2004-01-10: *UC*).

T is going to be modified as T_1 = (EN, Engineer, 7500, 2003-01-01:2003-10-31, 2004-01-10: *UC*) at *TTs* = 2005-03-10.

Step 1: T is deleted logically. The result is shown in Table 7.8.

Table 7.8 Relation in processing of modification (1)

Name	Dept	Title	Salary	Vs: Ve	Ts: Te
Black	EN	Engineer	7500	2003-01-01: *Now*	2003-01-10: 2004-01-09
Black	EN	Engineer	7500	2003-01-01: 2003-12-31	2004-01-10: 2005-03-09
Black	EN	Sr. Engineer	8500	2004-01-01: *Now*	2004-01-10: 2005-01-09
Black	EN	Sr. Engineer	8500	2003-11-01: 2004-12-31	2005-03-10: *UC*
Black	EN	Sr. Engineer	9000	2005-01-01: *Now*	2005-01-10: *UC*

Step 2: T_1 is inserted into the relation in Table 7.8 and the result is shown in Table 7.9.

Table 7.9 Relation in processing of modification (2)

Name	Dept	Title	Salary	Vs: Ve	Ts: Te
Black	EN	Engineer	7500	2003-01-01: *Now*	2003-01-10: 2004-01-09
Black	EN	Engineer	7500	2003-01-01: 2003-12-31	2004-01-10: 2005-03-09
Black	EN	Engineer	7500	2003-01-01: 2003-10-31	2004-01-10: *UC*
Black	EN	Sr. Engineer	8500	2004-01-01: *Now*	2004-01-10: 2005-01-09
Black	EN	Sr. Engineer	8500	2004-01-01: 2004-12-31	2005-03-10: *UC*
Black	EN	Sr. Engineer	9000	2005-01-01: *Now*	2005-01-10: *UC*

Step 3: The first three tuples constitute SG_1, the fourth and the fifth tuples constitute SG_2, and the last tuple constitutes SG_3. The valid time period in the last tuple in SG_1 is [2003-01-01, 2003-10-31] and in SG_2 is [2004-01-01, 2004-12-31]. These two periods do not intersect, which violates the condition "③" in the definition, and the modification should be propagated backward. The result is shown in Table 7.10.

Table 7.10 Deleted temporal relation propagating backward

Name	Dept	Title	Salary	Vs: Ve	Ts: Te
Black	EN	Engineer	7500	2003-01-01: *Now*	2003-01-10: 2004-01-09
Black	EN	Engineer	7500	2003-01-01: 2003-12-31	2004-01-10: 2005-03-09
Black	EN	Engineer	7500	2003-01-01: 2003-10-31	2005-03-10: *UC*
Black	EN	Sr. Engineer	8500	2004-01-01: *Now*	2004-01-10: 2005-01-09
Black	EN	Sr. Engineer	8500	2004-01-01: 2004-12-31	2005-01-10: 2005-03-09
Black	EN	Sr. Engineer	8500	2003-11-01: 2004-12-31	2005-03-10: *UC*
Black	EN	Sr. Engineer	9000	2005-01-01: *Now*	2005-01-10: *UC*

The operation of modification may be described formally as follows.

The tuple to be modified is the last tuple in TR. Without loss of generality, we only consider the case that MI consists of the last tuples of every SG, that is,

$$MI_j = \text{List}\{T_1, T_2, \cdots, T_{j0-1}, T_{j0}, T_{j0+1}, \cdots, T_{jp}\}$$

where $T_1, T_2, \cdots, T_{j0-1}, T_{j0}, T_{j0+1}, \cdots, T_{jp}$ are all the last tuples of every corresponding SG.

If the tuple $T_{j0} = (v(T_{j0}), VT(T_{j0}), TT(T_{j0}))$ in MI_j needs to be modified to $Tj_1 = (v(T_{j1}), VT(T_{j1}), TT(T_{j1}))$, the transformation Γ of the modification is defined as follows:

$$\Gamma_{T_{j0}}(MI_j) = \text{List}\{T_{j1}', T_{j2}', \cdots, T_{j0-1}', T_{j0}', Tj_1, T_{j0+1}', \cdots, T_{jp}'\}.$$

Case 1. If T_{j0} is "maximum" tuple in $\{VT(T_1), VT(T_2), \cdots, VT(T_{jp-1}), VT(T_{jp})\}$ by the relation"⊆", then

$v(T_{j0}') = v(T_{j0})$, $VT(T_{j0}') = VT(T_{j0})$, $TTs(T_{j0}') = TTs(T_{j0})$, $TTe(T_{j0}') = TTs(T_{j1}) - 1$; $T_k' = T_k(k \neq j0)$.

Case 2. If $\exists T_{jq} \in MI_j$, $T_{j0} ⊆ T_{jq}$, then

- for $k \neq j_q$, $T_k' = T_k$;
- for T_{j0}', $v(T_{j0}') = v(T_{j0})$, $VT(T_{j0}') = VT(T_{j0})$, $TTs(T_{j0}') = TTs(T_{j0})$, $TTe(T_{j0}') = TTs(T_{j1}) - 1$;
- for $k = j_q$, $v(T_{jq}') = v(T_{jq})$, $VTs(T_{jq}') = VTe(T_{j1}') + 1$, $VTe(T_{jq}') = VTe(T_{jq})$, $TT(T_{jq}') = TT(T_{jq})$.

7.4 Data Querying

The critical point of data querying with variables is the binding of *Now*. This procedure is related to its semantics and it is necessary to analyze the different cases of *Now*.

7.4.1 *Now* in Current Versions

7.4.1.1 Analysis on semantics of variable

The introduction of *Now* is caused by the indetermination of the ending point in periods. The ending point *VTe* for a valid time period [*VTs*, *VTe*] can be divided into the following cases:

- ***VTe* is a determinate instant.** In this case, *VTe* is a determinate time point. It indicates that the life span of the corresponding data either ends in the past (*VTe*<*CT*) or in the future (*VTe*<*CT*). Here, *CT* represents the current time.
- ***VTe* is an indeterminate instant.** In this case, *VTe* indicates that the valid time of data monotonously increases as time lapses. This indeterminacy means that the valid time of the data does not end until "present time" and it is unsure when it will end in the future.

It is an important issue to study this uncertainty of ending point of valid time. There are two methods to deal with this problem (Clifford et al. 1997).

1. Constant method

The uncertain ending points are treated as constants. There are also two implementing ways.

(1) Method of maximal value: For example, *VTe* is treated as the maximal time point of the system. The maximal time of IBM "DB2" is 8000 years.

(2) Method of varying constants: *VTe* is regarded as the "variable", where *VTe* is bound to present day *CT* along with the every day evolvement.

The method of maximal value is simple, however, it cannot reflect the meaning of "indetermination", and is contrary to people's day-to-day knowledge. The maximal time of the system is too big for the practical valid time. Meanwhile, it is difficult for us to distinguish one *VTe* from another *VTe*, because they are all the maximal time. Furthermore, the consistency of data may be affected when updating happens.

The method of varying constants can reflect the phenomenon of "indetermination", but it is also contrary to people's day-to-day knowledge because it does not need to nor is not possible to update the end points every day in time. The certain and uncertain *VTe* are all in the same "constant" form, however, it is necessary to distinguish the changing end points from constant ones in practical applications.

2. Variable method

VTe is treated as a true variable, such as *Now*, and a certain time value is assigned to it according to some rules when needed during operations. The implementing methods are as follows:

(1) Method based on present time semantics: According to the natural semantics of *Now*, that is "at present time", one can design a kind of mechanism to bind now to current time *CT* to realize the assignment of time values of *Now* (Clifford et al. 1997; Ye and Tang 2005; Stantic et al. 2004). This method can indicate "uncertainty", and it is quite intuitionistic that variational point is denoted by a variable. Furthermore, it avoids the deficiency brought by the usage of constants. In fact, since now is bound to the current time *CT*, this binding time is called now-relative time.

(2) Method based on the general semantics of *Now*: *Now* is endowed with many semantics according to different situations in the practical application.

In fact, the practical cases are not as simple as one may think when the temporal variables are introduced (Clifford et al. 1997; Torp et al. 2004; Stantic et al. 2004).

- **Semantic problem from concept of valid time:** When $VTe = Now$, if one takes the original semantics, the corresponding data is only valid to the current time and will be invalid in the next second. This case contradicts with ones common sense and brings inconvenience in practice.
- **Semantic problem when associated with transaction time:** There is a common hypothesis when using *Now*, that is, the time when data becomes "true" in real word is isochronous with the time that data is stored in the databases. This hypothesis obviously contradicts with the independence of the valid time *VT* and transaction time *TT*, and it is difficult to carry out this synchronization in the operation of temporal databases. There is a difference between the time point (*TTs*) of data being physically stored into database and the time point (*VTs*) of data logically becoming true in the real world. That is, usually $VTs \neq TTs$. There is a difference $\Delta = TTs - VTs \neq 0$.

If $VTs < TTs$, that is $\Delta > 0$, *Now* indicates "the past". If $TTs < VTs$, that is $\Delta < 0$, *Now* indicates "the future". Therefore, the introduction of variable *Now* not only deals with the uncertain time, but also shows some interesting and useful issues in the operation of the temporal databases.

Example 7.5 The temporal records are shown in Table 7.11 and Table 7.12.

Table 7.11 John's temporal recording

Name	Title	VTs: VTe
John	Peofessor	2006-03-01, *Now*

Table 7.12 Black's temporal recording

Name	Title	VTs: VTe
Black	Manager	2006-10-01, *Now*

With reference to Table 7.11, suppose the data in database is changed 3 days after they have been logically changed. If John resigned on 2006-08-20, his record physically changes in database on 2006-08-23. However, one needs to query the information of all employees at *CT*= 2006-08-22. If we simply bind *Now* to *CT*= 2006-08-22, we may obtain an incorrect result. If we bind *Now* to a certain time in the "past", such as "2006-08-19", we may get a suitable result. As mentioned above, the time that a fact becomes a record in database is usually later than it becomes true in real world, and in this case, the semantics of *Now* does not indicate the present but the past.

With reference to Table 7.12, if we need to query Black's case on 2006-12-20 at *CT*= 2006-08-01, it is not suitable to bind *Now* to the *CT*. We may consider binding *Now* to a certain day in the future. Assume the bargain period of Black is 3 years, it is reasonable to bind *Now* to 2009-09-31(*VTs*= 2006-10-01). In this case, *Now* has the semantics of some time in the future.

7.4.1.2 Uncertain querying

The uncertainty hypothesis on *VTe* discussed above has two senses. One is determinacy sense, which means, "having effect from some past moment to present"; the other one is the uncertainty, which means, "not able to determine when the validity ends". The meanings can cause different binding of values of *Now*, which would lead to the uncertainty of the query results.

(1) ***VTs<TTs*:** The difference $\delta = TTs - VTs > 0$ is in the lagging state. According to semantics analyzed in Section 7.4.1.1, variable *Now* needs to be bound to certain time in the past relative to current time *CT*, for example, bind it to $CT - \delta$. Different cases are listed below according to *QT* (query time):

- If $QT < VTs$, output the certain query result (else, no output).
- If $VTs \leqslant QTT < CT - \delta$, output the certain query result (if yes, output data)
- If $CT - \delta \leqslant QT < CT$, output the uncertain query result (if yes, output lowly uncertain data)
- If $CT \leqslant QT$, output the uncertain query result (if yes, output highly uncertain data)

The four cases discussed above are shown in Fig. 7.1

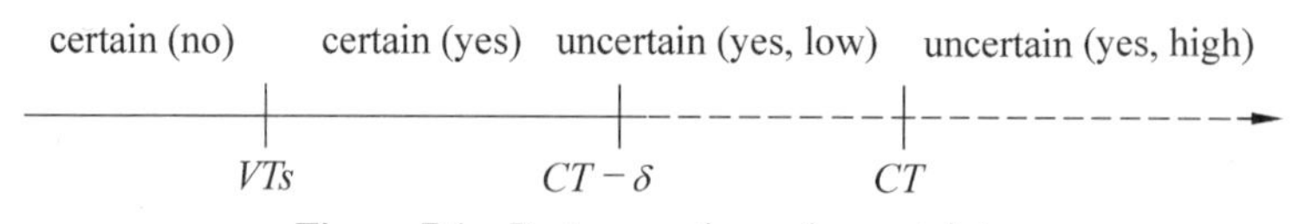

Figure 7.1 Past semantics and uncertainty

(2) ***TTs<VTs*:** We may define a predicted valid length Δ, and the corresponding valid period $VT = [VTs(P), VTe(P)] = [VTs(P), VTs(P) + \Delta]$. For instance, Δ indicates the employee's valid contract length, and $[VTs(P), VTs(P) + \Delta]$ indicates the contract period. According to the above semantic analysis, *Now* has semantics with "future time". In this case, it is rationally to bind *Now* to some future time such as *VTe*(*P*).

We can analyze it in the following cases according to the querying time QT:

- If $QT < VTs$, output the certain query result (else, no output).
- If $VTs \leqslant QT < CT - \delta$, output the certain query result (if yes, output data).
- If $VTs + \Delta < QT$, output the uncertain query result (if yes, output highly uncertain data).

The three cases are shown in Fig. 7.2.

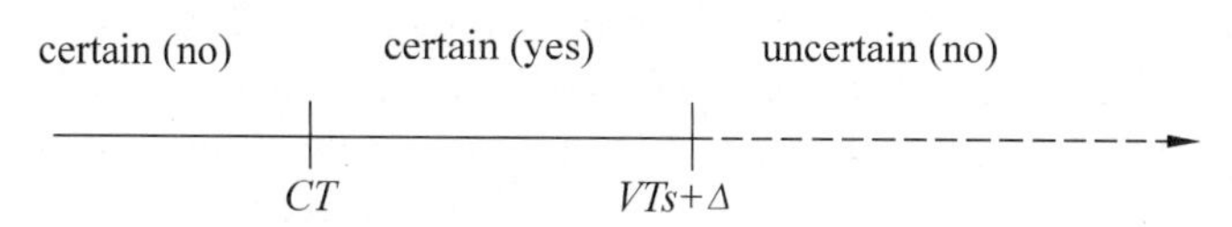

Figure 7.2 Future semantics and binding

7.4.2 *Now* in Non-Current Version

In the non-current version, it is more common that *Now* indicates the non-present semantics. Let $T_k \in SG$ and $VTe(T_k) = Now$, other tuples of SG be $T_1, \cdots, T_{k-1}, T_{k+1}, \cdots, T_m$. Their transaction time hold is $TT(T_1) \prec \cdots \prec TT(T_k) \prec \cdots \prec TT(T_m)$.

There are two cases: T_k is the maximal element of the SG or it is not.

7.4.2.1 T_k has succeeded tuple

In this case, there are succeeding records behind T_k. Let the maximal record be T_m. This case mainly happens when a new record is inserted. Then, *Now* is bound to the $VTe(T_m)$.

Example 7.6 There is an *MI* as shown in Table 7.13.

Table 7.13 Temporal records of Raul (1)

Name	Dept	Title	Salary	Vs: Ve	Ts: Te
Raul	CS	Assistant Prof.	7500	2005-01-01: *Now*	2005-01-10: 2006-01-09
Raul	CS	Assistant Prof.	7500	2005-01-01: 2005-12-31	2006-01-10: *UC*
Raul	CS	Prof.	8500	2006-01-01: *Now*	2006-01-10: *UC*

If we need to query Raul's records, which are at TT= [2005-09-01: 2005-11-30], VT= [2006-02-20: 2006-02-20] at CT= 2006-02-28, *Now* is not suitable to be bound to CT= 2006-02-28, or it will output the result as follows:

(Raul, CS, Assistant prof, 7500, 2006-02-20: 2006-02-20; 2005-09: 2005-11)

However, this contradicts with the real result:

(Raul, CS, prof, 8500, 2006-02-20: 2006-02-20; 2006-01-10: 2006-02-28)

In fact, *Now* should be determined by the $VTe = 2005\text{-}12\text{-}31$ of the largest succeeded record. For example, we may bind *Now* to *VTe*. Then the query result is *null*, which is consistent with the fact.

7.4.2.2 T_k has no succeeded tuple

It mainly happens when T_k is logically deleted and the non-current version record T_k has no subsequent record, that is, *Now* appears in the tense label of T_k. For example, a record about Raul is disabled for some reason. There are two cases.

Case 1: In this case, there is the succeeded group of the *SG*, to which T_k belongs. Let the last tuple of the group be T'. The instance of the case is shown in Table 7.14, where the first record denotes the tuple deleted, and the second denotes the current version T' reestablished. If we need to query the information of tuple Q, where

$$VT(Q) = [2005\text{-}03\text{-}01\text{: } 2005\text{-}09\text{-}30],\ TT(Q) = [2005\text{-}02\text{-}01\text{: } 2005\text{-}07\text{-}31]$$

Then *Now* could be bound to $TTs(T') - 1 = 2006\text{-}01\text{-}09$.

Table 7.14 Temporal records of Raul (2)

Name	Dept	Title	Salary	Vs: Ve	Ts: Te
Raul	CS	Assistant Prof.	7500	2005-01-01: *Now*	2005-01-10: 2006-01-09
Raul	CS	Assistant Prof.	8500	2005-01-01: *Now*	2006-01-10: *UC*

Case 2: The *SG*, to which T_1 belongs, has no succeeded *SG*, and T_1 is the maximal tuple in *MI*. The instance of this case is shown in Table 7.15 and *Now* can be bound to the current time *CT* of the query operation.

Table 7.15 Temporal recording of Raul (3)

Name	Dept	Title	Salary	Vs: Ve	Ts: Te
Raul	CS	Assistant Prof.	7500	2006-09-01: *Now*	2006-05-10: 2006-06-10

7.4.3 Temporal Querying Algorithms

7.4.3.1 Temporal joining

In addition to constraints of the common join conditions, there are some other constraints in temporal join.

Temporal Joinable Tuples: Tuple T_1 in the temporal relation TR_1 and tuple T_2 in temporal relation TR_2 are considered to be temporally joinable if the following conditions hold:

- T_1 and T_2 are in the same version.

- T_1 and T_2 can be joined as common relation tuple when the time stamps are removed.
- The valid time periods of T_1 and T_2 overlap.

In non-current version, *Now* is bound by the method described in Section 7.4.2. The non-temporal parts of the corresponding tuple are joined so that we obtain a new tuple for which the time stamp is the overlapping parts of the time stamps in T_1 and T_2. In current version, if *Now* is bound to the corresponding time value according to Section 7.4.1, then the overlapping part of the valid periods of the tuples that fulfill the join condition is the valid period of the tuple joined.

Let MI_1 and MI_2 be the main body instances, and T_1 and T_2 be the tuples in MI_1 and MI_2, respectively, where $T_1 = (v(T_1), VT(T_1), TT(T_1))$ and $T_2 = (v(T_2), VT(T_2), TT(T_2))$. In addition, Let MI_3 be the relation connected by MI_1 and MI_2.

If T_1 and T_2 satisfy following conditions:

- $TT(T_1) \subseteq TT(T_2) \vee TT(T_2) \subseteq TT(T_1)$.
- $VT(T_1) \cap VT(T_2) \neq \varnothing$.
- Non-temporal parts as common tuples are connectable.

Then we may define the temporal join operator as follows:

$$(T_1, T_2) = \{v(T_3), VT(T_3), TT(T_3)\}$$

where, $v(T_3) = v(T_1) \bowtie v(T_2)$, $VT(T_3) = VT(T_1) \cap VT(T_2)$, $TT(T_3) = \min(TT(T_1), TT(T_2))$.

Example 7.7 Two temporal relations MI_1 and MI_2 are shown in Table 7.16 and Table 7.17, respectively.

Table 7.16 Temporal relation MI_1

Name	Title	Vs: Ve	Ts: Te
Green	Manager	2003-01-01:2003-12-31	2005-01-10: *UC*
Green	Sr.Manager	2004-01-01: *Now*	2005-01-10: *UC*

Table 7.17 Temporal relation MI_2

Name	Salary	Dept	Vs: Ve	Ts: Te
Green	6800	RD	2004-01-01: 2004-12-31	2005-01-10: *UC*
Green	8000	RD	2005-01-01: *Now*	2005-01-10: *UC*

Here, the second tuple in MI_1 and the two tuples in MI_2 are temporally joinable. The overlapped parts of the valid time of these are "2004-01-01: 2004-12-31" and "2005-01-01: *Now*", respectively. We take them as the valid time periods of the tuples joined. The results are shown in Table 7.18.

Table 7.18 Temporal connection of MI_1 and MI_2

Name	Salary	Dept	Title	Vs: Ve	Ts: Te
Green	6800	RD	Sr. Manager	2004-01-01: 2004-12-31	2005-01-10: *UC*
Green	8000	RD	Sr. Manager	2005-01-01: *Now*	2005-01-10: *UC*

7.4.3.2 Temporal projecting

In projecting operations, there may be some tuples for which the non-temporal attributes are the same but the time stamps are different. These time stamps may need to be coalesced according to some rules, that is, to group the relevant time stamps. Firstly, the stamps are grouped according to the valid periods. Subsequently, the corresponding transaction times are merged based on the merging of valid time. Finally, they are grouped into new SGs by the new transaction time. In fact, the projecting operations are the ones that need to rebuild the snapshot equivalent group. There are two cases that need to be considered:

Case 1: If tuples in the projecting result with the same non-temporal attributes are in the same version, they do not need to be merged.

Case 2: If tuples in the projecting result with the same non-temporal attributes are in different versions, we should order their valid time stamps by their start points *VTs*, and obtain a valid period sequence. Then, from the smallest valid period, if the end point $VTe(t_i)$ of a certain $VT(T_i)$ fulfills the condition:

$$VTs(T_{i+1}) \leqslant VTe(T_i) \leqslant VTe(T_{i+1}),$$

then, we may merge $VT(T_i)$ and $VT(T_{i+1})$ to get a new valid period $VT(T_i) \cup VT(T_{i+1})$. Otherwise, the merge operation will not be executed. In the former case, if there is a temporal variable *Now* in tuple T_0 in non-current version and T_1 is the succeeded tuple of T_0, *Now* is bound to $VTs(T_1) - 1$. In addition, there exist two subcases:

- The valid time variable *Now* in non-current version usually does not denote current time but denotes past time. In this case, if there is no *Now* in ending point of the current-version but there is *Now* in the non-current version, the relative time periods cannot be merged even if there is overlapping in the valid time stamps.
- Because of the meaning of the temporal relation (Definition 7.5) and the original semantics of the variable *Now*, if *Now* appears in current-version, then we can always do the temporal merging.

During the process of the coalescing mentioned above, the transaction time is merged along with the coalescing of the valid periods.

Example 7.8 In the temporal relation shown in Table 7.19, if it is projected on the attributes "Name" and "Dept", according to Case 2, all time periods could be merged. Then, we get the relation projected in Table 7.19.

Table 7.19 Projecting on attributes "Name" and "Dept"

Name	Dept	Vs: Ve	Ts: Te
Black	EN	2003-01-01: *Now*	2003-01-10: *UC*

If the relation needs to be projected on "Title", the time stamps of "Engineer" is in "①" of Case 2, and we cannot merge them. "Sr.Engineer" is in "②"of Case 2, so we get the temporal relation shown in Table 7.20.

Table 7.20 Projection on "Title"

Title	Vs: Ve	Ts: Te
Engineer	2003-01-01: *Now*	2003-01-10: 2004-10-09
Engineer	2004-01-01: 2004-12-31	2005-01-10: *UC*
Sr. Engineer	2005-01-01: *Now*	2005-01-10: *UC*

For the same reason, if the relation is projected on "salary", the tuples with temporal values "7500", "8500" and "9000" are of the same version, which are in Case 1, so they do not need to be merged. The final relation is shown in Table 7.21.

Table 7.21 Projecting on "Salary"

Salary	Vs: Ve	Ts: Te
7500	2003-01-01: *Now*	2003-01-10: 2004-01-09
7500	2004-01-01: 2004-12-31	2004-01-10: *UC*
8500	2004-01-01: *Now*	2004-01-01: 2005-01-09
8500	2004-01-01: 2004-12-31	2005-01-10: *UC*
9000	2005-01-01: *Now*	2005-01-10: *UC*

We may obtain the operation on temporal projection as follows:

Step 1: Let MI_1 be the projection of MI_0 on the set of attributes A_0, which is denoted as List$\{SG_1, SG_2, \cdots, SG_q\}$. In every SG_k, If there is *Now* in it, *Now* is bound to "4.2" in non-current version or to "4.1" in current version. In this way, we obtain the valid time set of $SG_k(1 \leqslant k \leqslant q)$.

Step 2: The set obtained in Step 1 is divided into equivalent groups by the temporal connectional relation defined in Definition 7.5 in the valid time periods.

Step 3: The valid time is merged in every equivalent group from Step 2.

Step 4: The transaction time is grouped in the same way, for it synchronously lags or leads to the valid time.

It is noticeable that the data with the same non-temporal attribute values may form many different temporal tuples, since a temporal relation should satisfy the constraint 1NF.

The formal description of temporal projection is shown as follows:

Let $MI_0 = \text{List}\{SG_1(A), SG_2(A), \cdots, SG_m(A)\}$, A_0 be a subset of A, which contains only the non-temporal attributes, and the projecting transformation Π_{A_0} be defined as

$$\Pi_{A_0}(MI_0) = MI_1 = \text{List}\{SG_1(A_0), SG_2(A_0), \cdots, SG_q(A_0)\}$$

where, $SG_k = \{T_{k1}, T_{k2}, \cdots, T_{krk}\}$, and $v(T_{kj}) = v(SG_k)$, $VT(T_{kj}) = \cup_{T_{kji} \in E_{kj}} VT(T_{kji})$, $TT(T_{kj}) = \cup_{1 \leqslant j \leqslant p} VT(T_{kji})$. In this case, $E_{kj} = \{T_{kj1}, T_{kj2}, \cdots, T_{kjp}\}$ are the equivalent groups in SG_k by the temporal connection relation ($1 \leqslant k \leqslant m$, $1 \leqslant j \leqslant q$).

7.4.3.3 Temporal selecting

When there is selection on temporal element in querying, in order to determinate the location of the temporal element, *Now* needs to be bound by "4.1" in current version and by "4.2" in non-current version.

Let $MI_0 = \text{List}\{VT_i(TR_1), VT_i(TR_0), TT_i(TR_0)\}$ be a main body instance, CT and QT be the current time when the time query happens, and $VT(QT)$ and $TT(QT)$ be the valid time and transaction time of QT, respectively. We define the binding operator as follows:

- If $End(TTe(T_i)) \neq UC$
  - When $VTe(T_i) = Now$, $P(VTe(T_i)) = TTe(T_i) - \delta$, $P(TTe(T_i)) = TTe(T_i)$.
  - When $VTe(T_i) \neq Now$, $P(VTe(T_i)) = VTe(T_i)$, $P(TTe(T_i)) = TTe(T_i)$.
- If $TTe(T_i) = UC \wedge VT(T_i) < TTs(T_i) \leqslant CT$
  - When $VTe(T_i) = Now$, $P(VTe(T_i)) = CT - \delta$, $P(TTe(T_i)) = CT$.
  - When $VTe(T_i) \neq Now$, $P(VTe(T_i)) = VTe(T_i)$, $P(TTe(T_i)) = CT$.
- If $TTe(T_i) = UC \wedge TTs(T_i) \leqslant CT < (VTs(T_i))$, then $P(VTe(T_i)) = VTe(T_i)$, $P(TTe(T_i)) = VTe(T_i)$.

where, δ is the lagging of the transaction time related to valid time and P is ending point operator defined above. We may define the temporal selecting operator Γ as follows:

If $VT(QT) \subseteq [VTs(T_i) : P(VTe(T_i))$, $TTi(QT)] \subseteq [TTs(T_i)] : P(TTe(T_i))$, then $\Gamma(MI_0) = \text{List}\{v_i(TR_0), VT_i(QT), TT_i(QT)\}$, else $\Gamma(MI_0) = \varnothing$.

Example 7.9 With reference to Example 7.3, if we need to query the information, where $TT = [2003\text{-}06\text{-}30 : 2003\text{-}06\text{-}30]$ and $VT = [2003\text{-}05\text{-}25 : 2003\text{-}05\text{-}25]$, then *Now* in the first record T_0 can be bound to $TTs(T_1) - 1 = 2004\text{-}01\text{-}09$ since T_1 is the succeeded tuple of T_2. The valid period of query "2003-05-25" is in the period [2003-01-01 : 2004-01-09], and we can get the corresponding result.

If $CT = 2005\text{-}04\text{-}20$, we need to query the information in the last version, where valid period is [2005-02-15 : 2005-03-15]. Because the end point of the query is smaller than the CT-lagging $\delta = 2005\text{-}04\text{-}10$, *Now* is bound to CT in the fifth record in Fig. 7.6, that is, $CT = 2005\text{-}04\text{-}20$. Then we get the information needed. If the valid period VT is [2005-02-15 : 2005-04-15], since $VTe(QT) = 2005\text{-}04\text{-}15$ is larger than "CT-lagging δ" = "2005-04-10", *Now* should be bound to "2005-04-10", in this case, the query returns uncertain information.

We may obtain the following basic theorem according to the discussion in Section 7.3 and Section 7.4.

Theorem 7.1 (Closeness of Temporal Operations) The BRDM defined in Definition 7.5 is closed on the various temporal operations in Section 7.3 and Section 7.4.

References

[1] Bliujute R, Jensen CS, Saltenis S, Slivinskas G (**1998**) R-tree based indexing of now-relative bitemporal data. In: Proceedings of the 24th VLDB Conference, New York, USA, pp 345 – 356

[2] Bliujute R, Jensen CS, Saltenis S, Slivinskas G (**2000**) Light-weight indexing of bitemporal data. In: Proceedings of the 12th International Conference on Scientific and Statistical Database Management, Berlin, pp 125 – 138

[3] Clifford J, Dyreson CE, Isakowitz T, Jensen CS, Snodgrass RT (**1997**) On the semantics of "*Now*" in databases. ACM Transactions on Database Systems (TODS) 22(2)

[4] Clifford J, Dyreson CE, Snodgrass RT, Isakowitz T, Jensen CS (**2000**) Now, temporal database management. Dr.Techn. thesis by Christian S. Jensende, defended April 2000. 455 – 464. http://www.cs.auc.dk/~csj/Thesis/

[5] Dyreson CE (**2003**) Temporal coalescing with now granularity, and incomplete information. In: Proceedings of the 2003 ACM SIGMOD International Conference Management of Data, San Diego, California, ACM Press, pp 169 – 180

[6] Snodgrass RT (**1987**) The temporal query language TQuel. ACM Transactions on Database Systems, 12(2): 247 – 298

[7] Stantic B, Khanna S, Thornton J (**2004**) An efficient method for indexing now-relative bitemporal data. In: Proceedings of the 15th Australasian Database Conference (ADC2004), Dunedin, NZ. Conferences in Research and Practice in Information Technology, Vol. 27, pp 113 – 122

[8] Stantic B, Thornton J, Sattar A (**2003**) A novel approach to model NOW in temporal databases. In: Proceedings of 10th International Symposium on Temporal Representation and Reasoning and the 4th International Conference on Temporal Logic, pp 174 – 180

[9] Torp K, Jensen CS, Snodgrass RT (**2004**) Modification semantics in Now-relative databases. Information Systems (ACM) 29(8): 653 – 683

[10] Wang F, Zaniolo C (**2003a**) XBiT: An XML-based bitemporal data model. In: Proceedings of the 10th International Symposium on Temporal Representation and Reasoning and the 4th International Conference on Temporal Logic (TIME-ICTL'03), pp 1530 – 1311

[11] Wang F, Zaniolo C (**2003b**) Preserving and querying histories of XML-published relational database in xml. http://wis.cs.ucla.edu/publications/ecdm02.pd

[12] Ye XP, Tang Y (**2005**) Semantics on "Now" and calculus on temporal relations (in Chinese with English Abstract). Journal of Software 16(5): 838 – 846

Part Ⅲ Temporal Index Technologies

- Temporal Indexes Supporting Valid Time
- Indexes for Moving-Objects Data
- Temporal XML Index Schema

8 Temporal Indexes Supporting Valid Time

Hai Liu[1,2], Xiaoping Ye[1+], Ming Shi[2], and Boling Yang[2],

[1] Department of Computer Science, Sun Yat-sen University, Guangzhou 510275, P.R. China
[2] Computer School, South China Normal University, Guangzhou 510631, P.R. China

Abstract This chapter studies the temporal index technique based on valid time. Firstly, it introduces the current study on temporal index. Subsequently, it introduces temporal equivalence, temporal preorder and other basic conceptions, and studies their relative properties, which provide the necessary mathematical frame for the temporal index. In addition, it introduces the algorithm for temporal linear order branch by studying the properties of time interval number. Thereby, the temporal index – *TRdim* is established. After that, we study the incremental updating algorithm based on the query algorithm and the index. Finally, we design and implement a simulation experiment, whose result indicates the feasibility and validity of *TRdim*.

Keywords *temporal preorder, time interval number, query and incremental updating, simulation and evaluation*

8.1 Introduction

In the real world, every entity exists in a time environment, so it should contain relative temporal characteristics. The data stored in a database is the manifestation of an entity, so the time characteristic of the entity is needed. A database that can manifest and display the time characteristics effectively is called a temporal database. Normal databases, such as relational database, object database and XML database, implicitly express the time characteristics. The data in such databases can be regarded as the snapshot of the entity. In practical applications, such as e-commerce, e-government, and other novel data managements, however, database must manage the temporal information of the data explicitly. The data containing temporal labels is named temporal data. However, ordinary data managing technology cannot work effectively as the data quantity is very large, the data structure is very complex and the data operation involves many other subjects.

[+] Corresponding author: mcsyxp@mail.sysu.edu.cn

With the development of computer hardware technology, massive storage of the temporal data can be realized. However, as the temporal database management has not been developed completely, index technology becomes the key to implement high-speed query. From the materials we master, temporal index technology is centralized in temporal relational index and temporal XML index. This chapter introduces the temporal relational index based on valid time. The organization of this chapter is as follows. In Section 8.2, the current work on temporal relational indexes is discussed. From Section 8.3 to Section 8.5, our recent original work, a temporal relational data model, is presented. In Section 8.3, a temporal index structure with temporal equivalence and temporal inclusion is established. In Section 8.4, the query and updating algorithms based on the index are introduced. In Section 8.5, the simulation experiment and results evaluation of the index are provided.

8.2 Summary of Temporal Index

Time interval is the most basic label in practical applications. The time interval can be regarded as a set of all the points bound by two ending points. Hence, temporal index can be divided into two types according to the structure characteristics of the time label: one based on time ending point and the other based on the time interval.

The index based on interval ending points mainly contain interval tree (Edelsbrunner 1983), segment tree (Bentley 1977), external segment tree (EST) (Blankenagel and Guting 1994), time-index (Elmasri et al. 1990), MVB-Tree (Becker et al. 1996), and ST-tree (Gunadhi and Segev 1993). The two interval ending points can be translated into a point in some way. The main work on it contains MAP21 (Nascimento et al. 1996), priority search tree (McCreight 1985), and so on.

A valid way to study the index based on the time interval is to use high-dimensional indexing (spatial index) directly or to use the improved one based on the high-dimensional indexing. The work about it mainly contains GR-Tree (Bliujute et al. 1998), 4R-Tree (Bliujute et al. 2000).

The time label can be divided into valid time and transaction time. Considering the time label semantics, temporal index can be divided into single temporal index (based on transaction time or valid time) and bitemporal index (based on transaction time and valid time). The segment tree, external segment tree, time-index, MAP21 and others are based on the valid time. The MVB-Tree, AP-Tree (Gunadhi and Segev 1993), Snapshot Index (Kumar et al. 1995), HR-Tree (Historical R-Tree) (Nascimento and Silva 1998) are based on the transaction time. The main work on bitemporal index contains GR-Tree, 4R-Tree, 2LBIT (Nascimento et al. 1995), IVTT (Nascimento et al. 1995) and M-IVTT (Nascimento et al. 1996).

8.2.1 Temporal Index Based on Transaction Time

Single temporal index based on transaction time mainly contains HR-Tree, MVB-Tree, AP-Tree, and Snapshot Index (Kumar et al. 1995).

8.2.1.1 HR-tree

HR-tree (Historical R-tree) can reduce massive redundant data caused by transaction operations with incremental method. The basic idea is displayed in Fig. 8.1. There are three transaction time points T_1, T_2, T_3 in Fig. 8.1(a). From T_1 to T_2, the right child of A is updated. Therefore, all the nodes in T_1 except nodes A and R_1 can be reused and need not be updated. From T_2 to T_3, the middle child of C is updated, so all the nodes in T_2 except nodes C and R_2 need not be updated. The data of T_3 is displayed in Fig. 8.1(b).

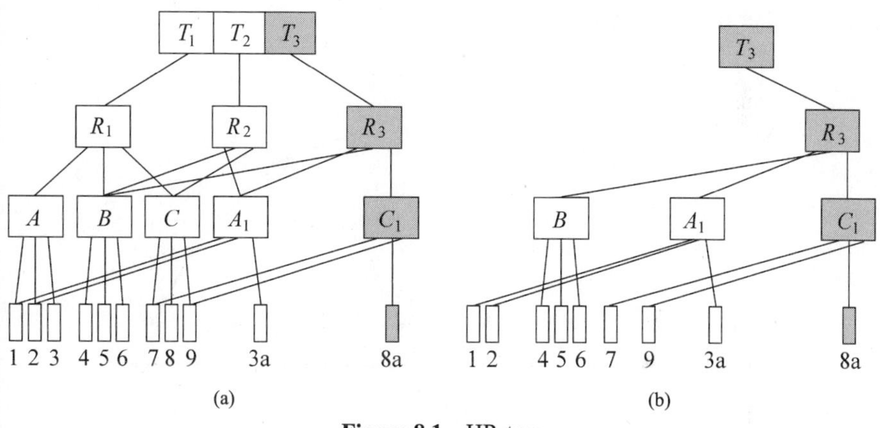

Figure 8.1 HR-tree

8.2.1.2 MVB-tree

MVB-tree (Multi-version B-tree) (Becker et al. 1996) provides a valid history version management method. It can be regarded as index based on transaction time, because the historical version will not be deleted and the new record is increases all the time. Its property is similar to B-tree and its basic structure B-tree. Its space complexity is $O(n/B)$. The time complexity of updating is $O(\log_B m)$, where m is the size of the updating part. The time complexity of query on the time point is $O(\log_B(n/B))$ and the time complexity of the query on the time point and the range of a value interval is $O(\log_B(n/B)+a/B)$, where a is the number of query results. It can be used generally. It can be used generally, as the basic structure of the index can be replaced by other index technologies, such as Cell-tree, and R-tree.

8.2.2 Index Based on Valid Time

8.2.2.1 ST-tree

Here, we take **ST-tree** (segment tree) as an example. A standard segment tree indexing n time intervals is a balanced binary tree, whose leaf nodes are the $2n$ time interval ending points. We define *range*(v) as the time interval of node v. If v is a leaf node whose time point is expressed by v, *range*(v) is interval $[v, v]$. If v is an internal node, the first child of the in-order traversal sequence of v is f, and the last is l, then *range*(v) is defined as $[f, l]$. Each node v of the segment tree contains a pointer pointing to another pair structure, which stores all the records whose valid time contains *range*(v) but does not contain *range*(*parent*(v)).

The space complexity of segment tree is $O(n\log n)$, the time complexity of building a segment tree is $O(n\log n)$, and the time complexity of point query, (selecting all the records whose valid time contains a given time point) is $O(\log n + k)$, where k is the number of the selected records.

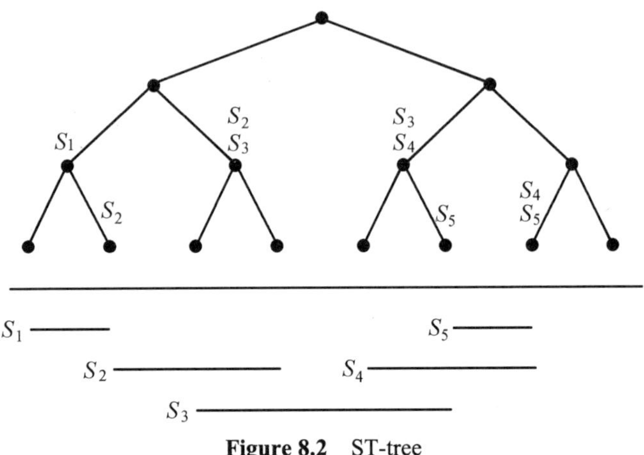

Figure 8.2 ST-tree

8.2.2.2 Interval tree

As **Interval tree** (Edelsbrunner 1983), we define x_{mid} as the median of end points of all the time intervals, S_l as the interval set in which the ending points of the time Interval are less than x_{mid}, S_r as the Interval set in which starting points are larger than x_{mid} and S_{mid} as the interval set in which all the intervals contain S_{mid}. Therefore, an Interval tree can be defined recursively as a binary tree, whose left sub-tree is the Interval tree of S_l, right sub-tree is the Interval tree of S_2 and root node is connected with two tables (L_l and L_r), which are the sorting results by the starting points of the intervals in S_{mid} and that by the ending points.

The space complexity of Interval tree is $O(n)$, time complexity of building an

Interval tree is $O(n\log n)$, and the time complexity of point query(selecting all the records whose valid time contains a given time point) is $O(\log n + k)$.

You can see an example of Interval tree in Fig. 8.3.

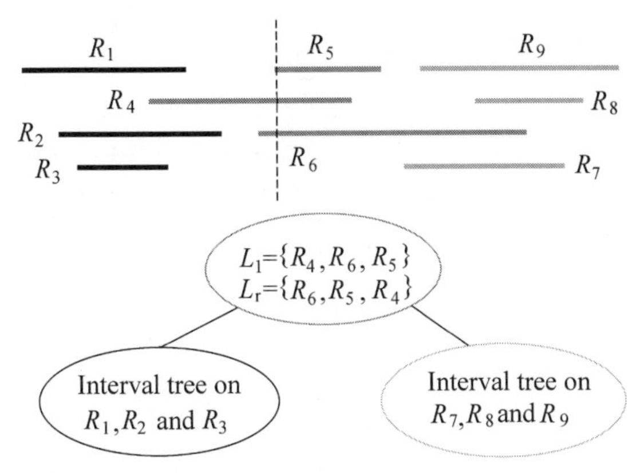

Figure 8.3 Interval tree

8.2.2.3 Time index

Time index tree is proposed by Ramez Elmasri and the others (1990). If B is the number of the records a block stores, S is the length of the query interval and s is the number of the query results, then the space complexity is $O(n^2/B)$, the time complexity of updating is $O(n/B)$, the time complexity of the query on the time point is $O(\log_B n + a/B)$ and the time complexity of the query on the time point and the range of a value interval is $O(\log_B n + s/B)$. The simplified idea and its structure is displayed in Fig. 8.4. However, as the space it uses is massive, increment method, time index+ and other improvements are proposed.

8.2.3 Bitemporal Index

Bitemporal index involves the valid time and the transaction time at the sametime. As valid time and transaction time can be treated as two cross dimensions, spatial index technology can be used as reference, which is the basic idea of the present work on bitemporal index.

8.2.3.1 2LBIT

2LBIT (2 Level Bitemporal Index Tree) is a visual bitemporal index, whose structure is displayed in Fig. 8.5. It is a two-layered structure. The upper layer is the transaction index, and each leaf node denotes a transaction time connected by many valid index trees. 2LBIT can be understood easily, and when you query a transaction time, it becomes very easy. However, as the data of the adjacent transaction nodes differ a little, it will waste storage space greatly as each transaction node stores the integrated valid indexes. Therefore, it is of little use in practice.

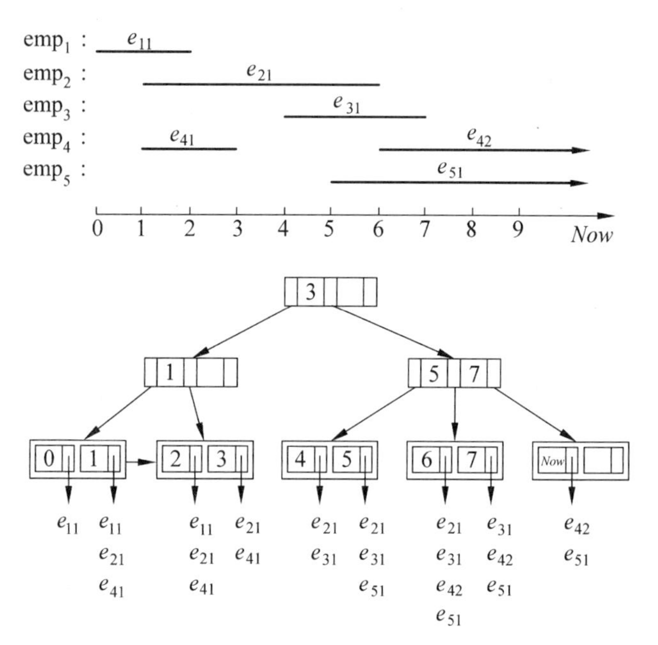

Figure 8.4 Time index

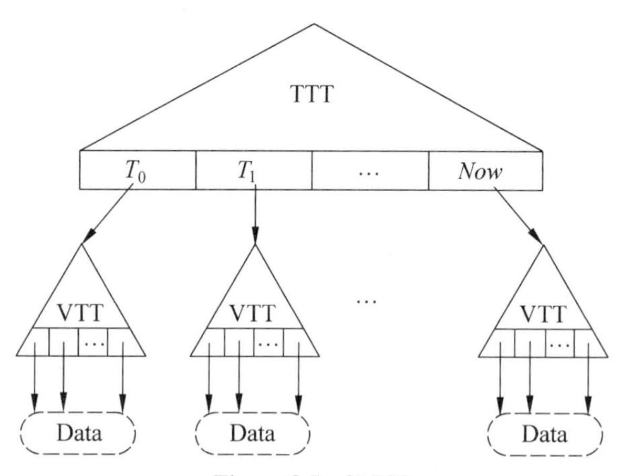

Figure 8.5 2LBIT

8.2.3.2 IVTT and M-IVTT

IVTT and M-IVTT are the improvements of 2LBIT. The structure of M-IVTT is displayed in Fig. 8.6. It is similar to that of 2LBIT, except for reducing valid indexes. Except for the present transaction node and some other nodes that contain the valid indexes, the others between them contain a patch (update from the present transaction node to the next). If you want to get the valid tree of a transaction node (assuming the transaction node contains no valid tree), you can first find the nearest transaction node that contains valid tree, then get the present valid tree by patching it sequentially. You can control the interval length of the valid tree dynamically according to space and property demands. If few indexes are demanded, the interval can be increased. However, if former transaction nodes are queried often, the interval should be reduced.

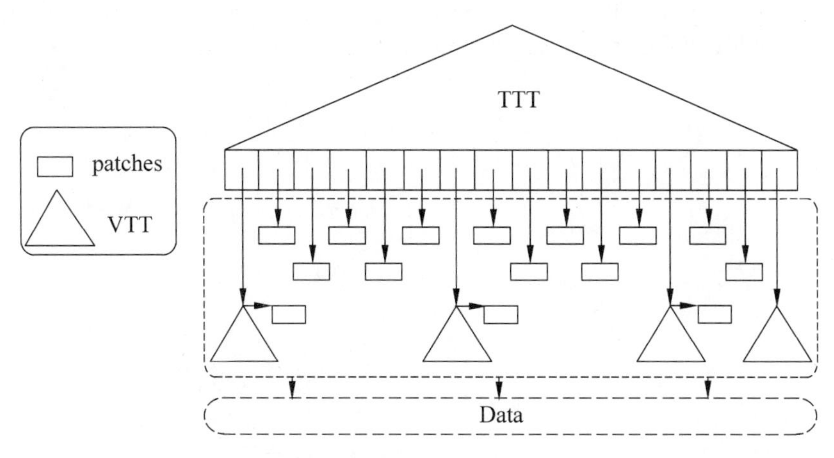

Figure 8.6 M-IVTT

8.2.3.3 GR-tree and 4R-tree

Some spatial index technologies, such as R-tree, KDB-tree are used in temporal index. However, it must be improved to process the temporal variables *Now* and *UC*, because spatial index does not consider the infinitely increasing object. The simplest way is to index the record whose valid time ends with *Now* and transaction time ends with *UC*, except for the main index. For example, we can build a B-Tree to index the start time of the records whose valid time ends with *Now*. If the end time of the record is determined, the record will be deleted from the B-Tree and added into the main index.

Here, we introduce 4R-tree simply. 4R-tree (Bliujute et al. 1999, Bliujute et al. 1998) allocates the data of an R-tree to 4R-tree or B-tree. Assuming the record is $\langle T_s, T_e, V_s, V_e, id \rangle$, where T_s and T_e are the starting and ending points of the transaction time, V_s and V_e are the starting and ending points of the valid time, and *id* is its label, we can determine the tree that it will be inserted into depending on whether T_e is *UC* and V_e is *Now*.

$$\langle T_s, T_e, V_s, V_e, id \rangle = \begin{cases} \langle T_s, T_s, V_s, V_s, id, R_1 \rangle, \text{if } T_e = UC \wedge V_e = Now \\ \langle T_s, T_s, V_s, V_e, id, R_2 \rangle, \text{if } T_e = UC \wedge V_e \neq Now \\ \langle T_s, T_e, V_s, V_s, id, R_3 \rangle, \text{if } T_e \neq UC \wedge V_e = Now \\ \langle T_s, T_e, V_s, V_e, id, R_4 \rangle, \text{if } T_e \neq UC \wedge V_e \neq Now \end{cases}$$

Correspondingly, you can convert the time internal query to a query on 4R-trees, whose basic idea is displayed in Fig. 8.7. In Fig. 8.7(a), if the right up corner of the query box is equal to the right up corner of a record and it is under the diagonal, the record will be output. In Fig. 8.7(b), if T_e of the query box is larger than the T_s of a record, and the valid time intersects the record's valid time, the record will be output. Figure 8.7(c) is similar to Fig. 8.7(b), but we must consider the rationality of the semantics. Figure 8.7(d) is the most former. If the query box intersects the rectangle of the record, the record will be output.

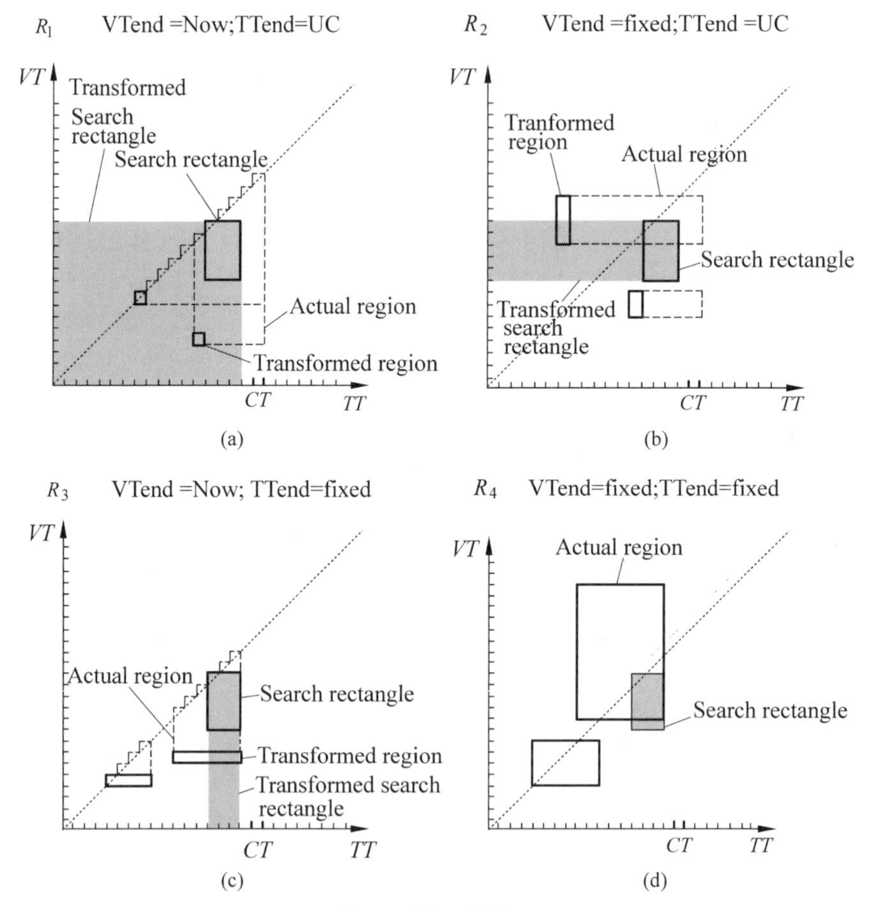

Figure 8.7 4R-tree

8.3 TRdim

Temporal node P is defined as a 2-tuple ($NT(P)$, $VT(P)$), where $NT(P)$ and $VT(P)$ represent non-temporal part and valid label, respectively. The equivalency of temporal nodes P_1 and P_2 is defined by $NT(P_1) = NT(P_2) \wedge VT(P_1) = VT(P_2)$. Often P and $VT(P)$ can be mixed and used if it causes no confusion, both representing valid time. The valid time period of P is denoted by $VT(P) = [VTs(P), VTe(P)]$, where $VTs(P)$ and $VTe(P)$ denote start and end points of $VT(P)$. We can denote $Ps = VTs(P)$, $Pe = VTe(P)$ or $Start(P) = VTs(P)$, $End(P) = VTe(P)$. In addition, we denote the valid span of temporal node P as $L(P)$, where $L(P) = Pe - Ps$. The set of all the temporal nodes are denoted as D and set of the nodes of a certain period Γ is denoted as $D(\Gamma)$ or Γ.

8.3.1 Relative Temporal Data Model

The temporal relation of this section is based on Snodgrass Representational Data Model (RDM).

In temporal relation TR, not all the attributes have relation with time. Here, we assume that A is an attribute of TR, whose values at different time periods are recorded by set Ω_A, and f is the map from the attribute value domain of A V_A to Ω_A. If $\forall x \in V_A$, $f(x) \cup \Omega_A$, A is time-independent attribute, otherwise A is time-dependent attribute. For example, in temporal relation (*name*, *id*, *title*, *dept*, *salary*, *VT*), *name* and *id* are regarded as the time-independent attributes, while *title*, *dept* and *salary* are treated as time-dependent attributes. TR can contain many time-independent attributes, which are relatively stable in a certain environment and will not change with time. Time-independent attribute is a type of non-temporal attribute with stricter limitation.

Definition 8.1 (Main body instances) If A is a time-independent attribute, we can import an equivalence relation "$\equiv$" on the tuples of TR via A. If T_1 and T_2 are temporal tuples, then $T_1 \equiv T_2 \Leftrightarrow \prod_A(T_1) = \prod_A(T_2)$, where $\prod_A(T)$ denotes the projection of T on the attribute A. Then, we can get equivalence class set $\{MI_1, MI_2, \cdots, MI_k\}$ based on "$\equiv$". A is called the main body attribute of TR, and equivalence class $MI_j (1 \leqslant j \leqslant k)$ is a main body instance of TR on A.

Definition 8.2 (*TRqdm*(*TR*)) The tree structure constructed as follows is called *TRqdm* and temporal relation tree model or tree model:

(1) The root of the tree is the name of the temporal relation data model.

(2) Inner nodes are the main body instances.

(3) The leaves are the temporal tuples of the corresponding main body instance.

Example 8.1 In the temporal relation, for the personnel placed in Table 8.1, *Name* is a time-independent attribute that can be used as the main body attribute.

Then there are five main body instances, respectively. In addition, *Code* denotes the serial number from the preorder travel of the corresponding *TRqdm*(*TR*).

Table 8.1 Temporal relation data instance personnel

Code		Name	Dept	Vs: Ve
2	3	Bob	Math	[3,7)
	4	Bob	Math	[7,16)
	5	Bob	Math	[16,*Now*)
6	7	White	IS	[4,6]
	8	White	IS	[6,20]
	9	White	CS	[20,25]
10	11	Raul	CS	[4,6]
	12	Raul	CS	[6,20]
	13	Raul	CS	[20,25]
14	15	John	CS	[3,7]
	16	John	CS	[7,*Now*]
17	18	Black	PHY	[3,7]
	19	Black	PHY	[7,16]
	20	Black	PHY	[16,*Now*)

The corresponding *TRqdm* of Table 8.1 is displayed in Fig. 8.8.

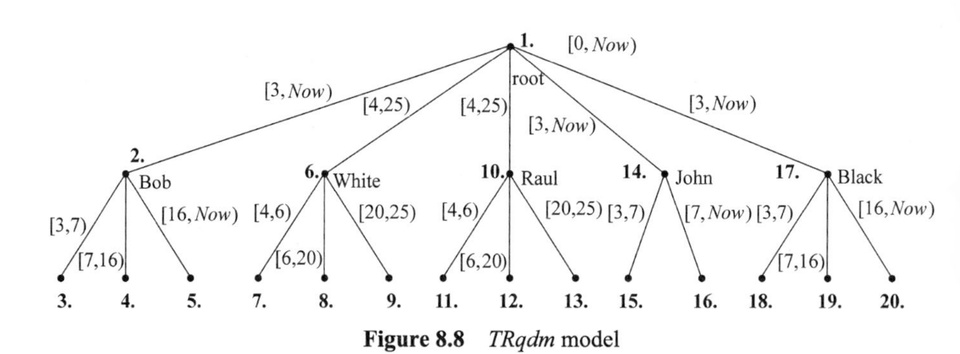

Figure 8.8 *TRqdm* model

8.3.2 Temporal Relation Index Model

In this section, we discuss temporal equivalence and temporal preorder relation as the basis of the temporal relation index.

8.3.2.1 Temporal equivalence and summary

Firstly, we assume that u and v are two temporal nodes. If $VT(u) = VT(v)$, then we say u, v satisfies time period equivalence relation denoted as $u \sim_T v$. Time period equivalence relation $\sim_T$ is a kind of equivalence relation that satisfies reflexivity, symmetry and transitivity.

Definition 8.3 (Temporal summary *TRsum*) Temporal summary, based on *TRqdm*, is recorded as $TRsum(TR) = \langle Nts(D), Vts(D), Ets(D) \rangle$. Here, the nodes are called summary nodes, its layer number is equal to that of *TRqdm* and its layer corresponds to the relative one of *TRqdm*. The *i*-th layer corresponds to the *i*-th layer of *TRqdm* and the root of *TRsum* is the name of the *TRqdm*.

① $Nts(D)$ is the node series number set from the preorder travel of the summary nodes, and $Vts(D)$ is the set of all the summary nodes in $TRsum(TR)$.

- The node of the first layer in $TRsum(TR)$ is the root.
- The second layer is the set of the main body nodes with the same time interval as that of the second layer in $TRqdm(TR)$, which are the equivalence $classes[u]$ based on the time period equivalence "~".
- The node of the third layer is the son of each main body equivalence class in the second layer, of which the data item is the tuple set of the main body with the same time interval.

② $Ets(D)$ is the set of the edges.

③ $LMAPsum(D) = \{LMAP_{[u]}(.) \mid [u] \in VTs(D)\}$ is the location map set for the tuples of the summary nodes in the third layer of $TRqdm(TR)$. For $W \in [U]$, $LMAP_{[U]}(W)$ maps it to its location index. If $[V]$ is the parent node of $[U]$, we sort the nodes x and w in $[V]$ and $[W]$ in ascending order according to their serial number $X_0, X_1, \cdots, X_k; W_0, W_1, \cdots, W_M$. If the parent of $W \in \{W_0, W_1, \cdots, W_M\}$ is X_{I_0}, then the location index of W is "I_0".

Temporal summary node $[u]$ is marked as 3-tuple: $[u] = (n_I([u]), VT([u]), LMAPsum([u]))$. Here, $n_1([u])$ is the serial number by preorder travel of $[u]$ in $TRsum(TR)$, $VT([u])$ is the time period of $[u]$ and $LMAPsum([u])$ is the location indexes of the nodes in $TRqdm(D)$.

From the above definitions, we know that $TRsum(TR)$ is the summary of $TRqdm(TR)$ based on the temporal property.

Example 8.2 Temporal summary of temporal relation data of Example 8.1 is displayed in Fig. 8.9 and the constitution of temporary summary nodes is shown in Table 8.2.

8.3.2.2 Temporal preorder and interval number

For the given set A, if a relation $\mathcal{R}$ on $A \times A$ satisfies reflexivity and transitivity, $\mathcal{R}$ is a preorder relation. If $\mathcal{R}$ is a preorder relation on D, then record D as Γ.

The essence of index is to get the necessary condition for it to query the data. The preorder relation set can provide this condition.

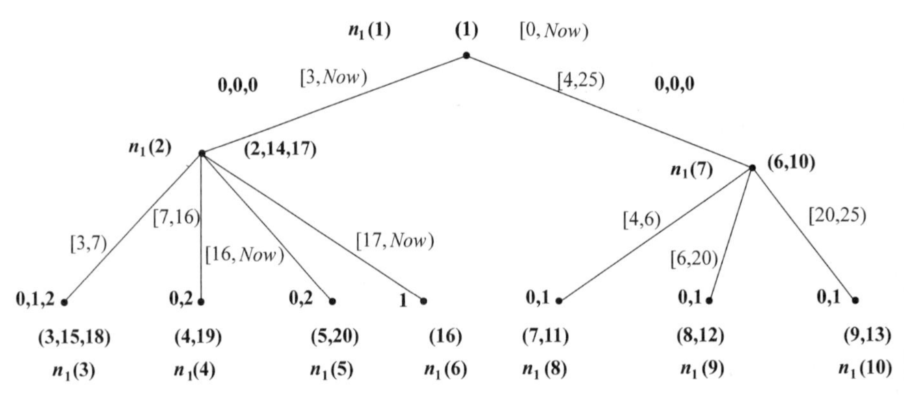

Figure 8.9 Temporal summary

Table 8.2 Summary nodes

Summary node	Period	Lable	Layer
$n_1(1) = (1)$,	[0,*Now*)	(0)	I
$n_1(2) = (2,14,17)$,	[3,*Now*)	(0,0,0)	II
$n_1(3) = (3,15,18)$,	[3,7)	(0,1,2)	III
$n_1(4) = (4,19)$,	[7,16)	(0,2)	III
$n_1(5) = (5,20)$,	[16,*Now*)	(0,2)	III
$n_1(6) = (16)$,	[17,*Now*)	(1)	III
$n_1(7) = (6,10)$,	[4,25)	(0,0)	II
$n_1(8) = (7,11)$,	[4,6)	(0,1)	III
$n_1(9) = (8,12)$,	[6,20)	(0,1)	III
$n_1(10) = (9,13)$	[20,25)	(0,1)	III

Definition 8.4 (Preorder and linear order branch) If $VT(u) \subseteq VT(v)$, then we say, v contains u temporally and records it as $u\mathcal{R}v$. If $VT(u) \not\subset VT(v)$, u, v are incompatible. Temporal containing relation is a preorder relation. For the preorder $\mathcal{R}$, we assume that $B \subseteq \mathcal{R}$, if $\forall \alpha, \beta \in B(\alpha \neq \beta), (\alpha\mathcal{R}\beta \vee \beta\mathcal{R}\alpha)$, then B is a linear order branch (*LOB*) of *R*. Here, we record the maximum and the minimum items of *B* as *maxB* and *minB*, respectively.

Here, we assume the ending point of time interval of Γ is an integer.

Definition 8.5 (Time interval number) Assuming α is the digit of the max end point of the time intervals of Γ, we define a map *F*(.) from Γ to *Z* (the set of positive integers): $\forall P \in \Gamma$. If $VT(P) = [Ps, Pe)$, then $F(P) = Ps * 10^{\alpha} + Pe = PsPe$. Here, *PsPe* is called the time interval number. The set of all the time interval numbers is recorded as $M(\Gamma)$, then $F(\Gamma) = M(\Gamma)$.

From the above definitions, we can get the basic properties of *F*(.) (basic property of map *F*(.)).

(1) $F(.)$ is bijective mapping

(2) if $P_1, P_2 \in \Gamma$, $P_1 \subseteq P_2 <=> V_s(P_1) = V_s(P_2)$, $F(P_2) \leqslant F(P_1)$ or $V_s(P_1) \neq V_s(P_2)$, $F(P_1) \leqslant F(P_2)$

(3) $P_1, P_2 \in \Gamma$, $(V_s(P_1) = V_s(P_2)) \wedge (F(P_2) \geqslant F(P_1))$ or $(V_e(P_1) = V_e(P_2)) \wedge F(P_1) \geqslant F(P_2) <=> P_1 \subseteq P_2$

Proof We can get the proof of (1) and (2) from (Nascimento and Dunham 1997). Here, we prove (3).

As provided in the problem, when $VTs(P_1) = VTs(P_2)$, $0 \leqslant F(P_1) - F(P_2) = (VTs(P_1) - VTs(P_2))10^{\alpha} + (VTe(P_1) - VTe(P_2)) = (VTe(P1) - VTe(P2))$.

Then, $0 \leqslant (VTe(P_1) - VTe(P_2))$, i.e., $VTe(P_1) \geqslant VTe(P_2)$, so $P_1 \subseteq P_2$.

When $VTs(P_1) \neq VTs(P_2)$, $0 \leqslant F(P_2) - F(P_1) = (VTs(P_2) - VTs(P_1))10^{\alpha} + (VTe(P_2) - VTe(P_1)) = (VTs(P_2) - VTs(P_1))10^{\alpha}$

Then, $0 \leqslant (VTe(P_2) - VTe(P_1))$, then $VTe(P_2) \geqslant VTe(P_1)$, so $P_1 \subseteq P_2$.QED

From (1), we treat time interval as time interval number, so $P_{ij} = [i,j) = ij$.

From (3), we define $P_1 \subseteq P_2$ as $F(P_1) \leqslant F(P_2)$.

Definition 8.6 (Temporal preorder matrix) Assuming that Γ is the set of the time interval preorder, i_1, i_m and j_1, j_n are the minimal and maximal starting points and the minimal and maximal ending points of Γ. then $[i, j) = ij(i_1 \leqslant i \leqslant i_m,\ j_1 \leqslant j \leqslant j_n)$. The triangle matrix displayed in Fig. 8.10 is named temporal preorder matrix and is recorded as $\Delta(\Gamma)$.

$$
\begin{array}{llll}
i_1 j_n, & i_2 j_n, & \ldots, & i_{m-1} j_n, \; i_m j_n \\
i_1 j_{n-1}, & i_2 j_{n-1}, & \ldots, & i_{m-1} j_{n-1} \\
\vdots & \vdots & \vdots & \\
i_1 j_2, & i_2 j_2 & & \\
i_1 j_1 & & &
\end{array}
$$

Figure 8.10 Temporal preorder $\Delta(\Gamma)$

For $i_0 j_0 \in \{ij\}(i_1 \leqslant i \leqslant i_m, j_1 \leqslant j \leqslant j_n)$, four sets are given in $\Delta(\Gamma)$ as displayed in Fig. 8.11:

$$LU(u_{i0j0}) = \{v_{ij} \mid i \leqslant i_0, j_0 \leqslant j\},\ RU(u_{i0j0}) = \{v_{ij} \mid i_0 \leqslant i, j_0 \leqslant j\}$$
$$LD(u_{i0j0}) = \{v_{ij} \mid i \leqslant i_0, j \leqslant j_0\},\ RD(u_{i0j0}) = \{v_{ij} \mid i_0 \leqslant i, j \leqslant j_0\}$$

In case of the above formulas, if "<" comes into existence, then they are named complementary submatrix and are recorded as *OLU*, *ORU*, *OLD* and *ORD*, respectively. If there exists two elements, $P_{ij} = ij$ and $P_{kq} = kq$, which satisfy $j = q \wedge (i = k - 1 \vee i = k + 1)$, then they are adjacent horizontally, and if $i = k \wedge (j = q - 1 \vee j = q + 1)$, they are adjacent vertically. The elements adjacent horizontally or vertically are connected by horizontal edge or vertical edge, respectively. The path between nodes u and v $Path(u,v)$ is defined by the shortest folded lines combined by horizontal and vertical edges. The length of $Path(u,v)$ is the number of the lines and the distance between u and v is the length of $Path(u,v)$. Assuming $E \subseteq \Delta(\Gamma)$, the nearest node of u_0 in E is defined by the node in E with the shortest length between u_0. For example, if E denotes the column to which u_0 belongs, the nearest node of u_0 is the nearest node above u_0. The Hass Map of preorder set Γ in $\Delta(\Gamma)$ is recorded as $H_\Delta(\Gamma)$.

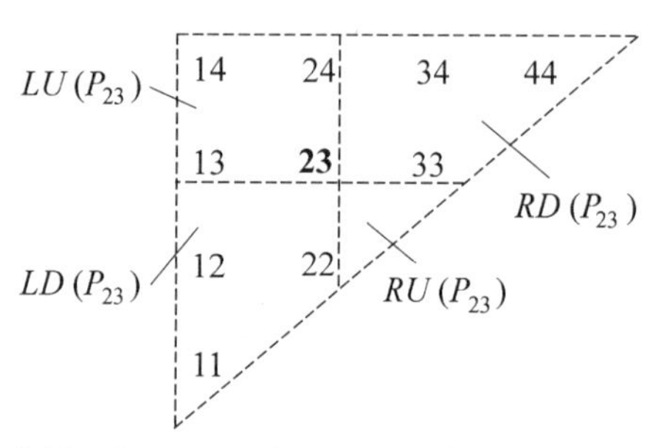

Figure 8.11 Open complementary submatrix relative to 23

Algorithm 8.1 (DFA)

Step 1: Constructs $H_\Delta(\Gamma)$, then does *DFA* search: from the first element in the first column, it searches down until the last element u_{i0} in this column.

Step 2: From u_{i0}, it turns to the adjacent element in the next column, then searches down from that element until the last element in that column, then turns to the adjacent element in the next column, recursively. In the end, it will search the element u_p and there is no other element of Γ except itself. Then, the nodes in the search path Γ construct a linear order branch L_1.

Step 3: Redoes *DFA* search in $\Gamma_1 = \Gamma \backslash L_1$ and it will get another linear order branch L_2, recursively. In the end, it will get all the linear order branches. The complexities of constructing $H_\Delta(\Gamma)$ and *DFA* are $O(n\log n)$ and $O(n^2)$, respectively.

Theorem 8.1 (The existence of the min linear order partition) The linear order partition (*LOP*) from *DFA* is the minimal linear order partition.

Proof Assuming that the linear order partition $\Delta_0 = \langle L_1, L_2, \cdots, L_n \rangle$ is from *DFA*, in which L_i is sorted by computing sequence, then $\forall a_{i0} \in L_i (1 < i \leqslant n)$, $a_{k0} \in L_{i-1}$, a_{i0} and a_{k0} are not compatible. In fact, we only need to prove that there exists an element in L_{i-1}, which is in $LD(a_{i0})$. Here, in reverse order, we assume no element of that kind exists, then L_{i-1} is in $OLU(a_{i0})$, then $ORD(\min L_{i-1})$ contains L_i, which contradicts with *LOB* property. Assuming that we get nodes $a_1, a_2, \cdots, a_n$, using the above method, $a_{i-1} \in OLD(a_i)(1<i \leqslant n)$.

That is, not every two nodes are compatible, so for any linear order partition, it contains a minimum of n *LOB*s.

Example 8.3 Assuming time interval set $\Gamma = \{14,12; 24,23,22; 33; 44\}$, then the solute process with *DFA* is displayed in Fig. 8.12 (here, we assume $\alpha = 1$).

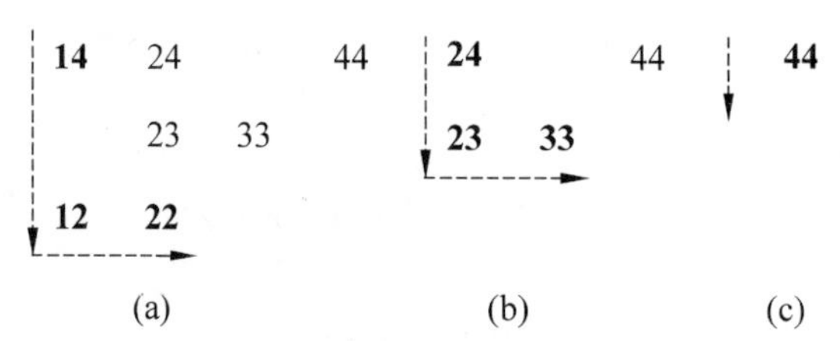

Figure 8.12 Linear order branch of $\Delta(\Gamma)$

8.3.2.3 Temporal relational index

We can obtain the temporal relational index model by creating the linear order branches on the set of (summary) nodes of the same layer of Temporal Summary through temporal preorder on set of periods *R*.

Definition 8.7 (Temporal Relational Index Model) Temporal Relational Index Model *TRdim* (Γ) is a tree structure. Its node is called (temporal) index node. *TRdim* (Γ) has the same number of layers as *TRsum*(Γ), and the corresponding layer is mutually corresponding, and the root is the name of *TRdim* (Γ). *TRdim* (Γ) can be described formally as the quadruple as follows:

$TRdim\ (TR) = (N_{ind}(D),\ VT_{ind}\ (\Gamma),\ E_{ind}\ (\Gamma),\ LMAP_{ind}\ (\Gamma))$

① $N_{ind}(\Gamma)$ is a set of node numbers of index elements resulting from pre-order traversal.

② $VT_{ind}\ (\Gamma)$ is a set of all (index) nodes.

The first layer in *TRdim* is just a single root, and this node is the name of *TRdim*(Γ).

The nodes in the second layer of *TRdim* are the *LOB*s on the corresponding summary nodes in the second layer of *TRsum*(Γ).

The nodes in the third layer are the *LOB*s on the corresponding summary nodes of each index node in the second layer of *TRdim* (Γ), $B(\min(B), \max(B))$ denotes a special *LOB*, where $\min(B), \max(B)$ means the minimum and maximum in *B*.

③ $E_{ind}\ (\Gamma)$ is a set of all edges in *TRdim* (Γ).

④ $LMAP_{ind}\ (\Gamma)$ is similar with $LMAP_{sum}$ in Definition 8.3 which describes the structural information of *TRdim* (*TR*).

The index node is the binary $B_i = (n_2(B_i), (maxB_i, minB_i))$, where $n_2(B_i)$ is the number obtained by pre-order traversal, and $maxB_i$ and $minB_i$ denote the maximum and minimum in *B*,respectively.

In the definition mentioned above, "①"shows that *TRdim* and *TRqdm* have a similar structure, or is homomorphism in some sense. "②" indicates the nodes in *TRqdm* are accumulation of nodes in the same layer in *TRdim.* ③ indicates the structure information of index nodes.

Example 8.4 Figure 8.13 shows the temporal index of relation data in Example 8.1

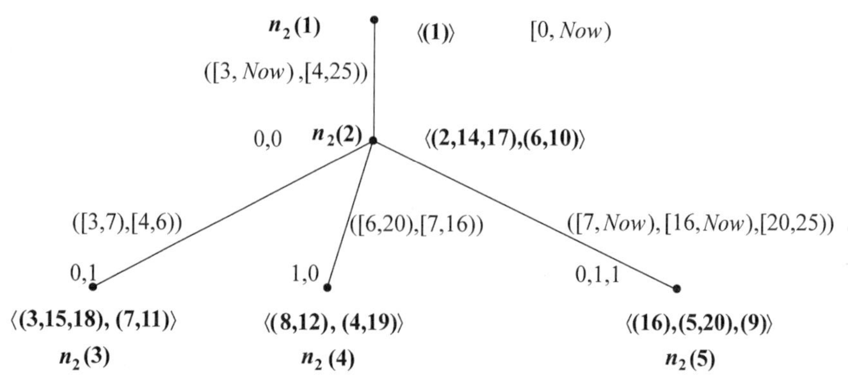

Figure 8.13 Temporal index

The information of index nodes is shown in Table 8.3.

Table 8.3 Index nodes

Nodes	Lable	LOB	Layer
$n_2(1) = \langle (1) \rangle$	(0)	[0,*Now*)	Ⅰ
$n_2(2) = \langle n_1(2), n_1(7) \rangle = (2,14,17), (6,10)$	(0,0)	[3,*Now*), [4,25)	Ⅱ
$n_2(3) = \langle n_1(3), n_1(8) \rangle = (3,15,18), (7,11)$	(0,1)	[3,7), [4,6)	Ⅲ
$n_2(4) = \langle n_1(9), n_1(4) \rangle = (8,12)(4,19)$	(1,0)	[6,20), [7,16)	Ⅲ
$n_2(5) = n_1(6), n_1(5), n_1(10) = (16), (5,20), (9)$	(0,1,1)	[7,*Now*), [16,*Now*), [20,25)	Ⅲ

8.4 Data Querying and Index Updating

This section discusses the temporal querying and the dynamic updating based on *TRdim*.

8.4.1 Index Querying

A temporal querying $Q_{t\ t} = \langle NT(Q_t), VT(Q_t) \rangle$ means that searching tuples $TTuple(TR) = (NT(TTuple), VT(TTuple))$ in *TR*, and meets $NT(Q_t) = NT(TTuple) \wedge VT(Q_t) \subseteq VT(TTuple)$.

Temporal relational querying often searches all the tuples that meet the temporal constraints. Only then do we obtain the superset of the result we need (if it exists). We can get the final result by implementing the general query on the superset. Temporal index *TRind* is mainly used in temporal searching.

Algorithm 8.2 (Index querying algorithm)

Step 1: Temporal disposal of data: Let $B_{i0} = (n(B_{i0}), maxB_{i0}, minB_{i0})$ be the leaf nodes of *TRdim*,

- If $VT(Q_t) \cap maxB_{i0} = \varnothing$, then no element in B_{i0} is a query result.
- If $VT(Q_t) \subseteq minB_{i0}$, then all elements in L_{i0} are possible query results.
- If neither ① nor ②, let the summary nodes in L_{i0} be $[u]_{i1}$, $[u]_{i2}$, $\cdots$, $[u]_{in}$ according to linear order. For array $\{VT([u]_{i1}), VT([u]_{i2}), \cdots, VT([u]_{in})\}$, we do a binary search on the start points to find the first period $VT([u]_{ij})$, which meets the condition: $Start(VT([u]_{ij})) \leqslant Start(VT(Q_t))$ and $Start(VT([u]_{i,j+1}))>Start(VT(Q_t))$. A subset of L_{i0} $\{VT([u]_{i1}), VT([u]_{i2}), \cdots, VT([u]_{ij})\}$ is obtained. Subsequently, another binary search is carried out on L_{i0} to find the first period $VT([u]_{ik})$, which meets the condition: $End(VT([u]_{ik})) \geqslant End(VT(Q_t))$ and $End(VT([u]_{i,k+1}))<$ $End(VT(Q_t))$. Then the elements in linear order segment $\{VT([u]_{i1}), VT([u]_{i2}), \cdots,$ $VT([u]_{ik})\}$ of L_{i0} are possible query results.

Step 2: Disposal of summary nodes: If we take the summary nodes of the *LOB* or segment out, then through the location mapping, we can get the main body instances and the set of tuples corresponding to the time period in the summary nodes, that is, the superset of the result we need. The tuples in this superset have the time stamp and the main body instances mark.

Step 3: Non-temporal disposal of data: Implement the non-temporal search on the superset and we can get the final query result.

The algorithm has time complexity of $O(\log n)$.

Example 8.5 With reference to the temporal relation in Example 8.1, there is a temporal querying $Q_t = \langle NT(Q_t), VT(Q_t) \rangle$ and $VT(Q_t) = [12,15)$.

(1) Temporal inquiry on data: With reference to Fig. 8.13, in the set of the leaf node of $Trind\{n_2(3), n_2(4), n_2(5), n_2(6)\}$, $maxT(n_2(3)) = [3,7) \cap VT(Q_t) = \varnothing$, is not the query result.

$minT(n_2(4)) = [7,16) \supseteq [12,15) = VT(Q_t)$, $n_2(4) = \langle n_1(9), n_1(4) \rangle$, is the possible inquiry result.

$maxT(n_2(5)) = [7,Now)$. If we carry out the binary search on the summary nodes of $n_2(5)$, $n_1(6)$ may be the possible answer.

(2) The constitution of instance-summary:

$n_2(4) = \langle n_1(9), n_1(4) \rangle$, its location mapping is (1,0), and its parent is $n_2(2) =$ $\langle n_1(2), n_1(7) \rangle$. Therefore, we have parent-children: $n_1(7)$-$n_1(9)$ and $n_1(2)$-$n_1(4)$, also we can get $n_1(2)$-$n_1(6)$.

$n_1(2) = (2,14,17)$, $n_1(7) = (6,10)$, and the location mapping of $n_1(4) = (4,19)$, $n_1(6) = (16)$, $n_1(9) = (8,12)$ are: (0,2),(1),(0,1), so we can get the instance-tuples: 2-4,17-19; 14-16,6-8; 10-12.

(3) Non-temporal disposal on data: We can get the final results by applying non-temporal query on the Instance-tuples: 2-4,17-19: 14-16,6-8: 10-12.

8.4.2 Index Updating

In *TRdim* framework, when updating data, firstly modification on the mathematic structure of *LOP* should be considered, then comes the structure change of *TRsum* and *TRdim*. In this part, we discuss inserting and deleting of the temporal nodes.

8.4.2.1 Nodes inserting

The inserting of new node could result in the split and merge of original *LOB*s. The reconstruction in *LOB*s should remain closed to *DFA*, that is, after reconstruction, the *LOP* should be the same as the one generated by *DFA* with these new data.

Theorem 8.2 (Reconstruction for *LOB*s based on inserting nodes) Reconstructed *LOP* can remain closed to *DFA* after inserting.

Proof There can be several cases after inserting a node u_0.

(1) $\Gamma \subseteq ORU(u_0)$ or $\Gamma \subseteq OLD(u_0) \cup ORD(u_0) \cup OLU(u_0)$, then u_0 constitutes a separate *LOB*.

(2) $\exists L_0(VT(u_0) \subseteq min(L_0) \vee max(L_0) \subseteq VT(u_0))$, then add $[u_0]$ to L_0.

(3) $\forall L \in H_0(\neg(VT(u_0) \subseteq min(L_i) \vee max(L_i) \subseteq VT(u_0)))$, let $D_0 = D \cup \{u_0\}$, then

① If there exists elements in $H_\Delta(\Gamma)$, which are in the column of u_0 and just above and below u_0 in the same *LOB*, then add u_0 to the *LOB*. If they are not in the same *LOB*, considering *DFA*, u_0 has nothing to do with the elements below it. The nearest elements above it and the nearest elements to the right of it are only considered.

- u_0 only has above nearest neighbor $v_i \in \Gamma$. Suppose $L(v_i) = \langle v_1, \cdots, v_i, v_{i+1}, \cdots, v_m \rangle$, and $v_{i+1} \in ORD(u_0)$, as shown in Fig. 8.14(a), then $LOB(u_0) = \langle v_1, \cdots, v_i, u_0, v_{i+1}, \cdots, v_m \rangle$.
- u_0 only has above nearest neighbor $v_i \in \Gamma$. $L(v_i) = \langle v_1, \cdots, v_i, v_{i+1}, \cdots, v_m \rangle$, and $v_{i+1} \in ORU(u_0)$, as shown in Fig. 8.14(b). Then we get $L(u_0) = \langle v_1, \cdots, v_i, u_0 \rangle$ and segment $\langle v_{i+1}, \cdots, v_m \rangle$. Take this segment as a virtual node with node value v_{i+1}, and insert it again. The reinsert may propagate to the last *LOB*.
- u_0 has both above neighbor $v_i \in \Gamma$ and right nearest neighbor $v_j \in \Gamma$. $L(v_i) = \langle v_1, \cdots, v_i, v_{i+1}, \cdots, v_{j-1}, v_j, \cdots, v_m \rangle$, as shown in Fig. 8.14(c), then we get $L(u_0) = \langle v_1, \cdots, v_i, u_0, v_j, \cdots, v_m \rangle$ and segment $\langle v_{i+1}, \cdots, v_m \rangle$. Take this segment as a virtual node with node value v_{i+1}, and insert it again.

② If u_0 consists of a new column along, suppose the right nearest neighbor of u_0 is v_i, $L(v_i) = \langle v_1, \cdots, v_{i-1}, v_i, \cdots, v_m \rangle$, as shown in Fig. 8.14(d), then we get $L(u_0) = \langle v_1, \cdots, v_i, u_0 \rangle$ and segment $\langle v_{i+1}, \cdots, v_m \rangle$. Take this segment as a virtual node and reinsert it as mentioned above.

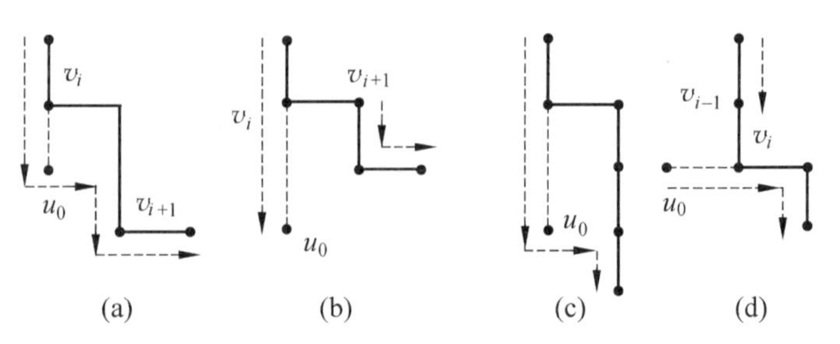

Figure 8.14 Reconstruction of *LOB* based on inserting the nodes

The above algorithm meets the constraints of *DFA* and since the *LOP* got by *DFA* is unique, then the theorem is obtained.

Algorithm 8.3 (insert node into *TRsum* and *TRdim*)

First, find the right position to insert u_0. Then, it can result in the following situations:

Step 1 (merge with existing summary node): If u_0 should be inserted into an existing summary node $[u_{i0}]$, update $[u_{i0}] = [u_{i0}] \cup \{u_0\}$, and $Label([u_{i0}]) = Label([u_{i0}]) \cup \{Label(u_0)\}$, then update the location mapping, add u_0 and its parent location information into the original $LMAP_{sum}([u_{i0}])$. *Now*, summary node $[u_{i0}]$ is also the summary node $[u_0]$ and *TRdim* needs no modification.

Step 2 (new summary node): If the insertion of u_0 needs creation of new summary node, then in *TRdim*:

- If $[u_0]$ needs to merge with existing index node L_{i0}, then update this $LOB = L_{i0} \cup \{[u_0]\}$, $Label(L_{i0}) = Label(L_{i0}) \cup \{Label([u_0])\}$, and update the location mapping, add $[u_0]$ and its parent location information into the original $LMAP_{tind}([L_0])$. *Now*, index node L_{i0} is also index node for u_0, and other part of index needs no modification.
- If insertion of $[u_0]$ needs creation of new index node, follow Theorem 8.2 to reconstruct index structure and update $LMAP_{tind}$ as well.

8.4.2.2 Nodes deleting

Theoretically, if we can find the summary nodes and index nodes, in which the node to be deleted exists, we only need to delete it. However, the summary nodes and index nodes have mathematical structures. We must consider the change of the structure in *TRsum* and *TRdim*.

Algorithm 8.4 The reconstructed *LOP* remains closed to DFA after deletion.

Suppose the *LOB* of u_0 is L_0, there can be several cases after deleting a node u_0.

- If there are elements both above and below u_0, then update *LOB* $L_0 = L_0 \backslash \{u_0\}$.
- If $L_0 = (u_1, \cdots, u_p, u_0)$ and the nearest node (down first) to u_p in $RD(u_p)$ is v_i, and $L(v_i) = \langle v_1, \cdots, v_{i-1}, v_i, \cdots, v_m \rangle$, then we get new *LOB* $L(v_i) = \langle u_1, \cdots, u_p, v_i, \cdots, v_m \rangle$ and linear order segment $\langle v_1, \cdots, v_{i-1} \rangle$, as shown in Fig. 8.15(a). Then find the nearest (down first) w_i node in $RD(v_{i-1})$ and repeat the above procedure.
- If $L(u_0) = (u_0, u_1, \cdots, u_p)$ and the nearest node (up first) to u_1 in $LU(u_1)$ is v_i, and $L(v_i) = \langle v_1, \cdots, v_i, v_{i+1}, \cdots, v_m \rangle$, then we get new *LOB* $L(v_i) = \langle v_1, \cdots, v_i, u_1, u_2, \cdots, u_p \rangle$ and linear order segment $\langle v_{i+1}, \cdots, v_m \rangle$, as shown in Fig. 8.15(b). Then find the nearest (up first) w_i node in $LU(v_{i+1})$ and repeat the above procedure.
- If $L(u_0) = (u_1, \cdots, u_{i-1}, u_0, u_{i+1}, \cdots, u_p)$ and the node nearest (above) to u_{i+1} in $LU(u_{i+1})$ is v_k and $L(v_k) = \langle v_1, \cdots, v_k, v_{k+1}, \cdots, v_m \rangle$, and the nearest node of $L(v_k)$ (below) to u_{i-1} in $RD(u_{i-1})$ is v_{j0}, then we get new *LOB* $L_0(v_k) = \langle u_1, \cdots, u_{i-1}, P, u_{i+1}, \cdots, u_p \rangle$, linear order segments $\langle v_1, \cdots, v_{j0-1} \rangle$ and $\langle v_{k+1}, \cdots, v_m \rangle$, as shown in Fig. 8.15(c). Then for those two linear order segments, find the nearest

(down first) node in $RD(v_{j0-1})$ and the nearest (up first) node in $LU(v_{k+1})$, then repeat the above procedure, respectively. This may propagate to the last *LOB*.

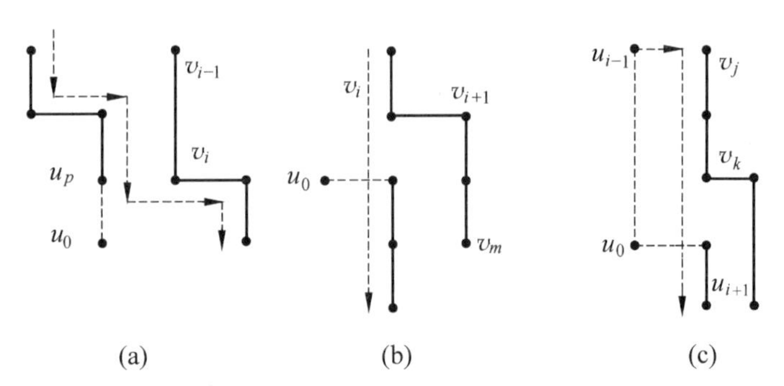

Figure 8.15 Reconstruction of the *LOB* based on deleted nodes

Algorithm 8.5 Delete node algorithm

Step 1 (find the node to be deleted): Find the node u_0 in *TRqdm* that needs to be deleted by query algorithm.

Step 2: If $[u_0]\backslash\{u_0\} \neq \varnothing$, the mathematical structure of *TRsum* and *TRdim* needs no modification, only structure information needs to be modified.

(1) Modification of structure information: Let the original summary node and the one after deletion be $[u_{i0}]$ and $[u_0]$, the original index node and the one after deletion be L_{i0} and L_0, then we have $Label([u_0]) = Label([u_{i0}]/\{u_0\})$, $Label(L_0) = Label(L_0/\{[u_{i0}]/[u_0]\})$. Delete the corresponding location mapping information in $LMAP_{sum}$ and $LMAP_{tind}$.

(2) Dealing descendents: If u_0 is an internal node in *TRqdm*, then all the descendents should also be deleted. We can directly delete the sub-graph in *TRsum* and *TRdim*.

Step 3: If $[u_0]\backslash\{u_0\} = \varnothing$, summary node $[u_0]$ needs to be deleted from *TRsum*, then:

(1) Reconstruct corresponding *Lob*s according to Theorem 8.2.

(2) Modification of structure information, semantic information and deletion of sub-graph is similar to Step 2.

The time complexity of Algorithm 8.5 is $O(\log n)$.

Example 8.6 We want to delete the node "2" as discussed in Example 8.1, and its children are "3","4","5", as shown in *TRsum*. Nodes "2", and "3","4","5" are in the summary nodes $n_1(2) = (2,14,17)$, $n_1(5) = (5,20)$, and its location mapping is (0,0,0), (0,1,2), (0,2) and (0,2). After deleting "2" and "3","4","5", the summary nodes become $n_1'(2) = (14,17)$, $n_1'(3) = (15,18)$, $n_1'(4) = (19)$ and $n_1'(5) = (20)$, and its location mapping is (0,0), (1,2), (2) and (2). *Now*, we only need to replace the original nodes with the new nodes and the other nodes need no modification. Then we can get *TRsum* after deleting nodes. As the summary nodes do not need to be cut down, the structure of *TRdim* does not need any modification. The corresponding index nodes change, which becomes $n_2'(2) = \langle (14, 17), (6, 10) \rangle$, $n_2'(3) = \langle (15,18),(7,11) \rangle$, $n_2'(4) = \langle (12,8),(19) \rangle$, $n_2'(5) = \langle (16),(20),(9) \rangle$, and its location mapping remains the same. The other index nodes do not need any modification.

If we want to delete the node "16" in example 8.1, where node "16" is in the summary node $n_1(6) = (16)$, we need to delete $n_1(6)$, which should be take away from *TRsum*. The index nodes, in which $n_1(6)$ exists is $n_2(5) = \langle (16),(5,20), (9,13) \rangle$. According to the *LOB* reconstruction algorithm based on deletion, after deleting node "16", we get the new index node $n_2'(5) = \langle (5,20),(9,13) \rangle$. We only need to replace $n_2(5)$ with $n_2'(5) = \langle (5,20),(9,13) \rangle$. The rest of the nodes remain the same and then we get the new index tree.

8.5 Simulation

The basic structure of temporal relation is the temporal relation table constituted by temporal tuples. Therefore, the querying mainly concerns the attribute value and tuple object. This thesis has designed two kinds of querying temporal data simulation test: the query only concerns temporal element and the query concerns both temporal and non-temporal element.

The test environment of simulation is: CPU: Cerlon 2.66GHz; memory: 768MB; operating system: Windows XP; development environment: Eclipse + JDK 1.5; development language: Java.

8.5.1 Index Constructing

When the number of nodes ranges from $5*10^4$ to $5*10^5$, the probability of having the same time is 0.5. When the main body instances range from 10^2, the average number of tuples in each of the main body instances is 10^2 and the span of average time period is 10^2. The results are shown in Figs. 8.16 and 8.17.

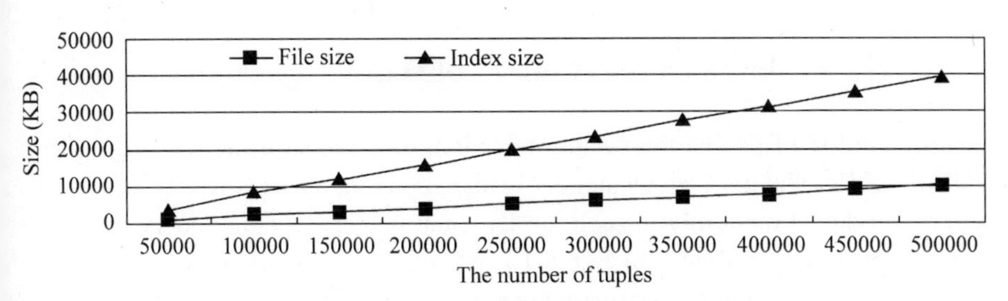

Figure 8.16 Space cost of constructing the index

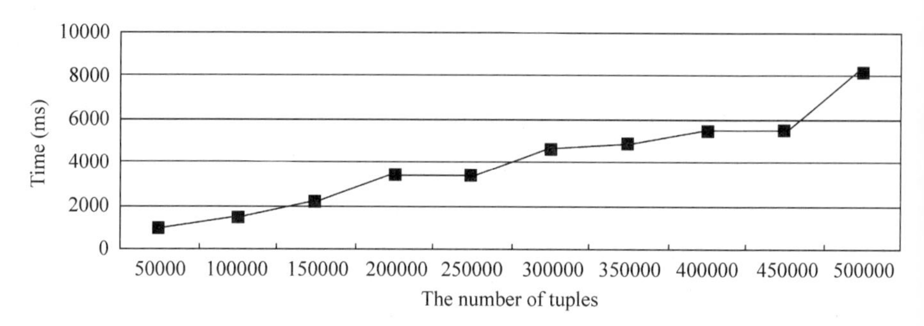

Figure 8.17 Time cost of constructing the index

8.5.2 Query Based on Probability

The query is the probability of having the same time ranges from 0.0 to 1.0. For each probability, the records produced randomly are $5*10^5$ and we have 10^2 main body instances. 10^3 query statements will be produced randomly. The result is shown in Fig. 8.18, where, the x-coordinate represents the probabilities with the same time ranges in given time set and the y-coordinate time (unit: millisecond). In the figuer, Erg: ergodic query; MAP21: query based on MAP21; *TRdim*: query based on *TRdim*.

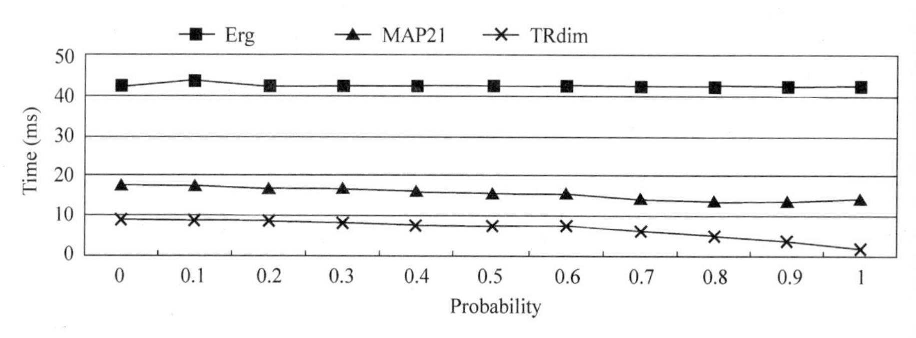

Figure 8.18 Query based on the change of the probability with the same time

8.5.3 Query Based on the Number of Data

The probability of having the same time is 0.5, the number of nodes ranges from $5*10^4$ to $5*10^5$, and the average number of main body instances is 10^2. 10^3 statements will be produced randomly in each implementation. The average query time cost is 10^3. The result is shown in Fig. 8.19, where, the x-coordinate represents the number of data and the y-coordinate the time (unit: millisecond).

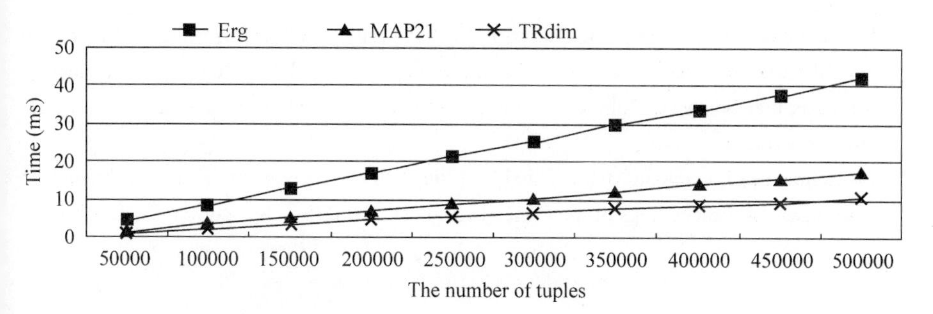

Figure 8.19 Query based on the number of data

References

[1] Becker B, Gschwind S, Ohler T, Seeger B, Widmayer P (**1996**) An asymptotically optimal multiversion B-Tree. LDB J 5(4): 264 – 275

[2] Bentley J (**1977**) Algorithms for Klee's rectangle problems. Computer Science Department, Carnegie Mellon University, Pittsburgh, PA

[3] Blankenagel G, Guting R (**1994**) External segment trees. Algorithmica 12(6): 498 – 532

[4] Bliujute R, Jensen CS, Saltenis S, Slivinskas G (**1999**) Light-weight indexing of general bitemporal data. Statistical and Scientific Database Management, pp 125 – 138

[5] Bliujute R, Jensen CS, Saltenis S, Slivinskas G (**1998**) R-Tree based indexing of now-relative bitemporal data. In: Proceedings of 24th International Conference on Very Large DataBases, New York, USA, pp 345 – 356

[6] Edelsbrunner H (**1983**) A new approach to rectangle intersections, Parts I & II. International J Comput Math 13: 209 – 229

[7] Elmasri R, Wuu GTJ, Kim YJ (**1990**) The Time Index: An access structure for temporal data. In: Proceedings of the 16th Very Large Databases Conference, San Francisco: Morgan Kaufmann Publishers Inc., pp1 – 12

[8] Gunadhi H, Segev A (**1985**) Efficient indexing methods for temporal relations. IEEE Transactions on Knowledge Data Engineering 5(3): 496 – 509

[9] Kumar A, Tsotras VJ, Faloutsos C (**1995**) Access methods for bitemporal databases. In: Clifford S, Tuzhlin A (eds) Proceedings of the International Workshop on Recent Advances in Temporal Databases (Zurich, Switzerland, Sept.), Springer-Verlag, New York, pp 235 – 254

[10] Mccreight EM (**1985**) Priority search trees. SIAM J Comput 14(2): 257 – 276

[11] Nascimento M, Dunham M (**1997**) Indexing valid time database via B+tree, the MAP21 approch. Technical Report CSE-97-08, Dallas, USA: School of Engineering and Applied Sciences, Southern Methodist University

[12] Nascimento M, Silva J (**1998**) Towards historical R-rrees. In: Proceedings of the 1998 ACM Symposium on Applied Computing, pp 235 – 240

[13] Nascimento M, Dunham MH, Elmasri R (**1995**) Analytical performance studies of the IVTT bitemporal access structure. Technical Report 95-CSE-19, School of Engineering and Applied Sciences, Southern Methodist University, http://www.seas.smu.edu/Papers/tr-96-cse-19.ps

[14] Nascimento M, Dunham MH, Elmasri R (**1996**) M-IVTT: a practical index for bitemporal databases. In: Proceedings of the Conference on DEXA (DEX '96, Zurich, Switzerland)

9 Indexes for Moving-Objects Data

Xiaoping Ye[1,2], Huan Guo[2], Xiongxiong Zhu[2] and Yidong Ja[2]

[1] Computer School, South China Normal University, Guangzhou 510631, P.R. China
[2] Department of Computer Science, Sun Yat-sen University, Guangzhou 510275, P.R. China

Abstract The basic characteristics of spatio-temporal data are the massive data and complex spatio-temporal relationship. Therefore, indexes are an effective method to implement spatio-temporal data's access. Moving-object data management is the main field of study in spatio-temporal management, which has strong application-driven and broad practical prospects. The first part of this chapter discusses the current moving data indexes, which indicates that the indexes of moving object data are the research frontier and hotspot in spatio-temporal databases. The second part of this chapter is our original work, which mainly proposes a moving-object data index technique based on time linear order. Current work usually deals with time dimension as a common spatial dimension, which is essentially a "pure" higher-dimensional spatial index. These schemas can use the existing special index technique directly, but cannot effectively reflect the basic differences between "time" and "space" from both theoretical principles and technical implementation. Thus, there may be greater possibilities for improving querying efficiency by certain novel indexing schemas. The index model in this chapter makes a proper separation between time and space, that is, discusses first the time and then the space, and then couples them in a suitable manner. Additionally, corresponding simulations are designed and implemented, which indicates that our index method has higher efficiency than TB-tree.

Keywords *moving-object index, time linear order, simulation test*

9.1 Introduction

Space and time are the basic factors of objective entities in the real world. In traditional database, because of restriction of the application range and the software-hardware environment of data management, the space and time characteristics of

---

mcsyxp@mail.sysu.edu.cn

data entities are usually represented as "implicit", that is, by default data taken as the current space position and time state. With an expansion and increase in depth of the application range of computers, we need to represent and manage the spatial and temporal information explicitly. Corresponding spatial and temporal data management technologies have also developed. Temporal databases and spatial databases have become the basic research field of modern database technology. Temporal database (TDB) first appeared in the late 1970s as a database system that can record and maintain the history of data. Its application requirement is personnel data management. TDB supports the management and operation of user-defined time, valid time and transaction time. Spatial database (SDB) is a database system that can describe, store, and manage spatial data and corresponding attribute data. SDB is a database with the GIS's development and application in 1980s, which can represent and manage spatial information, such as spatial objects' position, shape, size, distribution characteristic and spatial relations. SDB and TDB manage spatial and temporal information separately. However, in many important applications, the object's temporal and spatial information are interconnected. This needs a unified description and management and needs to integrate the object's temporal and spatial attributions. As such, the database system that studies both temporal and spatial factors appeared in the early 1990s. The corresponding spatio-temporal database (STDB), which includes temporal data and spatial data, and can deal with the temporal and spatial attributions of spatio-temporal objects, also appeared and obtained rapid development.

In the research of spatio-temporal management, trajectory management of moving objects got wide attention. The basic conceptions of moving objects were mainly proposed in 1997 by Sistla (Sistla et al. 1997). The trajectories of moving objects are represented as the data entities. This includes the dynamic attribute of the moving vector, which describes the current state of the moving object. The update of database can be implemented by changing the moving vectors. In this case, the same query statement will return different results at different times and can implement the query for some expected future time.

From then on, moving-objects management became the basic research field of spatio-temporal database. Moving objects consist of moving points and moving-region object, where both moving region object's shape and position change, for example, city's development, the region change from natural disaster, the change from some meteorological phenomenon. Moving-point object's shape does not change, only its position changes with time, which mainly applies to vehicle navigation (VNS), Location-based Service (LBS), Intelligent Transportation System (ITS), steamship navigation and the management of airplane line, and also some situation applications (e.g., the monitor of an aircraft carrier). The database system that manages moving objects and their position is called moving-objects database (MOD). In case of moving-point objects, only their positions change with time, so corresponding spatial attributes can be abstracted as a point in a plane. With the time change, the point position (x, y, z) in three dimensions forms a curve, which is called a trajectory. Therefore, the basic managed data units of MOD are moving-objects trajectories. Recently, with the development of mobile positioning technology and wireless communication technology, the development of mobile positioning devices, such as Vehicle Positioning Equipment, PDAs and mobile phones, the application domain has become wider. The application requirement has become more and more urgent. The corresponding research has become a hot spot of spatio-temporal database.

Moving-object database needs to manage large amounts of moving data. When dealing with query operation, traversal scanning will affect system performance. To improve query performance, research and development of efficient moving-data index method has become the basic approach of moving-data query. In order to use traditional spatial data indexing technique for reference, the corresponding methods mainly deal with the time dimension as another common spatial dimension. From the temporal point of the query information, the indexing techniques of moving objects can be divided into three classes: index based on historical trajectory information, index based on historical and current position information and index based on current and future position information.

1. Index to historical information

The method of description of moving object's historical position by recording the position of a moving object at any time is impossible since the position of a moving object is varied with the evolution of time. Two solutions have been taken to store moving objects, one is sampling and the other is function description. Under the condition that the moving objects do linear action between sampling points, sampling is mean to sample stored points for the historical position according to some constant time interval, so that we can describe motion approximately. Function description can be seen as taking the trajectories of moving objects as a time function, whose value will change only when the referred parameters (speed or direction) change. *Now* the index of a moving object's trajectory can be divided into three kinds: based on traditional space index, on multiple version structure and overlapping technique and index structure for trajectory-oriented query of moving objects.

The main methods based on traditional space index are RT-tree (Xu et al. 1990), 3D R-tree (Theodoridis et al. 1996), and STR-tree (Pfoser et al. 2000).

- **RT-tree** uses R-tree (Guttman 1984) as the accessing method of historical trajectory's spatial information and uses TSB-tree (Lomet and Salzberg 1989) as the accessing method of temporal information. Its spatial query performance is as good as standard R-tree. Being used for temporal query, TSB-tree has good query performance for time slice query. However, in some cases, it has to access the whole RT-tree when it is used for time window query.
- **3D R-tree** uses traditional R-tree to index moving data, which takes time dimension as a spatial dimension, but it does not consider the different characteristics between time and space. 3D R-tree will not work effectively if

it does not know the time interval of indexed historical trajectory before querying. Since 3D-tree is easy to implement, it can be used to index some spatio-temporal objects whose position and the range of motion do not change obviously or change a little. However, when MBR, in three-dimension, gets much bigger, it will bring plenty of overlapping areas. Therefore, the query efficiency will be reduced. Moreover, time slice query will no longer depend on the effective entity at query time, but will depend on all the historical entities.

- **STR-tree** is the extension of R-tree. It services well in query of the trajectories of moving objects. The main differences between R-tree and STR-tree are inserting and splitting algorithms. R-tree's insertion algorithm is based on minimum expansion strategy, while STR-tree considers not only the spatial contiguous but also the partial integrality of trajectory, which can keep line segments belonging to the same trajectory together as much as possible.

The indexed models based on multi-version and overlapping technique mainly include MR-tree (Xu et al. 1990), HR-tree (Nascimento and Silva 1998), HR+-tree (Tao and Papadias 2001a), and MV3R-tree (Tao and Papadias 2001b).

- The main idea of **MR-tree** and **HR-tree** is to reduce R-tree's storage burden at different time points. The common objects will no longer be stored in multi R-trees but will be pointed to the only stored node by pointers from different root nodes of successive R-trees. This method is effective for time slice query, but it does not perform well for time interval query. Its main defect is entity copy. For example, all the other nodes will be copied between two successive R-trees if only one node entity is changed at successive time points. There is a concrete algorithm and the discussion for detailed implementation in HR-tree. This indexed method has good performance in time slice query, but poor performance in time window query and it may bring a huge amount of copy in indexed tree when object changes. The main idea of HR-tree is to avoid reproduction of node in the index tree mentioned above. It allows the entities at different time stamps to be stored in the same node. Therefore, it has a good space utilization rate.
- The main idea introduced in the **MV3R-tree** is building two index trees, one is MVR-tree to deal with temporal query and the other is 3D R-tree to deal with query of longer time interval. The index tree that will be used for short time interval query is determined by their performance. Since MVR-tree and 3DR-tree use common leaf nodes, the storage overhead is reduced and the query performance is improved, but it increases the updating cost (two indexed tree need to be maintained).

The main modes used in moving object query are TB-tree (Pfoser et al. 2000), SETI (Chakka et al. 2003), and SEB-tree (Song and Roussopoulos, 2003).

- **TB-tree** is the extension of STR-tree, which strictly preserves integrality of trajectories. Every leaf node in TB-tree only contains segments belonging to the same trajectory. There is only inserting algorithm in TB-tree, since it just needs to record the object's trajectories. TB-tree is built from left to right, the leftmost is the first insertion node and the rightmost is the last one. In order to implement the management for trajectories, the ability of space distinguishing and nodes overlapping owned by R-tree is dropped in TB-tree. Therefore, TB-tree can put different segments in different nodes although they may be near in space positions. In consideration of the geographical locations, in case of TB-tree, the capability of the space distinguishing capability may decrease and the cost of the range query may increase. However, there is higher efficiency for "pure" spatio-temporal querying.

- **SETI** divides space dimension into static and non-overlapping cells. Every cell includes only line segments that are included in it. If there is one line segment spanning the cell boundary, it will be cut into two line segments and each of them is inserted into respective cells. At the same time, one temporal index is created for each cell. The advantage of SETI is that it has good range and time slice query performance, and better updating efficiency.
- **SEB-tree** has fast inserting and querying algorithm. The main design idea of SEB-tree is similar to that of SETI. It also divides space into sub-areas, which may overlap with each other. Each sub-area only considers the beginning and ending time stamps of moving objects, while using Hash method to map each object to a corresponding sub-area. The key difference between SEB and SETI is that there is no conception of trajectory in SEB, which only indexes spatial points.

2. Index to past and present information

The index methods focusing on historical trajectory-oriented queries assume that all the movements are known in advance, so only closed trajectories are stored. The current position of moving object cannot be queried, since it is not stored, but we need to query the current and historic position of moving object in many of the practical applications. With time evolving, moving object's current position keeps updating. Conventional index structures are difficult to be adapted effectively to this dynamic frequent update. So, building a spatio-temporal index structure that stores both historical and current information is challenging. Such index structures include 2 + 3 R-tree (Nascimento et al. 1999), 2-3 R-tree (Abdelguerfi et al. 2002), LUR-tree (Kwon et al. 2002), Bottom-up Updates (Lee et al. 2003), Hashing (Song and Roussopoulos 2001) and so on.

- **2 + 3 R-tree** is used to index the past and current information of moving objects. Its key idea is to establish two independent R-trees. One is used to index the current two-dimensional points and the other is used to index historical three-dimensional trajectory (one for temporal dimension, two for spatial dimension). The present moving object will be deleted from current two-dimensional R-tree and will be inserted into three-dimensional R-tree, once it has been updated. Both the R-trees may be queried according to different time queries.
- The basic idea of **2-3 R-tree** is the same as that of 2 + 3 R-tree. The main differences are: first, the three-dimensional R-tree in 2-3 R-tree indexes only

multi-dimensional points and not trajectories, which avoids forming plenty of dead space; second, 2-3 R-tree uses the bottom structure of TB-tree to deal with trajectory query, which has better query efficiency.

- **LUR-tree** only indexes the current position of spatio-temporal objects. The old entry will be deleted and a new entry will be inserted once the position of the object changes. The main idea is that the position of the moving object is modified only when the updated position is still in the original MBR. Once one object leaves its current MBR, then it will be deleted and a new object will be inserted, or when the object is not far away from its current MBR, then current MBR will be extended to include the new position of the object.
- **Bottom-up updates** extends the thought of LUR-tree by referencing the down-up updating methods of R-tree. Some down-up updating methods are studied to adapt the frequent update of moving objects. For example, we can enlarge current MBR to contain the updated position of moving object or put current object into its brother node. At the same time, a summary structure is introduced in order to avoid too many I/O operations caused by update and the search for its brother and parent nodes.
- In space-division-based indexing methods, the space is divided into several areas, which may overlap each other. Therefore, the indexed tree can only represent the approximate view of moving objects. To solve this uncertainty, in **Hashing** index method, a filter layer is introduced between the database and moving object, which includes the accurate location of moving objects. After a range query is transformed into a range set, all the entities in the range are query results, if the range is included in the query range. If the range intersects with the query range, then all the entities in this range are passed to the filter layer to look for the concrete entities satisfying the query range.

3. Index to present and future information

Indexing moving object's present and future positions is the research hotspot of MOD. To predict the future positions of moving objects, it is necessary to store extra information (e.g., the current velocity and direction) and build model object's motion by a time function (usually represented by linear function). For instance, in n-dimensional space, moving object can be simulated by the referenced position X $(x_1, x_2, \cdots, x_d)$ and velocity vector $v=(v_1, v_2, \cdots, v_d)$ at the referenced time t, where the value of time t in future is equal to $x_t=x_{\text{ref}}+(t-t_{\text{ref}})$. In the case of one-dimension space, the motion can be simulated by the linear equation $x_t=at+b$, where a, b are constants. The index methods for moving the object's present and future trajectory can be divided into index based on originally temporal domain, index transformation domain, parametric space domain. The last two domains are the main research fields at present. Index structures based on the present and future prediction of moving objects mainly include PMR-Quadtree (Tayeb et al. 1998), TPR-tree (Saltenis et al. 2000), TPR*-tree (Tao et al. 2003), NSI and PSI (Porkaew et al. 2001).

- **PMR-Quadtree** is a dynamic index method based on a type of distortion of Quadtree-tree. Initial state in PMR-Quadtree corresponds to single space, and then line segments are inserted into this space one by one. If the number of line segments is more than a given threshold, the space will be spitted. PMR-Quadtree organizes data dynamically in double rules: splitting rule and merging rule. The splitting rule is that if the number of line segments in one node is more than the threshold, then it will be divided into four parts when one more segment is inserted into this node. The merging rule is that if deleting one segment from a node, when the total number of segments in it and its brother nodes is less than or equal to the giving threshold, the node is merged with its brother node. The disadvantage of PMR-Quadtree is that it costs too much storage space and needs frequent updating. Moreover, one segment line may be stored in several nodes.
- **TPR-tree** is an index method based on R*-tree (Beckmann et al. 1990), which can effectively index the present position of moving object as well as predict the object's future position at some time. TPR may approximately predict object's position in the near future by the speed and direction of objects. Its updating method can make itself adjust the border of bounding rectangle to adapt the dynamic variation data set. TPR-tree is a better method to index moving objects, but its algorithm is too simple to be suitable in the situation that the tree's node will grow fast. With time evolving, the node's area will grow bigger and the overlap becomes serious, which will reduce the query efficiency. Thus, researchers have done some improvements on it, which are carried out based on TPR-tree, such as **TPR***-tree, STAR-tree (Procopiuc et al. 2002), R^{EXB}-tree (Saltenis and Jencen. 2002) and so on.
- **NSI** is similar to TPR-tree, which uses bounding rectangle of moving objects and the concrete definition is different from TPR-tree.
- **PSI** describes the movement of objects in the parameter's space consisting of motion parameters. Here, R-tree is used to index $(2d+1)$-dimension parameter space, which consists of referenced position, speed and time. The advantage of PSI is that it is easy to deal with historical query as long as the historical data is not deleted from index. The deficiency is that the object located in parameter's space may be different in actual space, for example, the objects getting away from each other in parameter's space may be closed. In addition, some complicated motion parameters, such as acceleration is not easy to be indicated in PSI.

9.2 Data Model for Moving Objects

The spatio-temporal modeling for moving objects is mainly based on general spatio-temporal model. Compared to other spatio-temporal objects, moving object's position always keeps changing. If limited to general spatio-temporal model, in order to keep the position valid at any time, it needs to update the corresponding data periodically. Therefore, moving object model needs to reflect the special properties of spatial-temporal data. Yeb and Cambray (1995) discussed the situation that spatial data frequently changes with time. They then introduced the concept of "behavioral time series", where every element includes one spatial object value, one time and one behavioral function. Here, the latter describes the evolvement of the next element in current element series. Sistla et al (1997) represents moving objects by introducing dynamic attribute, which includes one moving vector to describe the current state (e.g. speed and direction) of moving objects, changes the corresponding moving vector when the moving object change its state (e.g. changes its direction). In this model, one query will return all the spatial object values at different times. Its disadvantage is that it cannot represent and deal with all the historical information of one moving object.

In the domain of multimedia database, some researchers proposed representing moving objects by their trajectories (as discrete snapshots). Then one group of objects can be represented by one map, where the vertex represents a spatial object, and an edge represents spatio-temporal relationship between spatial objects. This method stresses the evolvement between moving objects and the global relationship, but ignores the change in their shapes. In the domain of computational geometry, researchers discussed moving spatial objects' (points) Voronoi map and how to maintain it when a set of points continuously change with time (Albers and Roos 1992). At present, MOD mainly uses the technique of abstracting moving objects' positions as the function of time. For example, by function $Location = f(t)$, system calculates moving object's position in any future time. Moving objects do not need to report their position periodically. Database is updated only when the deviation between the actual position and calculating position reaches or exceeds a certain threshold. This method is better to reduce the updating expenditure for database. The main work in this area includes that of Wolfson in Chicago Branch Campus of Illinois University and the MOST model proposed by his research group (Wolfson et al. 1999).

9.2.1 Data Model *Modm*

In order to efficiently store and manage moving objects' motion trajectories, system generally samples the position of moving objects in discrete time points and then represents the trajectories by curves consisting of a series of line segments connected to the sampling locations. In this chapter, the line segment connected by two adjacent sampling points is called *trajectory segment*, *segment* for short. The segment end points are called *update points*, and the valid time stamp of update point u is denoted as $VT(u)$. A trajectory consists of a line sequence, where lines are connected by start and end points. Here, data u's valid time is represented as $VT(u) = [VTs(u), VTe(u)]$, where $VTs(u)$, $VTe(u)$ are called the starting time point and ending time point of the valid time period $VT(u)$, respectively and they always satisfy that $VTs(u) \leqslant VTe(u)$. Here, $[VTs(u), VTe(u)]$ represents the set of all the time points that are greater than or equal to $VTs(u)$ and less than $VTe(u)$ under a given time granularity.

Definition 9.1 (Trajectory, Trajectory line segment, Trajectory fragment, Time constraints) The definitions of Trajectory, Trajectory line segment, Trajectory fragment, Time constraints are as follows:

- **Trajectory:** Trajectory $Trj = (Oid(Trj), Tid(Trj), \langle u_0, u_1, u_2, \cdots, u_k \rangle, VT(Trj))$, where $Oid(Trj)$ is the object identifier that contains trajectory Trj. $Tid(Trj)$ is the unique identifier. $\langle u_0, u_1, u_2, \cdots, u_k \rangle$ is a sequence of Trj's update points, where $u = (P(u), VT(u))$, $P(u) = (x(u), y(u))$ is a moving object $Oid(Trj)$'s position at time $VT(u)$, $VT(Trj) = [VTs(u_0), VTe(u_k)]$ is Trj's valid time stamp. In addition, if we denote S_i as the line segment that connects the adjacent update points $u_{i\text{-}1}$ and $u_i (1 \leqslant i \leqslant k)$, then Trj's update sequence $\langle u_0, u_1, u_2, \cdots, u_k \rangle$ can also be represented as the corresponding line sequence $\langle S_1, S_2, \cdots, S_k \rangle$.
- **Trajectory line segment:** Let trajectory line segment $S = (Tid(S), Sid(S), P(S), VT(S))$, where $Tid(S)$ is the trajectory identifier that contains S. $Sid(S)$ is the identifier of S, u_{i-1} and u_i are the starting point and ending point of S_i, respectively. $P(S) = (Ps(S), Pe(S))$, where $Ps(S) = P(u_{i-1}) = (x(u_{i-1}), y(u_{i-1}))$, $Pe(S) = P(u_i) = (x(u_i), y(u_i))$, separately expresses the spatial location coordinates of S's starting and ending points. If two line segments S_i and S_j satisfy $Pe(S_i) = Ps(S_j)$, then we can say that S_i and S_j are adjacent, where S_i is called the precursor of S_j and S_j is called the successor of S_i.
- **Trajectory fragment:** Trajectory fragment, which is a sequence of some continuous line segments of the same trajectory, is represented as $TP = \{Tid(TP), TPid(TP), \langle S_{i_0}, S_{i_{0+1}}, \cdots, S_{q_0} \rangle, VT(TP)\}$. Here, $Tid(TP)$ is the trajectory identifier of the trajectory, which contains TP. $TPid(TP)$ is the trajectory fragment identifier of TP, S_{k-1} and S_k $(i_0 \leqslant k \leqslant q_0)$ are adjacent, $VT(TP)$ is TP's valid time stamp. Suppose $S_{i_0} = \min(TP)$, $S_{q_0} = \max(TP)$. If two trajectory line segments TP_i and TP_k satisfy $\max(TP_i) = \min(TP_k)$, we say that TP_i and TP_k $(i_0 \leqslant k \leqslant m_0)$ are adjacent.
- **Time constraints:** Suppose u_{i-1} and u_i are two adjacent update points, then $VTe(u_{i-1}) \leqslant VTs(u_i)$, that is, the time parameters of line segments are monotonically increasing. Trajectory's time stamp is the union of all its segments' time stamps. Trajectory fragment's time stamp is the union of all its segments' time stamps. Moving object's time stamp is the union of all its trajectories' time stamps.

Definition 9.2 (Moving object data model, *Modm*) Moving object data model *Modm* is a tree structure with four levels:

- Root node level: The node in the 1st level is the only (virtual) root node.
- Moving object nodes level: Nodes in this level are the moving object nodes, which are represented as object prefix encoding, object's MBR, object's valid time, object ID.

- Trajectory level: Nodes in this level are the moving object's trajectory nodes, the incoming edge and outgoing edge.
- Trajectory fragment level: The nodes in the forth layer are the fragment nodes of each moving object. The incoming edges label the corresponding time stamp, the node is represented as prefix-code, line segment's MBR, valid time, trajectory's identifier.

Every node n_0 in *Modm* can be described as $(ID_j(n_0), MBR_j(n_0), VT_j(n_0), Pcode(n_0))$, $2 \leqslant j \leqslant 4$, where $Pcode(E)$, $VT(E)$, $MBR(E)$ and $ID(E)$ represent the prefix encoding, valid time, spatial information and object's identifier of data *E*, respectively. $j = 2, 3, 4$ indicates that corresponding nodes are moving objects, trajectory and trajectory fragment, respectively.

Example 9.1 Suppose there are three trajectories *Trj*(1), *Trj*(2) and *Trj*(3) formed by three moving objects, respectively, as shown in Fig. 9.1. The concrete information is shown in Table 9.1, and corresponding *Modm* mode is shown in Fig. 9.2, where the object's node layer is omitted for simplicity.

9.2.2 Temporal Summary

Definition 9.3 (The mode of temporal summary, *Motsum*) The temporal summary $Motsum(D) = (Nts(D), Vts(D), Ets(D), Lts(D))$ is a tree structure, whose nodes are called summary nodes, the *Motsum*(*D*) root nodes is the name of *Modm*(*D*), the (*i*-th)-layer of *Motsum*(*D*) corresponds the (*i*-th)-layer of *Modm*(*D*).

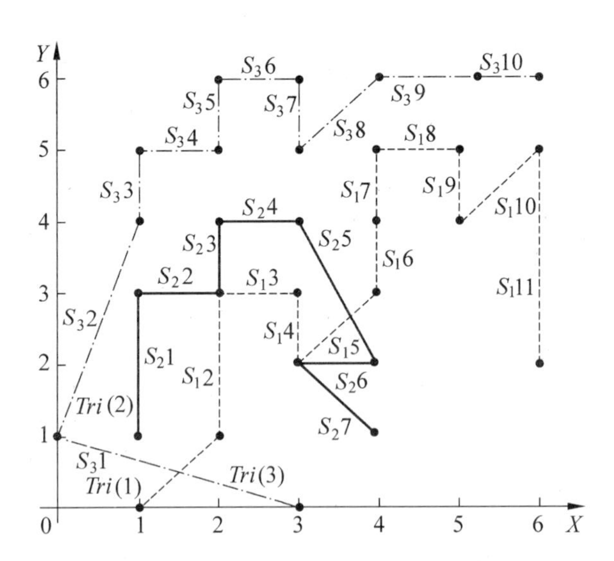

Figure 9.1 Trajectories of moving object

Table 9.1 Concrete information of trajectories

Trj	Segment					[*VTs*,*VTe*]	*Pcode*
	SID	Start (*x*,*y*)		End (*x*,*y*)			
		x	*y*	*x*	*y*		
Trj (1)	$S_1$1	1	0	2	1	[1,3]	000
	$S_1$2	2	1	2	3	[3,6]	001
	$S_1$3	2	3	3	3	[6,9]	002
	$S_1$4	3	3	3	2	[9,11]	003
	$S_1$5	3	2	4	3	[11,14]	004
	$S_1$6	4	3	4	4	[14,17]	005
	$S_1$7	4	4	4	5	[17,19]	006
	$S_1$8	4	5	5	5	[19,23]	007
	$S_1$9	5	5	5	4	[23,26]	008
	$S_1$10	5	4	6	5	[26,28]	009
	$S_1$11	6	5	6	2	[28,30]	0010
Trj (2)	$S_2$1	1	1	1	3	[1,6]	010
	$S_2$2	1	3	2	3	[6,9]	011
	$S_2$3	2	3	2	4	[9,14]	012
	$S_2$4	2	4	3	4	[14,17]	013
	$S_2$5	3	4	4	2	[17,19]	014
	$S_2$6	4	2	3	2	[19,23]	015
	$S_2$7	3	2	4	1	[23,30]	016
Trj (3)	$S_3$1	3	0	0	1	[2,3]	020
	$S_3$2	0	1	1	4	[3,9]	021
	$S_3$3	1	4	1	5	[9,11]	022
	$S_3$4	1	5	2	5	[11,14]	023
	$S_3$5	2	5	2	6	[14,19]	024
	$S_3$6	2	6	3	6	[19,23]	025
	$S_3$7	3	6	3	5	[23,24]	026
	$S_3$8	3	5	4	6	[24,26]	027
	$S_3$9	4	6	5	6	[26,28]	028
	$S_3$10	5	6	6	6	[28,30]	029

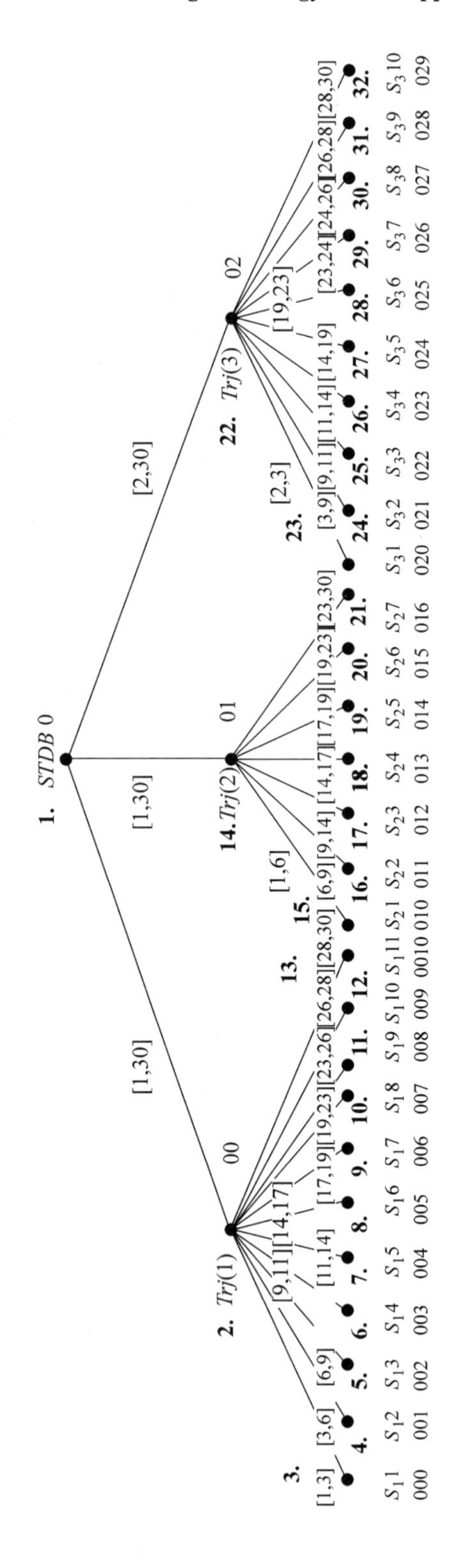

Figure 9.2 Data mode for moving objects

(1) *Nts*(*D*) is a prefix-code set of all the summary nodes in *Motsum*(*D*).

(2) *Vts*(*D*) is a set of all the summary nodes in *Motsum*(*D*).

- Node in the first level of *Motsum*(*D*) is root node.
- Nodes in the second level of *Motsum*(*D*) are made of the nodes on the second layer of *Modm*(*D*), which have the same time interval. In other words, it is the relation class based on the rule that they have the equal time interval.
- Node in the third level of *Motsum* is the set of children nodes of the nodes on the second layer of *Motsum*(*D*), which have the same time interval. In general, if we have the *i*-th layer summary nodes set $\{[u_i]\}$, for any $[v] \in \{[u]\}$, all temporal equivalence classes in the children of all *Modm* (*D*) nodes in $[v]$ are the children of $[v]$. The time stamps of $[v]$ are the ones of elements in $[v]$.

(3) *Ets*(*D*) is the set of edges under the condition of (2).

(4) *Idtsum*(*D*) is the set of all the elements' identifiers in all summary nodes of *Motsum*(*D*).

The temporal summary node can be represented as $[u] = (n_1([u]), VT([u]), Idtsum([u]))$, where $n_1([u])$ is the prefix encoding of $[u]$ in *Motsum*(*D*), and $VT([u])$ and $Idtsum([u])$ are the valid time and the set of identifiers of all the elements in $[u]$, respectively. In *Motsum*(*D*), the name of *Modm*(*D*) is on the first layer, the second layer is object ID, the third layer is the trajectory ID and the leaf-node layer is line segment ID.

Example 9.2 The temporal summary mode of the moving data in Example 9.1 is shown in Fig. 9.3 and the corresponding information is shown in Table 9.2.

Table 9.2 Data of summary nodes

No.	*VT*	*Dnodes*	$Pcode_{tsum}$	No.	*VT*	*Dnodes*	$Pcode_{tsum}$
$n_1(1)$	[1,30]	1	0	$n_1(14)$	[26,28]	12	0011
$n_1(2)$	[1,30]	2,13	00	$n_1(15)$	[28,30]	13	0012
$n_1(17)$	[2,30]	22	01	$n_1(16)$	[23,30]	21	0013
$n_1(3)$	[1,3]	3	000	$n_1(18)$	[2,3]	23	010
$n_1(4)$	[3,6]	4	001	$n_1(19)$	[3,9]	24	011
$n_1(5)$	[1,6]	15	002	$n_1(20)$	[9,11]	25	012
$n_1(6)$	[6,9]	5,16	003	$n_1(21)$	[11,14]	26	013
$n_1(7)$	[9,11]	6	004	$n_1(22)$	[14,19]	27	014
$n_1(8)$	[9,14]	17	005	$n_1(23)$	[19,23]	28	015
$n_1(9)$	[11,14]	7	006	$n_1(24)$	[23,24]	29	016
$n_1(10)$	[14,17]	8,18	007	$n_1(25)$	[24,26]	30	017
$n_1(11)$	[17,19]	9,19	008	$n_1 26)$	[26,28]	31	018
$n_1(12)$	[19,23]	10,20	009	$n_1(27)$	[28,30]	32	019
$n_1(13)$	[23,26]	11	0010				

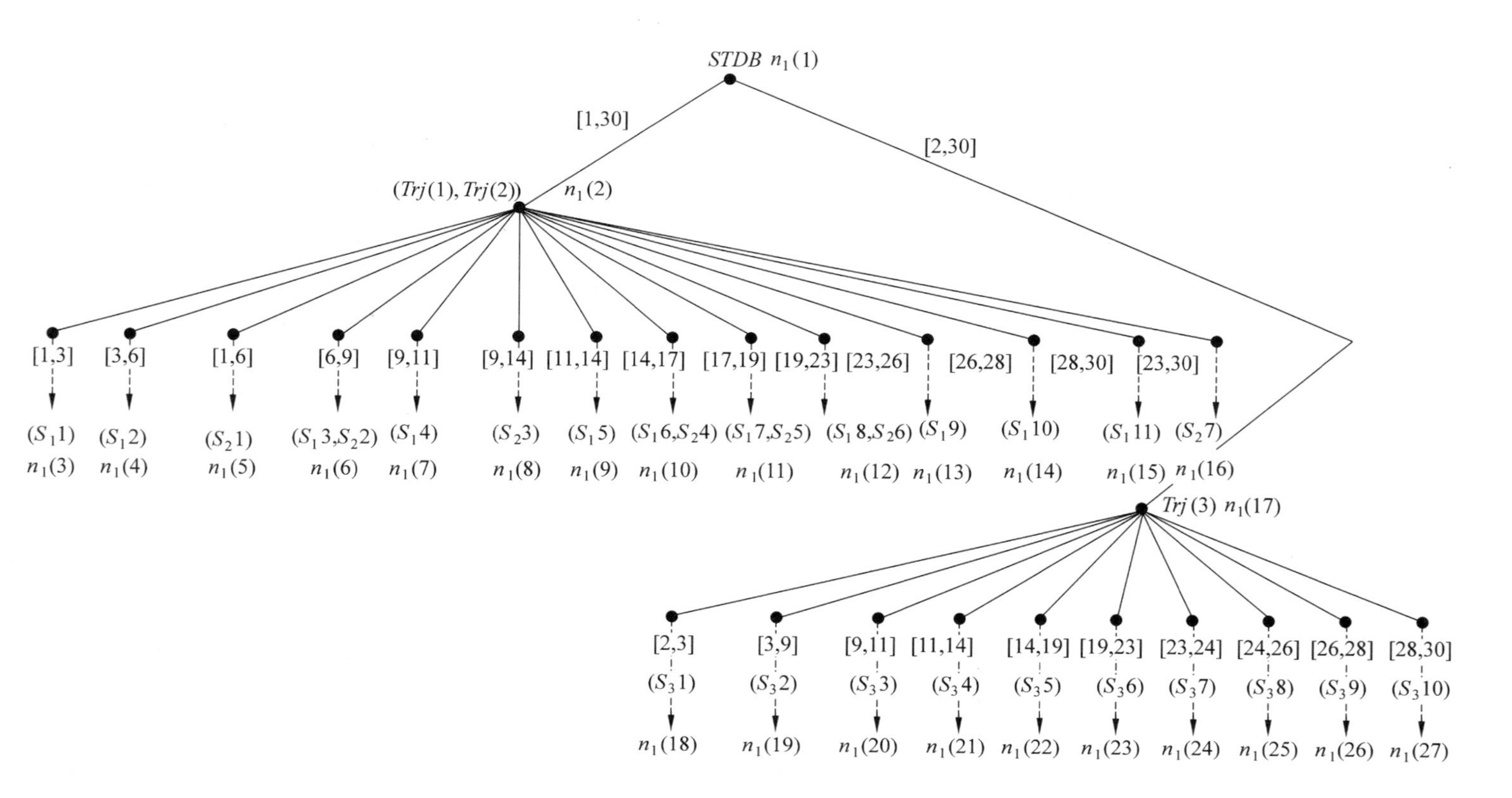

Figure 9.3 Mode for temporal summary

9.3 Index for Moving Object Data

9.3.1 Linear Order Division

To establish the temporal index structure, some basic mathematical relationships need to be discussed on the set of time periods.

9.3.1.1 Time-order matrix

Definition 9.4 (Time-order relation) Let D be the set of time periods. The corresponding temporal relationships are defined as follows:

(1) $\forall u, v \in \Gamma$: If their valid time $VT(u)$, $VT(v)$ satisfy $VT(u) \subseteq VT(v)$, we say that $u \leqslant v$. If $\neg(u \leqslant v \vee v \leqslant u)$, then u and v are called incompatible.

(2) $\forall u, v \in \Gamma$: Its valid times are represented as $VT(u) = [VTs(u), VTe(u)]$ and $VT(v) = [VTs(v), VTe(v)]$, respectively. If $VTs(u) < VTs(v)$ or $VTe(v) \leqslant VTe(u)$ when $VTs(u) = VTs(v)$, then the relationship "$\precsim$" between u and v is established, denoted as $u \precsim v$.

When $u \leqslant v$, we can also say that v temporal contains u, or u is temporally contained by v. In fact, relation "$\precsim$" sorts all the elements in Γ in ascending order by their starting time points. Elements that have the same starting time points are sorted in descending order by their ending time points. It can be verified that "$\leqslant$" is a pseudo-order, which satisfies reflexivity and transitivity. In this chapter, we also represent corresponding set of all the pseudo-orders by Γ. Relationship "$\precsim$" is a pseudo-linear order (linear order, for short), which satisfies reflexivity, transitivity, and totally ordered relation.

Definition 9.5 (Time-order matrix) Let D be the set of all time periods, i_1 and i_n be the minimum and the maximum starting time points, respectively, of all the time periods in Γ, j_1 and j_m are the minimum and the maximum ending time points, respectively. Then the time period $[i, j]$ can be denoted as ij $(i_1 \leqslant i \leqslant i_n, j_1 \leqslant j \leqslant j_m)$. The set of all the lattice points, defined by the set of time periods $\{i, j | i_1 \leqslant i \leqslant i_n, j_1 \leqslant j \leqslant j_m\}$ in the two dimensional space, where the x-axis and y-axis represent the starting and ending time points respectively, is called as time order matrix based on Γ, $TOM(\Gamma)$ for short, which is shown in Fig. 9.4.

$$\begin{array}{llll} i_1 j_m, & i_2 j_m, & \ldots, & i_{n-1} j_m, \; i_n j_m \\ i_1 j_{m-1}, & i_2 j_{m-1}, & \ldots, & i_{n-1} j_{m-1} \\ \vdots & \vdots & \vdots & \\ i_1 j_2, & i_2 j_2 & & \\ i_1 j_1 & & & \end{array}$$

Figure 9.4 Time order matrix

From the above definition, if we link the column of *TOM*(*D*) from top to bottom, left to right, we can get the pseudo-linear order of *D*.

$\forall u_{i_0 j_0} \in TOM(\Gamma)$, divide $TOM(\Gamma)$ into four areas: $UL(u_{i_0 j_0}) = \{v_{ij} \mid i \leqslant i_0, j_0 \leqslant j\}$, $UR(u_{i_0 j_0}) = \{v_{ij} \mid i_0 \leqslant i, j_0 \leqslant j\}$, $DL(u_{i_0 j_0}) = \{v_{ij} \mid i \leqslant i_0, j \leqslant j_0\}$ and $DR(u_{i_0 j_0}) = \{v_{ij} \mid i_0 \leqslant i, j \leqslant j_0\}$. If "<" holds above formulas, corresponding areas are "open" areas and are denoted by *OUL*, *OUR*, *ODL*, *ODR*. In $TOM(\Gamma)$, if u_{ij} and u_{kq} satisfy $j = q \wedge (i = k - 1 \vee k = i - 1)$, then u_{ij} and u_{kq} are horizontally adjacent. If they satisfy $i = k \wedge (j = q - 1 \vee q = j - 1)$, then they are vertically adjacent. Two adjacent nodes u and v are linked by horizontal and vertical line segments, denoted as $\langle u, v \rangle$. Let u, v be two elements in $TOM(\Gamma)$. If $\exists w_1, w_2, \cdots, w_m \in TOM(\Gamma)$ make the following three groups of elements adjacent: u and $w_1, \cdots, w_i$; $w_{i+1}, \cdots, w_{m-1}$ and w_m; w_m and v, then we say that $\langle u, w_1 \rangle$, $\langle w_1, w_2 \rangle$, $\cdots$, $\langle w_{m-1}, w_m \rangle$, $\langle w_m, v \rangle$ compose a path between u and v. The length of the path that connects u and v is the number of connected edges. The distance between u and v is the shortest length of the path connecting u and v, denoted as $d(u, v)$. The nearest neighbor of u_0 in $TOM(\Gamma)$'s subset Γ is the element whose distance to u_0 is minimum. The column and row that include u_0 are denoted as $column(u_0)$ and $row(u_0)$. If $column(u_0)$ does not include the elements above (below) u_0, we denote it as $column^U(u_0)$ ($column^D(u_0)$). If $row(u_0)$ does not include elements at the left (right) of u_0, we denote it as $row^L(u_0)$ ($row^R(u_0)$). u_0's nearest neighbors in $column^U(u_0)$ and $row^R(u_0)$ are called the upper-nearest neighbor point and right-nearest neighbor point, respectively. The Hass graph of pseudo-order set D in $TOM(D)$ is denoted as $H_{TOM}(D)$.

Example 9.3 The upper left regional *UL*(*23*) and the lower right regional *DR*(*23*) of temporal node "23" is shown in Fig. 9.5.

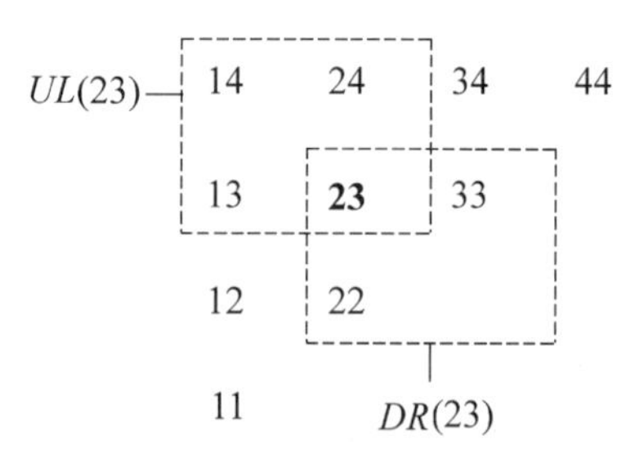

Figure 9.5 *UL*(23) and *DR*(23)

Theorem 9.1 (Pseudo-order and Time-order Matrix) $\forall u_0 \in H_{TM}(D)$, ① $u_0 \leqslant v_0 \Leftrightarrow v_0 \in UL(u_0)$; ② $v_0 \leqslant u_0 \Leftrightarrow v_0 \in DR(u_0)$; ③ u_0, v_0 are exclusive $\Leftrightarrow v_0 \in OUR(u_0) \vee v_0 \in ODL(u_0)$.

Proof ① Let $u_0 = [i_0, j_0)$, $v_0 = [k_0, l_0)$, $u_0 \leqslant v_0 \Leftrightarrow u_0 \subseteq v_0 \Leftrightarrow k_0 \leqslant i_0, j_0 \leqslant l_0 \Leftrightarrow v_0 \in UL(u_0)$;

② and ③ can be proved in the same way.

9.3.1.2 Linear order partition

Definition 9.6 (Linear order partition) Let $L \subseteq D$, $\forall u, v \in L$. If "$\leqslant$" satisfies totally ordered relation: $u \leqslant v \vee v \leqslant u \vee v = u$, then we can say that L is called a linear order branch in D, denoted as *LOB*. The greatest and the least element of *LOB* $\boldsymbol{L}$ are denoted as max $\boldsymbol{L}$ and min $\boldsymbol{L}$, respectively. If the *LOB* set Δ of D composes one partition of D, then Δ is defined as one linear order partition (*LOP*).

The time set D can have many *LOP*s. In terms of temporal index query efficiency, we need to consider the *LOP* with special characteristics.

Definition 9.7 (*Least-LOP*) Let Δ_0 be *LOP* in D. If, for any *LOP* Δ in D, such that $|\Delta_0| \leqslant |\Delta|$, we can say that Δ_0 is *LOP* in D, which has the least equivalence class, short for *Least-LOP*.

Algorithm 9.1 (Down-first *LOB*, *DFA*)

Input: time interval set $D = \{P_1, P_2, \cdots, P_n\}$

Output: *LOP*

Step 1: Sort all the time intervals in D. First, sort the time intervals by ascending start time. If start time is the same, then sort by descending end time. Let interval set be $D = \{P_1, P_2, \cdots, P_n\}$ after sorting, and let $i = 1$.

Step 2: If D is empty, algorithm stops. Let $LOB_i = \varnothing$. Begin with the first interval in the left of D, add them into the tail of LOB_i. Let $j = 2$. $Tail(LOB_i)$ represents the tail element of LOB_i.

Step 2.1: If $j \leqslant |D|$ and $P_j \subseteq Tail(LOB_i)$, then add P_j into the tail of LOB_i and delete P_j from D. If $j > |D|$, go to Step 2.3.

Step 2.2: If $j = j + 1$, go to Step 2.

Step 2.3: Output all elements of LOB_i, which compose a linear order branch. $i = i + 1$. Go to Step 2.

The algorithm above is equivalent to lower right search algorithm in $H_{TOM}(\Gamma)$. In fact, Step 1 is equivalent to build up $H_{TOM}(\Gamma)$, Step 2 is equivalent to lower right searching in $H_{TOM}(\Gamma)$. The left first element corresponds to the topmost element of the left column in $H_{TOM}(\Gamma)$. Searching rightwards to the tail corresponds to that search downwards until the tail of the column, then rightwards search at the same line (according to Theorem 9.1, every element contains those elements that only exist in $DR(P_i)$), then downwards search at the same column, and so on. The complete path found in every search is an *LOB*.

Theorem 9.2 (The *Least-LOP* obtained by *DFA*) The *LOP* obtained by *DFA* satisfies the minimum property of Definition 9.4.

Proof Let $\langle L_1, L_2, \ldots, L_n \rangle$ be the *LOP* obtained by *DFA*, which is sorted by the calculating order, then $\forall a_{i_0} \in L_i (1 < i \leqslant n)$, $\exists a_{k_0} \in L_{i-1}$, a_{i_0} and a_{k_0} are incompatible. This means that elements of L_{i-1} always exsts in $OLD(a_{i_0})$. If it does not, all elements in L_{i-1} have to be in $UL(a_{i_0})$. In this case, $a_{i_0} \in DR$ (min L_{i-1}), but $a_{i_0} \in L_i$. This is a contradiction. Then from L_n to L_1, we can obtain a sequence of these elements $a_n, a_{n-1}, \cdots, a_1$, which are all incompatible to each other. As a result, any *LOP* should have at least n *LOB*s.

Time complexity analysis for Algorithm 9.1: Step 1 sorting needs $O(n\log n)$; Step 2 contains two level loop. The worst situation of the inner loop needs to traverse all the elements, so it is $O(n)$. The worst situation of the outer loop also needs to loop n times, so it is $O(n^2)$ totally. The time complexity of Algorithm 9.1 is $O(n^2)$.

Suppose the *LOP* gained by *DFA* is $\{L_1, L_2, \cdots, L_n\}$, and all the *LOBs* are sorted by the calculated order. If $u_0 \in L_i$, then there exists an element in $L_{i-1}(i>1)$ lying in $UL(u_0)$, in which lies in the left-nearest neighbor of u_0 and an element in $L_{i+1}(i<n)$ lying on $DR(u_0)$, in which lies in the nearest down-right neighbor of u_0.

Example 9.4 Suppose there exists a time order matrix $TOM(\Gamma)$, as shown in Fig. 9.6.

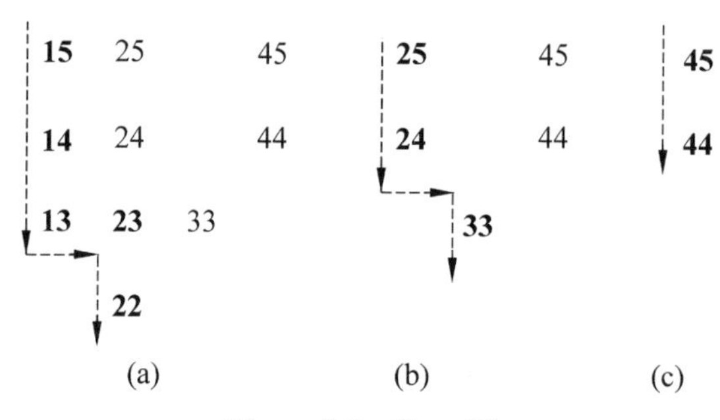

Figure 9.6 $H_{LOM}(D)$

Based on *DFA* calculation process:

From u_{15} downwards to u_{13}, from u_{13} rightwards to u_{23}, from u_{23} downwards u_{22}. Here, we can get $L_1 = \langle u_{15}, u_{14}, u_{13}, u_{23}, u_{22} \rangle$ as shown in Fig. 9.6(a).

In $D \backslash L_1$, from u_{25} downwards to u_{24}, from u_{24} rightwards and downwards to u_{33}, we can get *LOB*: $L_2 = \langle u_{25}, u_{24}, u_{33} \rangle$ as shown in Fig. 9.6(b).

In $D \backslash (L_1 \cup L_2)$, from u_{45} downwards to u_{44}, we can get *LOB*: $L_3 = \langle u_{45}, u_{44} \rangle$ as shown in Fig. 9.6(c).

Therefore, we can get all the *LOBs*: $L_1 = \langle u_{15}, u_{14}, u_{13}, u_{23}, u_{22} \rangle$, $L_2 = \langle u_{25}, u_{24}, u_{33} \rangle$, $L_3 = \langle u_{45}, u_{44} \rangle$.

9.3.2 Index Model *Modim*

Definition 9.8 (Moving objects data index model, *Modim*) Moving objects data index model $Modim(D) = (N_{ind}(D), V_{ind}(D), E_{ind}(D), L_{ind}(D), ID_{ind}(D))$, based on summary model $Motsum(D)$, is a kind of tree structure in which node is called index node. The number of levels of $Modim(D)$ is the same as that of $Motsum(D)$ and they are mutually corresponding. The root node is the name of $Motsum(D)$.

(1) $N_{ind}(D)$ is the set of pre-order codes of all the index nodes in *Modim*(*D*).

(2) $V_{ind}(D)$ is the set of all index nodes in *Modim*(*D*).

- The node in the first level of *Modim*(*D*) is a root node.
- The node in the second level of *Modim*(*D*) is the *LOB* obtained from the second level of *Modim*(*D*) by "$\leqslant$".
- The node in the third level of *Modim*(*D*) is the *LOB* obtained from every index node (summary node) in the second level by "$\leqslant$" in the son nodes of *Mosum*(*D*). Generally, when $i \geqslant 2$, if the set of temporal index nodes $\{L_k\}$ that is in the i-th level of *Modim*(*D*) is known, then the *LOB*s, which are obtained from the elements of L_k in the son nodes of *Modim*(*D*) by "$\leqslant$", compose the nodes of the $(i+1)$-th level of *Modim*(*D*).

(3) $E_{ind}(D)$ is the edge set under constraint of (2).

(4) $ID_{ind}(D)$ is the pre-order code set of index nodes.

Index node $L_0 = (n_2(L_0), \langle \max(L_0), \min(L_0) \rangle, ID_{ind}(D))$, in which $n_2(L_0)$ is pre-order code of L_0, $\min(L_0)$ and $\max(L_0)$ are the least and greatest element of L_0 respectively. $ID_{ind}(L_0)$ is the label set of all elements in L_0.

Example 9.5 The moving objects data model corresponding to Example 9.1 is shown in Fig. 9.7. Corresponding information for every index node is shown in Table 9.3.

Table 9.3 Data of index node

No.	*LOB*	*Snodes*	$Pcode_{ind}$
$n_2(1)$	[1,30]	$n_1(1)$	0
$n_2(2)$	[1,30],[2,30]	$n_1(2)$, $n_1(17)$	00
$n_2(3)$	[1,6],[1,3],[2,3]	$n_1(5)$, $n_1(3)$, $n_1(18)$	000
$n_2(4)$	[3,9],[3,6]	$n_1(19)$, $n_1(4)$	001
$n_2(5)$	[6,9]	$n_1(6)$	002
$n_2(6)$	[9,14],[9,13],[9,11]	$n_1(8)$, $n_1(7)$, $n_1(20)$	003
$n_2(7)$	[11,14]	$n_1(9)$, $n_1(21)$	004
$n_2(8)$	[14,19],[14,17]	$n_1(22)$, $n_1(10)$	005
$n_2(9)$	[17,19]	$n_1(11)$	006
$n_2(10)$	[19,23]	$n_1(23)$, $n_1(12)$	007
$n_2(11)$	[24,26]	$n_1(25)$	008
$n_2(12)$	[23,30],[23,26]	$n_1(16)$, $n_1(13)$, $n_1(25)$	009
$n_2(13)$	[26,28]	$n_1(26)$, $n_1(14)$	0010
$n_2(14)$	[28,30]	$n_1(27)$, $n_1(15)$	0011

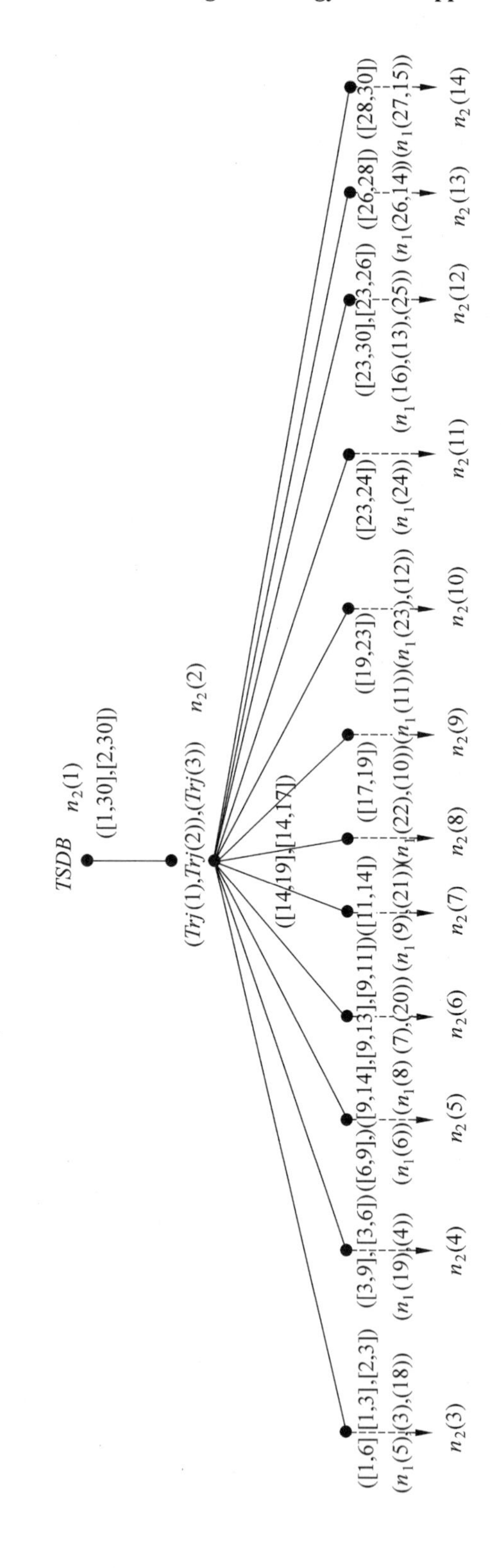

Figure 9.7 Index mode for moving object

9.4 Data Query

The querying schema Q of the moving objects can be described as $Q=(ID(Q), VT(Q), R(Q))$, where $ID(Q)$ is the queried moving object's identifier, $VT(Q) = [VTs, VTe)$, $R(Q) = [(x_1, y_1), (x_2, y_2)]$.

Depending on whether there is any specific query object, queries can be divided into two types:

(1) Query with no specific object: In this case, $Q=(VT(Q), R(Q))$. Its semantic is to query all trajectory (segment), which intersect with spatial query frame $R(Q)$ in $VT(Q)$.

(2) Query with specific object: In this case, $Q=(ID(Q), VT(Q), R(Q))$. Its semantic is to query all trajectory (segment) of query moving object $ID(Q)$, which intersects with spatial query frame $R(Q)$ in $VT(Q)$.

In addition, the trajectory query only needs to be located at the third level (index level of trajectory) of *Modim*, while the trajectory segment needs to be located at the leaf level (trajectory segment level).

Without loss of generality, this chapter only considers the algorithm based on track segment query with no specific moving object and the algorithm based on trajectory query with specific moving object.

Algorithm 9.2 (trajectory segment query)

Let the query with moving object $Q=(VT(Q), R(Q))$.

Step 1 (index node query): Localize at the leaf level of *Modim*.

(1) Temporal filtrate: Let the valid time *LOB* set corresponding to all index nodes be $\{L_{i_0}\}$, (simply, L_{i_0} is represented as spatial index node). Let $L_{i_0} \in \{L_{i_0}\}$ and $L_{i_0} = (n_{i_0}(j), \langle \max(L_{i_0}), \min(L_{i_0})\rangle)$ be index node, in which $n_{i_0}(j)$, $\langle \max(L_{i_0}), \min(L_{i_0})\rangle$ represents pre-order code, the greatest and the least time intervals, respectively. Let all index nodes be sorted by ascending start time, descending by end time (when start time equal each other), so that we can get $L_{i_1}, L_{i_2}, \ldots, L_{i_n}$. If a subset in array $\{L_{i_1}, L_{i_2}, \ldots, L_{i_n}\}$ holds $\max(L_{i_i}) \cap VT(Q) \neq \varnothing$, we can assume that the son set is $\{L_{i_1}, L_{i_2}, \ldots, L_{i_n}\}$ itself. Begin with L_{i_1}, binary search: First, execute binary search on the end time of $\{\max(L_{i_i})\}$, find the first index node L_{i_1} that holds $VTs(Q) \leqslant VTe(\max(L_{i_i}))$. Then execute the binary search on the start time of $\{\max(L_{i_i})\}$, get the last index node L_{i_n}, which holds $VTe(Q) \geqslant VTs(\max(L_{i_i}))$. Then all the index nodes between L_{i_1} and L_{i_n} are the result of temporal filtrating.

(2) Time query: As we can know from (1), all the candidate index nodes hold $VT(L) \cap \max(L_{i_0}) = \varnothing$, that is, all the candidate index nodes must have index entity that satisfies the time constraint. Time query is shown as follows:

① If $VT(L) \cap \min(L_{i_0}) \neq \varnothing$, then the time intervals in the elements of *LOB* node L_{i_0} are index query results.

② If it is not the case ①, let the summary nodes in L_{i_0} be sorted by time containment relation $[u]_{i1}$, $[u]_{i2}$, ⋯, $[u]_{in}$ (that is, $[u]_{i1} \subseteq [u]_{i2} \subseteq \cdots \subseteq [u]_{in}$):

- If $VTe(Q) < VTs$ (min(L_{i_0}), in array $\{VT([u]_{i1}), VT([u]_{i2}), \cdots, VT([u]_{in})\}$, then execute a binary search on the start points of the time intervals. Find the first time interval that holds $VT([u]_{ij})$: $VTs([u]_{ij}) < VTe(Q)$, then the elements corresponding to the segment of time sequence branch $LOBs = \{VT([u]_{i1}), VT([u]_{i2}), \cdots, VT([u]_{ij})\}$, are temporal query results.
- If VTe (min(L_{i_0})) $< VTs(Q)$, in array $\{VT([u]_{i1}), VT([u]_{i2}), \cdots, VT([u]_{in})\}$, execute a binary search on the end points of the time intervals. Find the first time interval that holds $VT([u]_{ij})$: $VTe([u]_{ij}) > VTs(Q)$, then the elements corresponding to the segment of time sequence branch $LOBs = \{VT([u]_{i1}), VT([u]_{i2}), \cdots, VT([u]_{ij})\}$, are temporal query results.

Step 2 (summary nodes backtracking) Let *LOB* or *LOB*s be the result after Step 1, and $L_{k_0} = \{[u]_{k1}, [u]_{k2}, \cdots, [u]_{km}\}$ be the data nodes obtained from summary nodes.

Step 3 (data nodes spatial query) Get the projections in x-axis and y-axis of the query rectangle, respectively, $x_{R(Q)} = (x_{R(Q)}.\text{low}, x_{R(Q)}.\text{high}) \wedge x_{R(Q)}.\text{low} \leqslant x_{R(Q)}.\text{high}$, $y_{R(Q)} = (y_{R(Q)}.\text{low}, y_{R(Q)}.\text{high}) \wedge y_{R(Q)}.\text{low} \leqslant y_{R(Q)}.\text{high}$. Spatially filtrate the data nodes in summary nodes that are obtained from Step 2 by query rectangle, specific process as follows:

- Project all data nodes (line segments) to x-axis, get the projection $\{x_{pi} \mid x_{pi} = (x_{pi}.\text{low}, x_{pi}.\text{high})$ on x-axis, and $x_{pi}.\text{low} \leqslant x_{pi}.\text{high}\}$, execute the same binary search like '(1)' in Step 1 on $\{x_{pi}\}$ and $x_{R(Q)}$, get the candidate data node set $\{S_{i_i}\}$.
- Project the data of candidate node set $\{S_{i_i}\}$ to y-axis, get the projection $\{y_{pi} \mid y_{pi} = (y_{pi}.\text{low}, y_{pi}.\text{high}$) on y-axis, and $y_{pi}.\text{low} \leqslant y_{pi}.\text{high}\}$, execute the same binary search like '(1)' in Step 1 on $\{y_{pi}\}$ and $y_{R(Q)}$, get the time spatial query result.

Step 4 (trajectory segment assembly) Execute the necessary assembly on the query result (trajectory segment), that is obtained from Step 3 by pre-order code of data nodes, then get the final query result.

Algorithm 9.2′ (trajectory query algorithm)

Let trajectory query with moving object $Q = (ID(Q), VT(Q), R(Q))$.

Step 1 (ID filtrate): Localize at the second level of *Modim*. First, filtrate the nodes on this level by *ID(Q)*, then select the index nodes that are contained in *ID* set.

Step 2 (based on index query): This step is the same as "Step 1~Step 3" in Algorithm 9.2.

Step 3 (data node query): Identify those data nodes (trajectory *Trj*) that are obtained from Step 1 by *ID(Q)*, that is, check whether they hold $ID(Q) = ID(Trj)$ or not, to get the final query result.

Example 9.6 Consider the moving object data instance in Example 9.1.

(1) Track segment query

Let there be a query $Q = (VT(Q), R(Q))$, $VT(Q) = [6,20)$, $R(Q) = [2,1; 4,6]$.

Step 1 (index node query): Localize at the leaf level of *Modim*.

① Temporal filtering

$\max(L_{i_0})$ corresponding to every leaf node L_{i_0} are: $\max(n_2(3)) = [1,6]$, $\max(n_2(4)) = [3,9]$, $\max(n_2(5)) = [6,9]$, $\max(n_2(6)) = [9,14]$, $\max(n_2(7)) = [11,14]$, $\max(n_2(8)) = [14,19]$, $\max(n_2(9)) = [17,19]$, $\max(n_2(10)) = [19,23]$, $\max(n_2(11)) = [24,26]$, $\max(n_2(12)) = [23,30]$, $\max(n_2(13)) = [26,28]$, $\max(n_2(14)) = [28,30]$. The sort result is $n_2(3)$, $n_2(4)$, $n_2(5)$, $n_2(6)$, $n_2(7)$, $n_2(8)$, $n_2(9)$, $n_2(10)$, $n_2(12)$, $n_2(11)$, $n_2(13)$, $n_2(14)$. According to the binary search in Algorithm 9.2, the first index node holds $VTs(Q) \leqslant VTe(\max(L_{ii}))$ is $n_2(4)$; the last index node holds $VTe(Q) \geqslant VTs(\max(L_{ii}))$ is $n_2(10)$. Therefore, the temporal filtrate result is $n_2(4)$, $n_2(5)$, $n_2(6)$, $n_2(7)$, $n_2(8)$, $n_2(9)$, $n_2(10)$.

② Temporal query

We can get $n_1(19)$ after binary search on $n_2(4)$. The minimum time interval of $n_2(5)$, $n_2(6)$, $n_2(7)$, $n_2(8)$, $n_2(9)$, $n_2(10)$ intersect with $VT(Q)$. Therefore, $n_1(6)$, $n_1(8)$, $n_1(7)$, $n_1(20)$, $n_1(9)$, $n_1(21)$, $n_1(22)$, $n_1(10)$, $n_1(11)$, $n_1(23)$, $n_1(12)$, all the summary nodes in $n_2(5)$, $n_2(6)$, $n_2(7)$, $n_2(8)$, $n_2(9)$, $n_2(10)$ are candidates for query results.

Step 2 (summary node query): Backtrack to the third level of *Mosum*. We can check the nodes and get the data nodes, S_32, S_13, S_24, S_23, S_14, S_33, S_15, S_34, S_35, S_16, S_17, S_25, S_36, S_18, S_26, that are contained by summary nodes $n_1(19)$, $n_1(6)$, $n_1(8)$, $n_1(7)$, $n_1(20)$, $n_1(9)$, $n_1(21)$, $n_1(22)$, $n_1(10)$, $n_1(11)$, $n_1(23)$, $n_1(12)$.

Step 3 (data node query): Query rectangle $R(Q) = [2,1; 4,6]$, with projections on x-axis and y-axis are $x_{R(Q)} = (2,4)$, $y_{R(Q)} = (1.6)$. Backtrack to the third level of *Modm*. We can get trajectory segments: $S_32 = (0,1,1,4)$, $S_13 = (2,3,3,3)$, $S_22 = (1,3,2,3)$, $S_23 = (2,3,2,4)$, $S_14 = (3,3,3,2)$, $S_33 = (1,4,1,5)$, $S_15 = (3,2,4,3)$, $S_34 = (1,5,2,5)$, $S_35 = (2,5,2,6)$, $S_16 = (4,3,4,4)$, $S_24 = (2,4,3,4)$, $S_17 = (4,4,4,5)$, $S_25 = (3,4,4,2)$, $S_36 = (2,3,3,6)$, $S_18 = (4,5,5,5)$, $S_26 = (4,2,3,2)$.

- The result of the projection of every candidate line segment on x-axis after sort is: {(0,1), (1,1), (1,2), (1,2), (2,2), (2,2), (2,3), (2,3), (2,3), (3,3), (3,4), (3,4), (3,4), (4,4), (4,4), (4,5)}. According to the binary search in Algorithm 9.2, the first projection interval that holds $x_{R(Q)}.\text{low} \leqslant x_{pi}.\text{high}$ is (2,2). The last projection interval that holds $x_{R(Q)}.\text{high} \geqslant x_{pi}.\text{low}$ is (4,5). So the candidate result set is: {(2,2), (2,2), (2,3), (2,3), (2,3), (3,3), (3,4), (3,4), (3,4), (4,4), (4,4)}. The relevant candidate line segments are {S_23, S_35, S_13, S_24, S_36, S_14, S_15, S_25, S_26, S_16, S_17}.
- The result of the projection of every candidate line segment on y-axis after sort is: {(2,2), (2,3), (2,3), (2,4), (3,3), (3,4), (3,4), (3,6), (4,4), (4,5), (5,6)}. According to the binary search in Algorithm 9.2, the first projection interval that holds $y_{R(Q)}.\text{low} \leqslant y_{pi}.\text{high}$ is (2,2). The last projection interval that holds

$y_{R(Q)}$.high$\geqslant y_{pi}$.low is (5,6). So the spatio-temporal query result set is: {(2,2), (2,2), (2,3), (2,3), (2,3), (3,3), (3,4), (3,4), (3,4), (4,4), (4,4)}. The relevant candidate line segments are $\{S_23, S_35, S_13, S_24, S_36, S_14, S_15, S_25, S_26, S_16, S_17\}$.

Step 4 (trajectory segment assembly): The pre-order code of S_13, S_23, S_14, S_15, S_35, S_16, S_24, S_17, S_25, S_36, S_26 are 002, 012, 003, 004, 024, 005, 013, 006, 014, 025, 015, respectively. Therefore, we can see that: S_13, S_14, S_15, S_16, S_17 are trajectory segments that belong to the same trajectory, S_23, S_24, S_25, S_26 are trajectory segments that belong to the same trajectory, S_35, S_36 are trajectory segments that belong to the same trajectory. Therefore, the query result is: $\{\{S_13, S_14, S_15, S_16, S_17\}, \{S_23, S_24, S_25, S_26\}, \{S_35, S_36\}\}$.

(2)Trajectory query

Let there be a query Q=($VT(Q)$, $R(Q)$), in which $VT(Q)$=[6,20), $R(Q)$= [2,1; 4,6].

Step 1 (index node query): Localize at the second level of *Modim*(*D*).

Because there is only one node $n_2(2)$ in the second level of *Modim*(*D*); [2,30] and [6,20], the minimum time span of $n_2(2)$ intersect. Then $n_1(2)$, $n_1(17)$, which are the index entity of $n_2(2)$, are candidate query results.

Step 2 (summary node query): Backtrack to the data nodes *Trj*(1),*Trj*(2), *Trj*(3), which are contained by the summary node $n_1(2)$, $n_1(17)$ in the second level of *Mosum*(*D*).

Step 3 (data node query): Query rectangle $R(Q)$= [2,1; 4,6], for which projections on x-axis and y-axis are $x_{R(Q)}$ = (2,4), $y_{R(Q)}$ = (1.6). Backtrack to the second level of *Modm*(*D*). We can get $MBR_{Trj(1)}$ = (1,0,6,2), $MBR_{Trj(2)}$ = (1,1,4,1) and $MBR_{Trj(3)}$ = (3,0,6,6):

- The result of the projections of every candidate line segment on x-axis after the sort is {(1,4), (1,6), (3,6)}. According to the binary search in Algorithm 9.2, the first projection interval that holds $x_{R(Q)}$.low$\leqslant x_{pi}$.high is (1,4), the last projection interval that holds $x_{R(Q)}$.high$\geqslant x_{pi}$.low is (3,6). Therefore, the candidate result set is {(1,4), (1,6), (3,6)}. The relevant candidate line set is {*Trj*(1), *Trj*(2), *Trj*(3)}.
- The result of the projection of every candidate line segment on y-axis after sorting is {(0,2), (0,6), (1,1)}. According to the binary search in Algorithm 9.2, the first projection interval that holds $y_{R(Q)}$.low$\leqslant y_{pi}$.high is (0,2), the last projection interval that holds $y_{R(Q)}$.high$\geqslant y_{pi}$.low is (1,1). Therefore, the spatio-temporal query result set is {(0,2), (0,6), (1,1)}. The trajectory set is {*Trj*(1),*Trj*(2),*Trj*(3)}.
- The spatio-temporal query result set is {*Trj*(1),*Trj*(2),*Trj*(3)}.

9.5 Index Update

Data update of moving objects mainly contains inserting trajectory and trajectory segment, as well as deleting object and trajectory. The following discussion is about *Modim* update and, for brevity, the discussion is carried out with an example.

Without loss of generality, we can assume what is to be inserted is a trajectory segment $S_{i0} = \{Tid(S_{i0}), TPid(S_{i0}), VT(S_{i0})\}$.

Example 9.7 Figure 9.3 is the temporal index graph *Modim*(*D*) of moving object. In this graph, the corresponding time interval to the summary node $n_1(i)$ of index node $n_2(2)$ is [2,12], in which son nodes of *Motsum*(*D*) are $n_1(i)_1, \cdots, n_1(i)_k$ and there is only one trajectory Trj_i contained by $n_1(i)$. If we insert a line segment (Trj_i, *Sid*, x_s, y_s, x_e, y_e, 12, 22), the trajectory adds a line segment with a label named *Sid*, start point (x_s, y_s), end point (x_e, y_e), and span (12,22). Then the time interval of Trj_i should be [2,22]. It has to delete summary node $n_1(i)$ from [3,12]. As there is no summary node whose time interval is [2,22] in *Motsum*(*D*), it needs to create a new summary node $n_1(Ti)$ and insert $n_1(Ti)$ into an appropriate *LOB* in *Modim*(*D*).

- Delete

Delete Trj_i from $n_1(i)$: At the same time, it needs to delete the son nodes of $n_1(i)$ from $n_1(i)$. Based on this point, it needs deleting from corresponding *LOB*. Then the *LOBs* corresponding to index node $n_2(2)$ and $n_2(k+1)$ are respectively ([1,20], [2,20], [2,18], [2,12], [3,12], [3,8], [3,4]) and ([3,23], [3,18], [3,16], [3,14], [4,14], [4,12], [4,8]). Here, we need to restructure *LOB*. After analysis, it belongs to the case shown in Fig. 8.15, that is, it needs to merge the branch segment ([3,18], [3,16], [3,14]) of $n_2(k+1)$ into $n_2(2)$. The summary nodes corresponding to that branch segment are $n_1(j), \cdots, n_1(j_k)$, in which the son nodes in *Motsum*(*D*) are $n_1(j)_1, \cdots, n_1(j)_n, \cdots, n_1(j_k)_1, \cdots, n_1(j_k)_m$. At the same time, it needs to delete $n_1(j)_1, \cdots, n_1(j)_n, \cdots, n_1(j_k)_1, \cdots, n_1(j_k)_m$ from the son nodes of $n_2(k+1)$ and insert them into the son nodes of $n_2(2)$. Then the *LOB*s corresponding to $n_2(2)$ and $n_2(k+1)$ are ([1,20], [2,20], [2,18], [3,18], [3,16], [3,14], [3,12], [3,8], [3,4]) and ([3,23], [4,14], [4,12], [4,8]), respectively.

- Insert

Insert into $n_1(Ti)$: After analysis, it belongs to the case shown in Fig. 8.15. Then it needs to split $n_2(k+1)$ into two *LOB*s, ([2,22], [4,14],[4,12],[4,8] and [3,23]).

The delete and insert of child nodes are the same as above. The final result is shown in Fig. 9.10 after insert segment (Trj_i, *Sid*, x_s, y_s, x_e, y_e, 12, 22) in Fig. 9.11.

It is available to assume that the trajectory $Trj_0 = (Oid(Trj_0), Tid(Trj_0), \langle S_1, S_2, \cdots, S_k \rangle, VT(Trj_0))$ needs to be deleted.

- Update summary node: Find the summary node where the trajectory segments in Trj_0 and Trj_0 lie. Delete those trajectory segments that are among the summary nodes in Trj_0 and Trj_0. If the relevant summary node is not empty after delete, then the structure of *Motsum* remains the same, and MBR of the relevant summary node is restructured. If the relevant summary node is empty after delete, then delete the relevant node in *Motsum*.

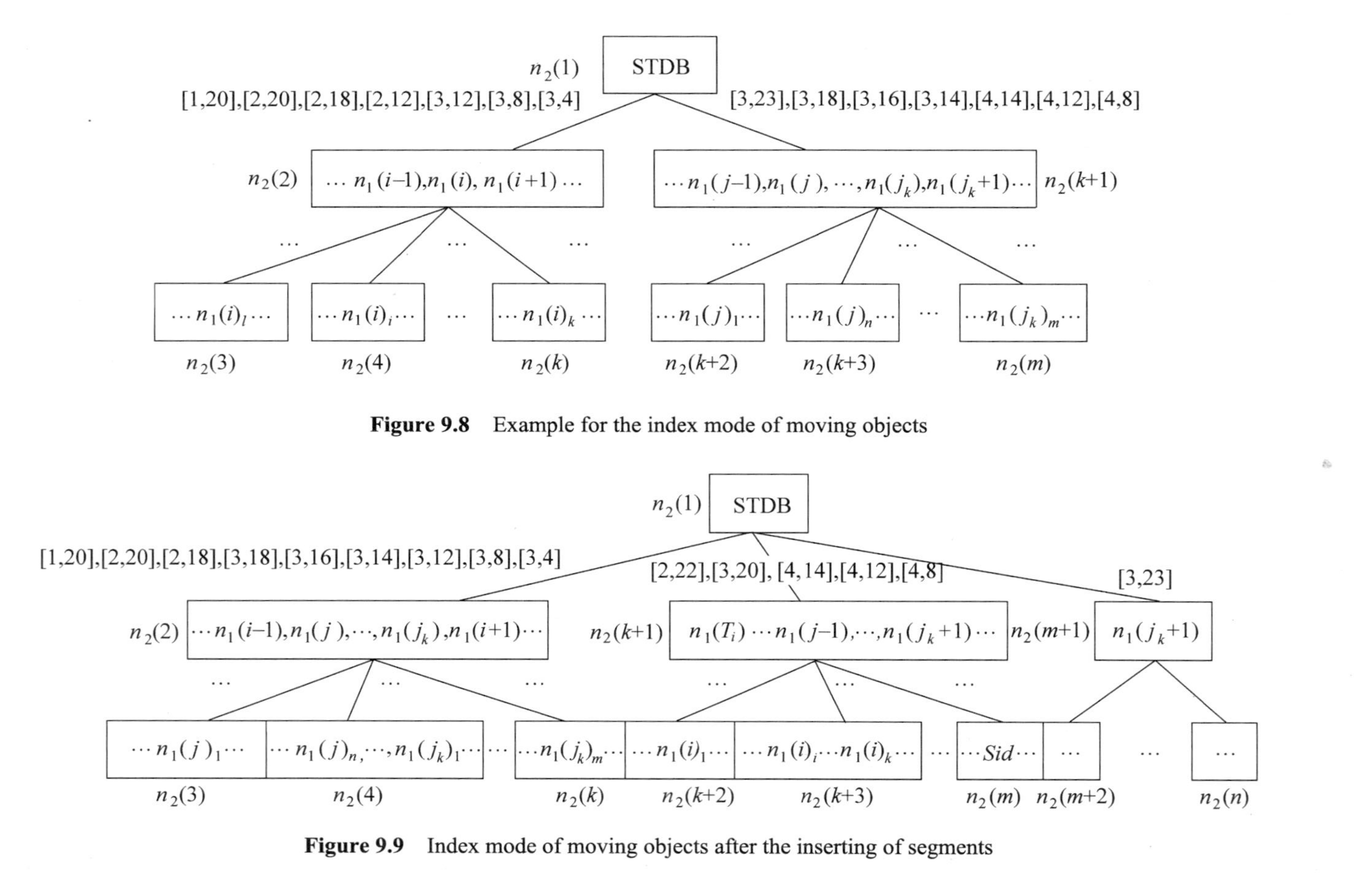

Figure 9.8 Example for the index mode of moving objects

Figure 9.9 Index mode of moving objects after the inserting of segments

- Update index node: If the relevant summary node is empty after deleting trajectory segments from Trj_0 and Trj_0, then the structure of *Motsum* remains the same, but MBR of relevant index nodes need to change. Else, it needs to delete the summary node where the trajectory segments in Trj_0 and Trj_0 are present.

The simulations are omitted because of limited space and if necessary, you can contact us. However, the related results show that the index schema is feasible and efficient.

References

[1] Abdelguerfi M, Givaudan J, Shaw K, Ladner R (**2002**) The 2-3 TR-tree, a trajectory-oriented index structure for fully evolving valid-time spatio-temporal datasets. In: Proceedings of the ACM workshop on Adv in Geographic Info Sys, ACM GIS, pp 29 – 34

[2] Albers G, Roos T (**1992**) Voronoi diagrams of moving points in higher dimensional spaces. In: Proceedings of the 3rd Scandinavian Workshop on Algorithm Theory

[3] Beckmann N, Kriegel H-P, Schneider R, Seeger B (**1990**) The R*-tree: an efficient and robust access method for points and rectangles. In: Proceeding of SIGMOD Conference, pp 322 – 331

[4] Chakka VP, Everspaugh A, Patel JM (**2003**) Indexing large trajectory data sets with SETI. In: Proceedings of the Conference on Innovative Data Systems Research, CIDR, Asilomar, CA

[5] Guttman A (**1984**) R-Trees: a dynamic index structure for spatial searching. In Proceedings of the ACM International Conference on Management of Data (SIGMOD), pp 47 – 57

[6] Kwon D, Lee S, Lee S (**2002**) Indexing the current positions of moving objects using the lazy update R-tree. In: Mobile Data Management, MDM, pp 113 – 120

[7] Lee M, Hsu W, Jensen C, Cui B, Teo K (**2003**) Supporting frequent updates in R-trees: a bottom-up approach. In: Proceedings of the International Conference on Very Large Data Bases (VLDB)

[8] Lomet DB, Salzberg B (**1989**) Access methods for multiversion data. In: Proceedings of the ACM International Conference on Management of Data (SIGMOD), pp 315 – 324

[9] Nascimento M, Silva R (**1998**) Towards historical R-trees. In: Proceedings of the ACM Symposium on Applied Computing (SAC). pp 235 – 240

[10] Nascimento M, Silva R, Theodoridis Y (**1999**) Evaluation of access structures for discretely moving points. In: Proceedings of the International Workshop on Spatio-Temporal Database Management (STDBM), pp 171 – 188

[11] Pfoser D, Jensen CS, Theodoridis Y (**2000**) Novel approaches in query processing for moving object trajectories. In: Proceedings of the International Conference on Very Large Data Bases (VLDB), pp 395 – 406

[12] Porkaew K, Lazaridis I, Mehrotra S (**2001**) Querying mobile objects in spatio-temporal databases. In: Proceedings of the International Symposium on Advances in Spatial and Temporal Databases (SSTD), pp 59 – 78

[13] Procopiuc CM, Agarwal PK, Har-Peled S (**2002**) STAR-tree: an efficient self-adjusting index for moving objects. In: Proceedings of the Workshop on Alg Eng and Experimentation, (ALENEX), pp 178 – 193

[14] Saltenis S, Jensen CS (**2002**) Indexing of moving objects for location-based services. In: Proceedings of the International Conference on Data Engineering (ICDE)

[15] Saltenis S, Jensen CS, Leutenegger ST, Lopez MA (**2000**) Indexing the positions of continuously moving objects. In: Proceedings of the ACM International Conference on Management of Data (SIGMOD), pp 331 – 342

[16] Sistla AP, Wolfson O, Chamberlain S, Dao S (**1997**) Modeling and querying moving objects. In: A1 Gray, P1 Larson (eds) Proceedings of the 13th International Conference on Data Engineering (ICDE'97)

[17] Song Z, Roussopoulos N (**2001**) Hashing moving objects. In: Mobile Data Management, pp 161 – 172

[18] Song Z, Roussopoulos N (**2003**) SEB-tree: an approach to index continuously moving objects. In: Mobile Data Management (MDM), pp 340 – 344

[19] Tao Y, Papadias D (**2001a**) Efficient historical R-trees. In: Proceedings of the International Conference on Scientific and Statistical Database Management (SSDBM), pp 223 – 232

[20] Tao Y, Papadias D (**2001b**) MV3R-tree: a spatio-temporal access method for timestamp and interval queries. In: Proceedings of the International Conference on Very Large Data Bases (VLDB), pp 431 – 440

[21] Tao Y, Papadias D, Sun J (**2003**) The TPR*-tree: an optimized spatio-temporal access method for predictive queries. In: Proceedings of the International Conference on Very Large Data Bases (VLDB)

[22] Tayeb J, Ulusoy O, Wolfson O (1998) A quadtree-based dynamic attribute indexing method. The Computer Journal 41(3): 185 – 200

[23] Theodoridis Y, Silva R, Nascimento M (**1999**) On the generation of spatiotemporal datasets. In: Proceedings of the 6th International Symposium on Spatial Databases, pp 147 – 164

[24] Theodoridis Y, Vazirgiannis M, Sellis T (**1996**) Spatio-temporal indexing for large multimedia applications. In: Proceedings of the IEEE Conference on Multimedia Computing and Systems (ICMCS)

[25] Wolfson O, Sistla AP, Chamberlain S, Yesha Y (**1999**) Updating and querying database that track mobile units. Distributed and Parallel Databases 7(3): 257 – 387

[26] Xu X, Han J, Lu W (**1990**) RT-tree: an improved R-tree indexing structure for temporal spatial databases. In: Proceedings of the International Symposium on Spatial Data Handling (SDH), pp 1040 – 1049

[27] Yeh TS, Cambray B (**1995**) Modeling highly variable spatio-temporal data. In: Porceedings of the 6th Australiasian Database Conference, pp 221 – 230

10 Temporal XML Index Schema

Xiaoping Ye[1,2], Junjie Luo[2], and Gongfu Zhong[2]

[1] Computer School, South China Normal University, Guangzhou 510631, P.R. China
[2] Department of Computer Science, Sun Yat-sen University, Guangzhou 510275, P.R. China

Abstract Because of the characteristics of temporal constraints, there may be a relationship between structural and temporal information in temporal XML. This chapter focuses on this and studies a temporal XML index based on linear order. Firstly, it discusses linear order partition, which provides a mathematical framework to the temporal index. Secondly, it researches algorithms on linear order branches, which are crucial in creating the index. Thirdly, it studies the querying paths based on temporal XML. Finally, it implements simulations with large number of testing data whose results indicate that the index is feasible and effective.

Keywords *linear order, temporal summary, index and querying, simulation and comparison*

10.1 Introduction

For databases, effective and efficient queries depend on two fundamental factors: one is the query language with strong expressing abilities, the other is querying processors that can work efficiently. Recently Xpath in W3C and XQuery may be considered to be query languages for XML, which are similar to SQL serving as in relational database. Because of massive data and high query frequency for XML, especially semi-structured characteristics in XML, the index schema for XML become the most crucial query technology in XML query processor.

As a type of data, XML can be modeled as a directed graph, which may be transformed into a directed tree. XML query is usually navigational query based on paths expressions. In the case of directed tree, the query can be classified into **absolute path** (root node is starting node) query and **relative path** (non-root node is starting node) query. It can also be classified as **value query** (ending node is leaf node) and **structure query** (ending node is non-leaf node). In

mcsyxp@mail.sysu.edu.cn

addition, it can be classified into **simple path query** (only search single path) and **branching path query** (need to search some sub-tree) according to practical issues. There are some querying methods in the light of the classification mentioned above: **summary, coding, summary coding.**

(1) Indexes based on **summary:** The main idea is that the paths starting from the root in XML consist of an indexing node, which satisfies some conditions (for example, they have the same semantic label), that is, an indexing node represents some paths in XML. Indexes based on summary are usually utilized directly in the case of graph, and it advances in the simple path query, but not in branching path query.

(2) Indexes based on **coding**: In this way, XML data are coded and the indexes corresponding to certain codes are built. There are some coding methods, such as vector coding and interval coding. Coding indexes can not only handle simple path but also branching path. However, the nodes need to be ordered and if it is not, it is not good enough to be utilized directly in a graph.

(3) Indexes based on **summary coding**: XML data is summarized in structure at first. The elements in the structure summary nodes are coded. One of them is Ctree, which can work efficiently in the case of experiments.

With the development of network, people gradually focus on semi-structured data with temporal constraints. Temporal XML adds temporal label (valid and transaction time) to attributes in elements, so there are three fundamental elements: temporal constraints, semantic labels and structure information. For our knowledge, the current work can be divided as follows:

- **Data modeling**: De Capitani (2002) considered an accredited visiting model to access the temporal XML documents, but it focuses on the mechanism of accreditation instead of the temporal characteristics (Goldman and Widom (1997)) studied a temporal XML model based on XPath. Dyreson presented an approach that supports the transaction time by extending XPath (Zou et al. 2004). Wang and Zaniolo (2003) discussed the problem of XML document historical version in web data warehouse research. Here, some continual versions of XML document are represented to a single document by XQuery. In addition, these authors dealt with the valid and transaction time (bitemporal) by four special attributes: Vstart, Vend, Tstart, and Tend (Wang and Zaniolo 2004). Wang et al. utilized the approaches similar to ones in (Wang and Zaniolo 2003, 2004) to process and query the historical XML data. Here, historical XML data is mapped into the documents called H-documents (Wang et al. 2005a, 2005b).
- **Querying language**: First, it is presented in temporal XML query language research. Gao. et al. presented a new querying language XQuery that supports the valid time querying by extending Xquery. The basic idea is transforming XQuery into Xquery by a given middleware and then implementing temporal querying through XQuery Engine (Amagasa et al. 2000, Dyreson et al. 1999). Milo and Suciu as well as Flavio and Alejandro, built the model based on TXPath, which is almost accorded with W3C, and extended the temporal

querying functions in XPath by defining the novel rules for the grammar and semantics of TXPath (Milo and Suciu 1999, Flavio and Alejandro 2008). In the aspect of temporal query, they defined value query and return value form, where returned result is the set of time point period binary tuple.

- **Indexes:** Alejandro et al. presented a temporal XML indexing schema based on the transaction time, which combines path information with temporal information to form a continual path. However, the indexing structure is rather complex and cannot deal with updates (Mendelzon et al. 2004). Alberto et al. (2004) proposed a temporal XML graphical model related to the transaction time. This discusses the reference of nodes and some related issues resulting from the reference, such as the consistency and so on. However, they had not handled the problem completely. In addition, it researched the update to the index (Vaisman et al. 2007). Campo and Vaisman (2006) discussed the resemble issues too. Ye et al. (2007) presented an indexing framework based on temporal connection and inclusion relations, and studied a tree-based XML indexing schema TXIDM. The indexing schema implements the value and paths querying, but the complexity of space and time are so high that the efficiency of the system may be decreased with data increasing.

In this chapter, index of temporal XML based on linear order is studied. Contributions of this chapter are as follows: the temporal information (temporal constraint) and structure information (ancestor node and offspring node, parent-child and left-right structure) in XML are researched, and the temporal information may contain some certain structure information. Vice versa, the structure information also contains some temporal information. Therefore, based on linear order, the indexing schema may be proposed in the suitable frame. In this way, it is possible to couple the temporal and structural information and implement the temporal and structural querying. This chapter is organized as follows: In Section 10.2, linear order is discussed and corresponding algorithms are given by introducing linear order matrix. In Section 10.3, temporal XML index model (*Txind*) is studied by means of temporal summary. In Section 10.4, XML value query and path query algorithms based on *Txind* are researched. In Section 10.5, simulations are designed and implemented and the results indicate that the index is effective and efficient.

10.2 Linear-Order Relation

A temporal data node is usually denoted as ($NT(u)$, $VT(u)$), where $NT(u)$ and $VT(u)$ represent non-temporal parts and temporal parts to P, respectively. The equivalence of P_1 and P_2 is defined as $NT(u_1) = NT(u_2) \wedge VT(u_1) = VT(u_2)$. To avoid misunderstanding, there is no difference in the use of labels P and $VT(P)$ to represent valid time stamp. The valid time stamp $VT(P)$ is a time period: $VT(u) = [VTs(u), VTe(u))$, where $VTs(u)$ and $VTe(u)$ are the starting point and ending point in $VT(P)$, respectively. The chapter also denotes them as $us = VTs(u), ue = VTe(u)$. If $VTs(u)$ and $VTe(u)$ are integers i and j, respectively, they also denote $VT(u) = [VTs(u), VTe(u))$ as "i,j". Without loss of generality, all temporal period points are positive integers. Meanwhile, D usually represents all temporal nodes or time periods corresponding to temporal nodes.

10.2.1 Linear-Order Matrix

In order to describe and research the relationship of time periods, it is needed for us to build some time sequence relations and introduce linear order matrix concept.

Definition 10.1 (Temporal relation) Let D be a period set. We define the temporal relation in D as follows:

① $\forall u,v \in D, VT(u) \subseteq VT(v)$, u and v have relation of "$\leqslant$", denoted as $u \leqslant v$, if $\neg(u \leqslant v \vee v \leqslant u)$, u, v are not compatible.

② $\forall u,v \in D, VT(u) = [VTs(u), VTe(u))$, $VT(v) = [VTs(v), VTe(v))$, if $VTs(u) < VTs(v)$, or $VTs(u) = VTs(v)$, and $VTe(v) \leqslant VTe(u)$, u and v have the relation "$\precsim$" denoted as $u \precsim v$.

If $u \leqslant v$, v is said to temporally contain u or u is contained in v, but time sequence relation "$\precsim$" is the increasing (ascending) sort of element u in D according to "starting point" $VTs(u)$.

It can be proved that "$\leqslant$" is the pseudo-order relation. It satisfies reflexivity and transitivity. D represents pseudo-order set in this chapter. In addition, "$\precsim$" satisfies reflexivity, transitivity and pseudo-linear order.

Definition 10.2 (Preorder matrixes to periods) Let D be a time period set, i_1, i_m be the minimal and maximal starting points, respectively, and j_1, j_n be the minimal and maximal ending points in all periods of D, respectively. Denote $[i,j)$ as $ij(i_1 \leqslant i \leqslant i_m, j_1 \leqslant j \leqslant j_n)$. Let the lateral axis and vertical axis in plane represent starting point and ending point, respectively. The matrix shown in Fig. 10.1 is called a temporal preorder matrix of D and is denoted as $LOM(D)$.

$$
\begin{array}{lllll}
i_1 j_m, & i_2 j_m, & \ldots, & i_{n-1} j_m, & i_n j_m \\
i_1 j_{m-1}, & i_2 j_{m-1}, & \ldots, & i_{n-1} j_{m-1} & \\
\vdots & \vdots & \vdots & & \\
i_1 j_2, & i_2 j_2 & & & \\
i_1 j_1 & & & &
\end{array}
$$

Figure 10.1 Linear matrix $LOM(D)$

For $u_{i_0 j_0} \in LOM(D)$, $u_{i_0 j_0}$ divide $LOM(D)$ into four areas: $UL(u_{i_0 j_0}) = \{v_{ij} \mid i \leqslant i_0, j_0 \leqslant j\}$, $UR(u_{i_0 j_0}) = \{v_{ij} \mid i_0 \leqslant i, j_0 \leqslant j\}$, $DL(u_{i_0 j_0}) = \{v_{ij} \mid i \leqslant i_0, j \leqslant j_0\}$ and $DR(u_{i_0 j_0}) = \{v_{ij} \mid i_0 \leqslant i, j \leqslant j_0 j\}$, as shown in Fig. 10.2. If only "<" is established, corresponding areas are denoted as *OUL*, *OUR*, *ODL* and *ODR*, respectively. In *LOM*(*D*), if u_{ij}, u_{kq} satisfies $j=q\wedge(i=k-1\vee k=i+1)$, we call u_{ij} and u_{kq} level adjacent. If it satisfies $i=k\wedge(j=q-1\vee q=j+1)$, we call u_{ij} and u_{kq} vertically adjacent. If the adjacent nodes u, v (let $v\prec u$) have the edge, which is level or vertical from u to v, it is denoted as $\langle u,v\rangle$. Let u, $v\in LOM(D)$. If $w_1, w_2, \cdots, w_m\in LOM(D)$, which satisfies u and $w_1, \cdots, w_i$ and w_{i+1} are adjacent, w_{m-1} and w_m, w_m and v are adjacent, respectively, $\langle u,w_1\rangle,\langle w_1,w_2\rangle,\cdots,\langle w_{m-1},w_m\rangle,\langle w_m,v\rangle$ is the path between u, v. The length of path connecting u and v is the number of contained edges in the path. The distance of u, v is the length of "shortest" path connecting u and v, denoted as $d(u,v)$. The nearest node from u_0 in the *LOM*(*D*) is defined as the node that has the shortest distance with u_0 in D. The column and row of u_0 in *LOM*(*D*) are $column(u_0)$ and $line(u_0)$. Upper half column and lower half column that do not contain u_0 in $column(u_0)$ are denoted as $column^U(u_0)$ and $column^D(u_0)$, respectively. Left half row and right half row that do not contain u_0 in $line(u_0)$ are denoted as $line^L(u_0)$ and $line^R(u_0)$, respectively. u_0 in $column^U(u_0)$ and $line^R(u_0)$ are called most upper near node and most right near node of u_0 in D, respectively. Hass Graph of time set D in *LOM*(*D*) is denoted as $H_{LOM}(D)$.

Theorem 10.1 (Preorder and preorder matrix) $\forall u_0\in H_{TM}(D)$, ① $u_0\leqslant v_0\Leftrightarrow v_0\in UL(u_0)$; ② $v_0\leqslant u_0\Leftrightarrow \in v_0\in DR(u_0)$; ③ u_0,v_0 are incompatible $\Leftrightarrow v_0\in UR(u_0)\vee v_0\in DL(u_0)$.

Proof ① let $u_0=[i_0,j_0)$, $v_0=[k_0,l_0)$, $u_0\leqslant v_0\Leftrightarrow u_0\subseteq v_0\Leftrightarrow k_0\leqslant i_0$, $j_0\leqslant l_0,\Leftrightarrow v_0\in UL(u_0)$. ② and ③ can be proved in the same way as ①.

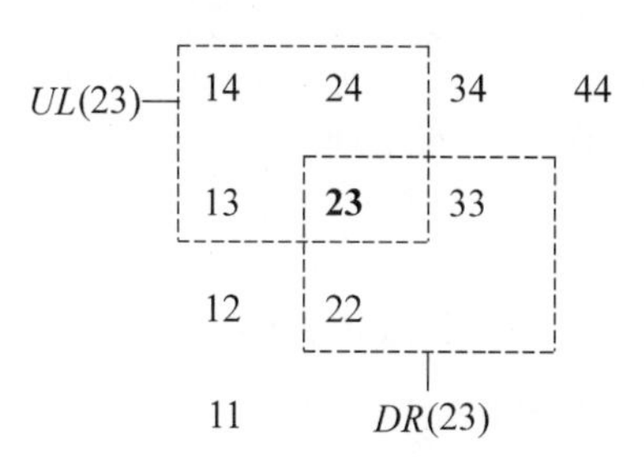

Figure 10.2 *UL*(23) and *DR*(23)

10.2.2 Linear-Order Equivalence Relation

Definition 10.3 (Linear equivalence relation) Let $L\subseteq D, \forall u,v\in L$. If pseudo-order "$\leqslant$" satisfies $u\leqslant v\vee v\leqslant u\vee v=u$, call L the linear order branch (*LOB*) in D. The greatest element and the least element are denoted as *maxL* and *minL*, respectively. If A is the partition of D in set *LOB,* call A linear order partition (*LOP*) in D. The equivalence relation is determined by a temporal order equivalence relation (*TOER*). $\varDelta$ represents temporal equivalence relation and corresponding temporal equivalence set (linear-order partition). At the same time, linear order branch is also called temporal equivalence in this chapter.

Time set D can have multiple linear-order partitions, so we have multiple linear equivalence relations. It needs to take some special features of relation into account, from the efficiency angle of temporal indexing query.

Definition 10.4 (Least and longest linear equivalence relation) Let $\varDelta_0$ be the linear-order equivalence relation in D. If every linear-order equivalence $\varDelta$ in D holds $|\varDelta_0| \leqslant |\varDelta|$, where $\varDelta_0$ is called the least equivalence in D, denoted as *Least-TOER*. Let L_0 be the *LOB* in $\varDelta_0$, if every *LOB* L holds $|L_0| \geqslant |L|$, L_0 is called the "longest" *LOB* in $\varDelta_0$. Let $\varDelta_0 = (L_1, L_2, \cdots, L_p)$ be the linear-order equivalence relation. If L_1 is the longest in D, L_2 is the longest in $D\backslash\{L_1\}, \cdots, L_p$ is the longest in $D\backslash\{L_1, \cdots, L_{p-1}\}$, $\varDelta_0$ is called the "longest" equivalence class of *Longest-TOER*.

Theorem 10.2 (*Least-TOER* existing) Pseudo-order set D has the least linear equivalence relation.

Proof According to *DFA*, the linear-order branching of linear-order set D may be gained.

First, $H_{LOM}(D)$ is built, then search in the light of *DFA* from the first element u_0 in the first column of $H_{LOM}(D)$ to down side, until the last element u_{i0} of D. We search from u_{i0} to right adjacent column, then down side, until the last element in this column, then move from this element to the right adjacent column. In this way, we can get an element u_p in D. There is no element in D except u_p. All nodes of the search path in D compose a linear-order branching L_1. Then in $D_1 = D\backslash L_1$, repeat the *DFA* search, get the linear-order branching. In $D_2 = D_1\backslash L_2$, repeat, until all linear-order branching $\langle L_1, L_2, \cdots, L_n\rangle$ is gained. The complexity of building $H_{LOM}(D)$ and *DFA* are $O(n\log n)$ and $O(n^2)$, respectively. If L_i and L_j are gained from *DFA,* then $ODR(minL_i) \cap L_j = \varnothing\ (i<j)$.

Let the linear partition $\varDelta_0 = \langle L_1, L_2, \cdots, L_n\rangle$ be obtained from *DFA,* L_i be the equivalence class obtained by calculating *DFA*. $\forall a_{i_0} \in L_i (1<i\leqslant n), \exists a_{k_0} \in L_{i-1}$, so that a_{i_0} and a_{k_0} are incompatible. In fact, it only needs to show that $\exists a_{k_0} \in L_{i-1} \wedge a_{k_0} \in ODL(a_{i_0})$. When $L_{i-1} \cap ODL(a_{i_0} = \varnothing)$, then $L_{i-1} \subseteq OUR(a_{i_0}) \cup ODR(a_{i_0}) \cup ODL(a_{i_0})$. In the light of *DFA*, $L_{i-1} \cap OUR(a_{i_0}) = \varnothing, L_{i-1} \cap ODR = \varnothing$, so $L_{i-1} \subseteq OUL(a_{i_0})$. *Now,* $ODR(minL_{i-1})$ contains L_i, but it contradicts $ODR(minL_{i-1}) \cap L_i = \varnothing$. In this way, we get the nodes denoted as $a_1, a_2, \ldots, a_n$. In fact, for $a_{i-1} \in ODL(a_i)$ $(1<i\leqslant n)$, there should be an element a_i that is incompatible with a_{i-1}. Therefore, any linear-order partition $\varDelta$ has at least n *LOB*s, that is, $|\varDelta_0| \leqslant |\varDelta|$.

Theorem 10.3 (*Longest-TOER* existing) Pseudo-order set D has the feature of linear-order equivalence relation of "longest" equivalence class.

Proof Let D be a temporal pseudo-order set. Build D's *LOB*s according to temporal total order algorithm, *TOA*.

First, execute the sort based on pseudo-order matrix. From the first column in $H_{TM}(D)$ to every column, gain a permutation: $A[1], A[2], \cdots, A[m]$. Second, build corresponding target set. Give the indicator $B[i_0]$ and $C[i_0]$ to the element i_0, after sorting. The initial indicators are $B[1]=1$, $C[1]=0$. For $B(i_0-1)$, $C(i_0-1)$, then $B[i_0]=B[i_1]+1$, $B[i_1]=max\{B[i] \| A[i_0] \subseteq A[i], 1 \leqslant i \leqslant i_0-1\}$, but $C(i_0-1)=A[i_1]$. Then gain the linear-order branching. The complexity of the algorithm is $O(n^2)$. The longest linear-order equivalence can be ensured by *TOA*.

Example 10.1 Let $H_{LOM}(D)$ be shown in Fig. 10.3

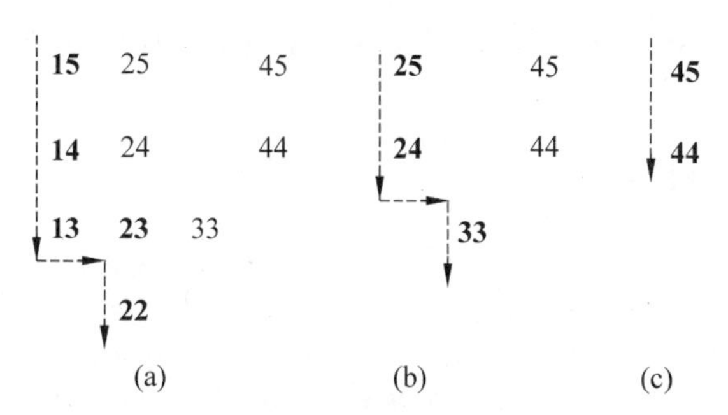

Figure 10.3 $H_{LOM}(D)$

The calculating course based on *DFA* may be as follows:

A[1]	A[2]	A[3]	A[4]	A[5]	A[6]	A[7]	A[8]	A[9]	A[10]
15	14	13	25	24	23	22	33	45	44

$B[1]=1$, $B[2]=2$, $B[3]=2$, $B[4]=1$, $B[5]=2$, $B[6]=4$, $B[7]=5$, $B[8]=5$, $B[9]=1$, $B[10]=2$

$C[1]=0$, $C[2]=1$, $C[3]=2$, $C[4]=1$, $C[5]=2$, $C[6]=3$, $C[7]=6$, $C[8]=6$, $C[9]=0$, $C[10]=9$

From u_{15} down to u_{13}, from u_{13} right to u_{23}, from u_{23} down to u_{22}, get $L_1=\langle u_{15}, u_{14}, u_{13}, u_{23}, u_{22} \rangle$, as shown in Fig. 10.3(a).

In $D \backslash \{L_1\}$, from u_{25} down to u_{24}, from u_{24} right and down to u_{33}, get *LOB*: $L_2=\langle u_{25}, u_{24}, u_{33} \rangle$, as shown in Fig. 10.3(b).

In $D \backslash (L_1 \cup L_2)$, from u_{45} down to u_{44}, get *LOB*: $L_3=\langle u_{45}, u_{44} \rangle$, as shown in Fig. 10.3(c).

In this way, we get all *LOB*s: $L_1=\langle u_{15}, u_{14}, u_{13}, u_{23}, u_{22} \rangle$, $L_2=\langle u_{25}, u_{24}, u_{33} \rangle$, $L_3=\langle u_{45}, u_{44} \rangle$.

In Example 10.1, the *LOP* gained by applying *DFA* or *TOA,* which satisfies *Least- TOER* = *Longest- TOER*, *Least- TOER* and *Longest- TOER* are two different equivalence relations. They do not contain each other. The corresponding processings are described in Fig. 10.4.

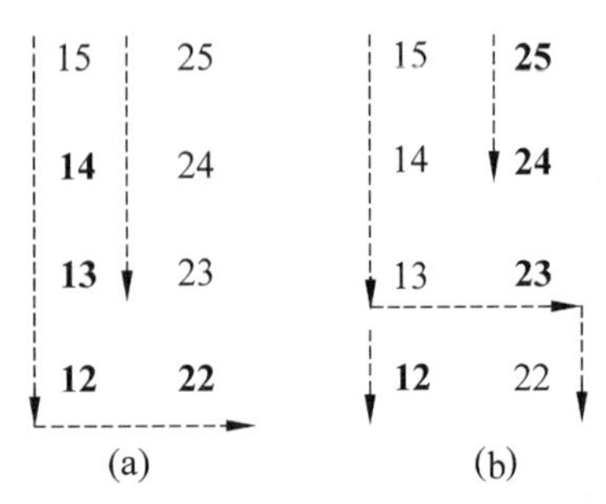

Figure 10.4 *Least-TOER* and *longest-TOER* are mutually not contained in each other

10.3 Temporal Summary and Temporal Indexing

10.3.1 Data Model

There are two ways to build temporal XML query data model. The first way is by denoting the node of valid time as an import edge. The second way is by building the order pair, which contains node and valid time. No matter which way is used, the model gained is a directed graph. In order to separate the levels, we gradually transform the graph to a tree. The temporal XML query data model is tree structure in this chapter.

Definition 10.5 (Temporal XML querying data model *TXqdm(D)*) Let D be the temporal XML document data. The temporal XML querying data model of an XML document D is a tree-structure:

$TXqdm(D) = \langle N(D), V(D), E(D), VT(D), L(D) \rangle$, where

- $N(D)$ is a set of the serial numbers for all elements in $TXqdm(D)$, and the serial number of node u is denoted as $n(u)$ or n_i.
- $V(D)$ is the set of all nodes in $TXqdm(D)$. There are four classes: root node r, element nodes n_e, attribute nodes n_t and value nodes n_v. If an element only has value nodes, we call it simple element node, which is denoted as n_{je}. In this chapter, the attribute node refers to simple element nodes.
- $E(D)$ is the set of all the edges (n_i,n_j) in $TXqdm(D)$. There are two classes in $E(D)$: value edge $e_v \Leftrightarrow n_j = n_v$ and non-value edge $e_{nv} \Leftrightarrow n_j \neq n_v$.
- $VT(D)$ is the multi-set of all the time stamp of the nodes in $TXqdm(D)$.
- $L(D)$ is the multi-set of all the semantic labels of the nodes in $TXqdm(D)$.
- $u \in V(D)$, $u = (n(u), VT(u), L(u))$, where $n(u)$ is the serial number of u, $VT(u)$ is the valid time of u, $L(u)$ is the node label. The elements in $V(D)$ are called the temporal nodes of $TXqdm(D)$.

Definition 10.6 (Temporal constraint) The nodes of $TXqdm(D)$ must satisfy the following temporal constraints:

- **Time period of child node cannot be larger than parent node.** If v is the parent node of u, $VT(u) \subseteq VT(v)$.

- **In one period the value of node is unique.** Let $n_0 \in V(D) - \{r\}$, which is a simple element node, n_{i0} and n_{j0} are the children value nodes of n_0, $VT(n_{i0}) \cap VT(n_{j0}) = \varnothing$. If n_0 only has a leaf (value) node n_{i0}, then $VT(n_0) = VT(n_{i0})$.
- **Brother nodes time period are adjacent or intersecting.** Let n_{i0} and n_{j0} be brother nodes, and $VT(n_{i0}) \leqslant sVT(n_{j0})$. $VTs(n_{j0}) \leqslant VTe(n_{i0}) \vee VTe(n_{i0}) < VTs(n_{j0})$, if "=" is satisfied, call n_{i0} and n_{j0} adjacent.

Example 10.2 An instance of temporal XML data is shown in Fig. 10.5.

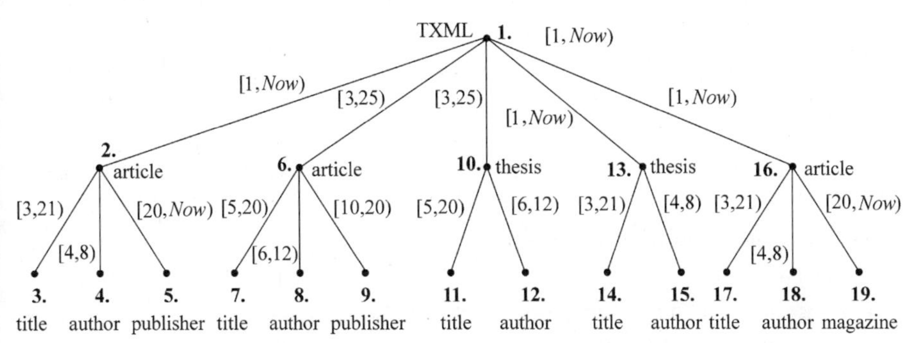

Figure 10.5 Instance of temporal XML data

Level 1: TXML (1);
Level 2: *article* (2,6,16); *thesis* (10,13);
Level 3: *title*(3,7,11,14,17); *author* (4,8,12,15,18); *publisher* (5,9); *magazine*(19)

10.3.2 Temporal Summary

Let u, v be two temporal nodes. If $VT(u) = VT(v)$, u, v are said to be temporally equivalent. Temporal equivalence is the equivalence relation of time period set D. We can do the classification according to that.

Definition 10.7 (Temporal summary *TXsum*) Temporal summary $TXsum(D)$ is defined as $TXsum(D) = (NTs(D), VTs(D), ETs(D), LTs(D), LMAPsum(D))$, which is a tree structure. $TXsum(D)$'s root is the name of $TXqdm(D)$, the number of its level is the same as $TXqdm(D)$.

- $NTs(D)$ is a set of series of numbers of all summary nodes in $TXsum(D)$ by go-ahead sequence.
- $VTs(D)$ is the set of all summary nodes in $TXsum(D)$. The node in the first level of $TXsum(D)$ is the root. A node in the second level of $TXsum(D)$ consists of the nodes in the second level, which have the same valid time periods, that is, the equivalence class $[u]$. The third level nodes are the equivalence classes, which compose every summary node in the second level in $TXqdm(D)$ according to time equivalence.

- $ETs(D)$ is the set of the edges with the constraints in $VTs(D)$.
- $LTs(D)$ is the set of the semantic labels of all summary nodes in $TXsum(D)$, where $LTs(.)$ is a mapping from $VTs(D)$ to $LTs(D)$: $\forall\,[u]\in VTs(D), LTs[u]=\{Lable(v)\,|\,v\in[u]\}$.
- $LMAPsum(D)=\{LMAP_{[u]}(.)\,|\,[u]\in VTs(D)\}$ is the set of the located mapping for the summary nodes, where $\forall w\in[u]$, $LMAP_{[u]}(w)$ maps w into its located guide codes: assuming $[v]$ is the parent node of $[u]$ in $TXsum(D)$, ranking all x in $[v]$ by ascending order as $x_0, x_1, \cdots, x_k$. Similarly, ranking all w in $[u]$ as: $w_0, w_1, \cdots, w_{\mathrm{m}}$. If $w\in\{w_0, w_1, \cdots, w_m\}$ and its parent node in $[v]$ is x_{i0}, then the located guide code of w is "i_0".

A temporal summary node $[u]$ is denoted as $[u]=n_1([u]), VT([u]), LTs([u]), LMAPsum[u]$, where $n_1[u]$ is the series of numbers of $[u]$ in $TXqdm(D)$ by go-ahead sequence, $VT[u]$ is the valid time of $[u]$, $LTs([u])$ is the set of semantic labels of all the nodes in $[u]$, $LMAPsum[u]$ is the located guide codes of $[u]$ in $TXqdm(D)$.

The temporal summary $TXsum(D)$ is an abstraction or summary for temporal characteristics in $TXqdm(D)$ according to the definition mentioned above. The following theorem shows that there is a connection between the temporal paths in $TXqdm(D)$ and the temporal summary path in $TXsum(D)$.

Theorem 10.4 (Property of temporal summary) ① Let B be the child node of A in $TXqdm(D)$, then $[B]$ is the child node of $[A]$ in $TXsum(D)$. ② If "$\cdots A(VT(A))\backslash\backslash B(VT(B))\cdots$" is a path of $TXqdm(D)$, then "$\cdots[A]\ (VT[A])\backslash\backslash [B](VT[B])\cdots$" is a path in $TXsum(D)$ also.

Proof "①" in the theorem can be obtained from Definition 10.6 and the nodes in $TXsum(D)$ will satisfy the same temporal constrains that the nodes in $TXqdm(D)$ satisfy. Thus the conclusion is that "②" is obtained from "①"of this theorem and Definition 10.6.

The mapping for a path in $TXqdm(D)$ to the corresponding path in $TXsum(D)$ corresponds 1-1, which keeps the temporal constraints and the route relation.

Level 1: $n_1(1)=(1)$, $I(n_1(1))=[1,Now)$, $Label(n_1(1))=\{\mathrm{TXML}\}$, $LMAP_{n_1(1)}(w)=(1)$;

Level 2: $n_1(2)=(2,13,16)$, $I(n_1(2))=[1,Now)$, $Label(n_1(2))=\{\text{article, thesis}\}$, $LMAP_{n_1(2)}(w)=(0,0,0)$; $n_1(6)=(6,10)$, $I(n_1(6))=[3,25)$, $Label\ (n_l(6))=\{\text{article, thesis}\}$, $LMAP_{n_1(6)}(w)=(0,0)$;

Level 3: $n_1(3)=(3,14,17)$, $I(n_1(3))=[3,21)$, $Label(n_1(3))=\{\text{title}\}$, $LMAP_{n_1(3)}(w)=(0,1,2)$; $n_1(4)=(4,15,18)$, $I(n_1(4))=[4,8)$, $Label\ (n_1(4))=\{\text{author}\}$, $LMAP_{n_1(4)}(w)=(0,1,2)$; $n_1(5)=(5,19)$, $I(n_1(5))=[20,Now)$, $Label\ (n_1(5))=\{\text{publisher, magazine}\}$, $LMAP_{n_1(5)}(w)=(0,\ 2)$; $n_1(7)=(7,11)$, $I(n_1(7))=[5,20)$, $Label\ (n_1(7))=\{\text{title}\}$, $LMAP_{n_1(7)}(w)=(0,1)$; $n_1(8)=(8,12)$, $I(n_1(8))=[6,12)$, $Label\ (n_1(8))=\{\text{author}\}$, $LMAP_{n_1(8)}(w)=(0,1)$; $n_1(9)=(9)$, $I(n_1(9))=[10,20)$, $Label\ (n_1(9))=\{\text{magazine}\}$, $LMAP_{n_1(9)}(w)=(0,1)$.

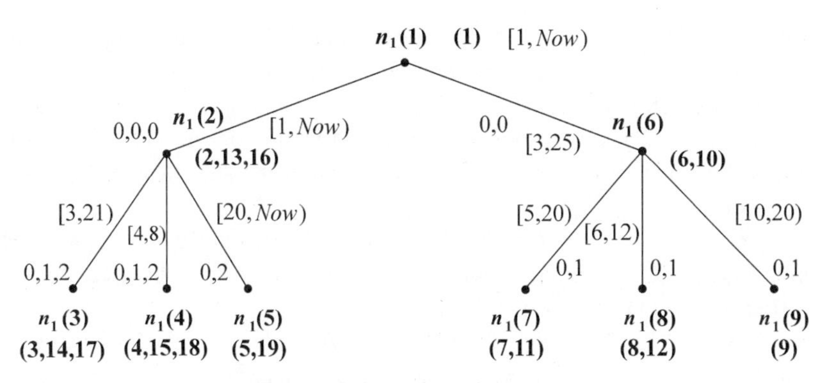

Figure 10.6 Temporal summary

10.3.3 Temporal Indexing

Definition 10.8 (Temporal XML indexing schema) The temporal XML indexing schema is defined as follows: $TXind\ (D) = (N_{ind}(D), V_{ind}\ (D), E_{ind}\ (D), L_{ind}(D), LMAP_{ind}(D))$, where $TXind(D)$ is a tree structure. The nodes are called temporal indexing nodes. The level is the same between $TXind(D)$ and $TXsum(D)$, corresponding to the same level. The root node is the name of the $TXsum(D)$.

① $N_{ind}(D)$ is the set of all nodes in $TXind(D)$. The node in the first level of $TXind\ (D)$ is the root.

② $V_{ind}(D)$ is the set of all nodes in $TXind(D)$. The root node is in the first level of $TXind(D)$. The nodes of second level are linear-order branching gained by temporal pseudo-order summary nodes in the second level of $TXsum(D)$. The nodes of the third level are the linear-order branching gained by element (summary nodes) of every indexing nodes of level two according to temporal pseudo-order. Where $i \geqslant 3$, if we get the set of temporal indexing nodes in the i-th level of $TXind(D)$, then the nodes in the $(i+1)$-th level consist of linear-order branching of nodes in L_k according to "$\leqslant$".

③ $E_{ind}\ (D)$ is the edge set constrained in "②".

④ $Lti(D)$ is the label set of all nodes in $TXind\ (D)$. $L_{ind}\ (.)$ is the mapping from $V_{ti}(D)$ to $L_{ind}(D)$: $\forall\ L_0 \in V_{ind}\ (D), L_{ind}\ (L_0) = \{Label([w]) |\ [w] \in L_0\}$.

⑤ $LMAP_{ind}(D) = \{LMAP\ L_0(.) | L_0 \in V_{ind}(D)\}$ is the location mapping set of the element in indexing nodes, $\forall\ [w] \in L_0, LMAP\ L_0\ ([w])$ will be mapped as location target. Let L_1 be the father node of L_0. Let the summary nodes $[x]$ and $[y]$ in L_1 and L_0 be sorted in ascending order $[x]_0, [x]_1, \cdots, [x]_k$; $[y]_0, [y]_1, \cdots, [y]_m$. If the father node of $[y] \in \{[y]_0, [y]_1, \cdots, [y]_m\}$ is in $TXsum(D)$, then the location target of $[y]$ is "j_0".

Indexing nodes $L_0 = (n_2(L_0), \langle\ min(L_0), max(L_0)\ \rangle, Label(L_0))$, $n_2(L_0)$ is the number of indexing node, $min(L_0)$, $max(L_0)$ are the maximal and minimal elements of L_0, $Label(L_0)$ is the semantic label of indexing node.

According to Definition 10.8, "①" indicates that *TXind*(*D*) and *TXqdm*(*D*) have the similar structure, that is, they are homomorphism in some degree. ① indicates that the node set in every level of *TXind*(*D*) is an aggregation of corresponding node set in *TXqdm*(*D*), which shows the indexing feature of *TXind*(*D*). ④ indicates the semantic feature of indexing node, ⑤ describes the structure information of indexing node.

Example 10.3 Temporal index of temporal XML data in Example 10.2 is shown in Fig. 10.7. The semantic label set of indexing node is shown as follows.

Level 1: $n_2(1)=(1)$,$Lable(n_2(1))=$ {TXML};

Level 2: $n_2(2)=(n_1(2), n_1(6))=((2,13,16),(6,10))$, $Lable(n_2(1))=$ {article, thesis};

Level 3: $n_2(3)=(n_1(3), n_1(4))=((3,14,17),(4,15,18))$, $Lable(n_2(3))=${title, author}; $n_2(4)=(n_1(5))=((5,19))$, $Lable(n_2(3))=$ {publisher}; $n_2(5)=(n_1(7), n_1(8))=((7,17),(8,12))$, $Lable(n_2(4))=$ {title, author}; $n_2(6)=(n_1(9))=((9))$, $Lable(n_2(5))=$ {magazine}.

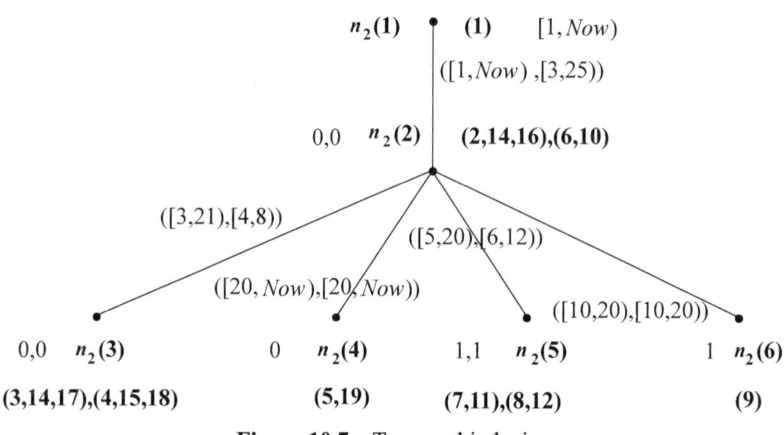

Figure 10.7 Temporal indexing

10.4 Data Query

Definition 10.9 (Temporal Path and Temporal Query) The temporal path in *TXqdm*(*D*) is denoted as $L_t=\langle Path(L_t), VT(L_t), Sem(L_t)\rangle$.

- $Path(L_t)$ is the sequence of the m_0 temporal nodes based on the Xpath in the *TXqdm*(*D*), the *k*-th node is noted as $L_t(k)$.
- $VT(L_t)$ is the sequence of $VT(L_t(k))$: $(VT(L_t(1)), VT(L_t(2)),\ldots,VT(L_t(m_0))$, where $VT(L_t(k))$ is the valid time constraint of $L_t(k)(1\leqslant k\leqslant m_0)$, which is usually shown as a temporal predicate.
- $Sem(L_t)$ is the set of all the semantics tags in L_t.

$L_t(1)$and $L_t(m_0)$ are called the start point and end point of L_t. If $L_t(1)=root$, then L_t is called the temporal absolute path, else, the temporal relative path.

10.4.1 Query Based on Absolute Paths

Let the basic query be TQ_{abs}: / $A[VT(A)]$. In $TXqdm(D)$, $A=(VT(A), Lable(A))$, where $VT(A)$ and $Lable(A)$ indicate the series of numbers, valid time and semantic label of A, respectively.

Algorithm 10.1 (Absolute path query based on *TXind*(*D*))

Step 1 (Getting levels): The level of A in $TXqdm(D)$ is confirmed (calculated) by the querying of the related temporal information and we assume that A is located on the i_A-th level in $TXqdm(D)$.

Step 2

(1) Querying to indexing nodes: Let $\{L_{i0}\}$ be the set of all linear order branches on the i_A-th level in $TXind(D)$ and $L_{i0}=(n_{i0}(j), max(L_{i0}), min(L_{i0}), Lable(L_{i0}))$ be the indexing node, where $n_{i0}(j)$, $max(L_{i0})$, $min(L_{i0})$ and $Lable(L_{i0})$ are the series of numbers, the maximal and minimal periods in the branch and the set of the semantic labels of the indexing node, respectively.

① If $VT(A)\cap max(L_{i0})=\varnothing$, all the elements in L_{i0} are not the querying results based on the index.

② If $VT(A)\subseteq min(L_{i0})$, all the elements in L_{i0} are the querying results based on the index.

③ If neither of the above statements are true, arrange the summary nodes in R_{i0} as $[u]_{i1}, [u]_{i2}, \cdots, [u]_{in}$, ranked by the starting points of their periods. The subset in $\{[u]_{i1}, [u]_{i2}, \cdots, [u]_{in}\}$, satisfying $VTs([u]_{i1})\leqslant VTs(A)$ is selected. Assume that $\{[u]_{i1}, [u]_{i2}, \cdots, [u]_{in}\}$ itself is such a subset. Beginning with $[u]_{i1}$, if $VTe(A)\leqslant VTe[u]_{i1}$, then operate by ①, else consider B_{i2} and so on, until a segment in L_{i0} is obtained, which is a branch itself.

(2) Semantic selection of indexing nodes: The semantic selection is carried out in the set $\{R_{i0}\}$ resulting from the querying on the indexing nodes. If $L_A\in Slable(R_{i0})$, indexing node R_{i0} is held, else it is deleted.

Step 3 (Process of data nodes): Let L_{i0} be the LOB acquired by Step 2, $u\in L_{i0}$, do the selection of u with $Lable(A)$, then the result will be obtained.

The complexity of the algorithm is $O(\log n)$.

10.4.2 Query Based on Relative Paths

Suppose there is a basic query with relative path TQ_{rel}: $A[VT(A)]\backslash B[VT(B)]$, similar to Algorithm 10.1. This is presented as follows:

Algorithm 10.2 (Basic query with relative path)

Step 1 (Getting the levels): The levels i_A and i_B, where A and B are located, are calculated by the given querying information, such as semantic tag.

Step 2 (Querying on indexing nodes):

(1) Querying on the temporal index: The querying of the indexing node is done

on the i_A-th and i_B-th levels in *TXind*(*D*) and the result will be the indexing node sets $\{n_2(A)\}$ and $\{n_2(B)\}$.

(2) Semantic selection of indexing nodes: The indexing nodes resulting from Step 2(1) are selected. If $Lable(A) \in Lable(n_2(A))(Lable(B) \in Lable(n_2(B)))$, then $\{n_2(A)\}$ ($\{n_2(B)\}$) will be kept. Else, the related indexing nodes will be dropped. There is no harm in assuming that the result is $\{n_2(A)\}$ ($\{n_2(B)\}$).

Step 3 (Disposal of the summary node): $L_{i0} = n_2(B)$, L_{j0} is L_{i0}'s parent node, $[u]_0 \in L_{i0} \wedge (Lable(A) \in Lable([u]_0))$. Find the parent node $[v]_0$ in L_{j0} according to the location index of $[u]_0$. Let L_{k0} be L_{j0}'s parent node and find the parent node in L_{j0} according to the location index of $[v]_0$, and so on. Repeat this process and backtrack to the i_A -th level, We will get the related node $[w]_0$ and the summary path $[w]_0/\cdots/[v]_0/[u]_0$. If $[w]_0 \in n_2(A)$, then the path will be kept, else the path will be dropped. In the same way, if $Lable(A) \in Lable([w]_0)$, the path will be kept, else the path will be dropped. The path obtained in this way is $[w]_0/\cdots/[v]_0/[u]_0$.

Step 4 (Disposal of the data nodes): $[u]_0 \in L_{i0} \wedge (Lable(A) \in Lable([u]_0))$. Find the parent node v_0 of u_0 according to the location index of u_0, and perform the same process on the node v_0. Repeat this process on the node w_0 in the i_A-th level and we will get path $w_0/\cdots/v_0/u_0$ in *TXqdm*. By checking whether $w_0 \in [w]_0 \wedge (Lable(B) = Lable(w_0))$, we can get the query result.

The complexity of the algorithm is $O(\log n)$.

Example 10.4 Let the temporal querying be "thesis\author[6,7)" on the XML datum in Example 10.2, where A = thesis, B = author.

(1) Getting the levels: A is in the second level and B is in the third level, calculated by the semantic labels of A and B. Since A = thesis has no temporal constraints, it is enough to carry out the temporal indexing querying for B.

(2) Querying on indexing nodes

(i) In the third level of *TXind*(*D*), $I(B) = [6,7) \subseteq [4,8) = min(n_2(3))$, $VT(B) = [6,7)$ $[6,12) = min(n_2(5))$, $max(n_2(4)) = [20, Now)$ and $max(n_2(6)) = [10,20)$, do not intersect with $[6,7) = VT(B)$, so $\{n_2(3), n_2(5)\}$ is the indexing nodes set for searching.

(ii) Semantic selection for indexing nodes: $Lable(B)$ = author $\in$ {{title, author}} = $Lable(n_2(3)) = Lable(n_2(5))$, so $n_2(3)$ and $n_2(5)$ are the indexing nodes set for searching.

(3) Disposing of the summary nodes

$n_2(3) = ((n_1(3), n_1(4)) = ((3,14,17),(4,15,18))$, location index(0,0), $Lable(n_2(3)) = Lable(n_1(3), n_1(4))$ = {title,author}, $Lable(B)$ = author $\in Lable(n_1(4))$ = {author}; $n_2(5) = (n_1(7), n_1(8)) = ((7,17)(8,12))$, location index(1,1), $Lable(n_2(5)) = Lable(n_1(7), n_1(8))$ = {title, author}, $Lable(B)$ = author $\in Lable(n_1(8))$ = {author}; $n_2(3)$ and $n_2(5)$ are parent nodes; $n_2(2) = (n_1(2), n_1(6)) = ((2,13,16),(6,10))$, $Lable(n_2(2)) = Lable(n_1(2), n_1(6))$ = {article,thesis}, $Lable(A)$ = thesis $\in Lable(n_l(2)) = Lable(n_1(6))$ = {article, thesis}.

Therefore, we can get the summary path $n_1(2)/n_1(4)$, $n_1(6)/n_1(8)$.

(4) Disposing the data nodes: In $n_1(2) = (2,13,16)$, only $Lable(13)$ = thesis. In

$n_1(4)$ = (4,15,18), location index is (0,1,2), *Lable*(4) = *Lable*(15) = author. Therefore, the acquired temporal index is 16/15.

In $n_1(6)$ = (6,10), only *Lable*(10) = thesis and in $n_1(8)$ = (8,12), location index is (0,1), *Lable*(8) = *Lable*(12) = author. The needed temporal index is 10/12. Therefore, 13/15 and 10/12 are the querying results.

10.5 Simulation and Evaluation

We design and implement relevant simulation experiment to verify the feasibility and validity of *TXind*(*D*).

10.5.1 Environment and Data Design

The experiment environment is WinXP SP2, Java, Celeron CPU at 2.4 GHz, memory with capacity of 512 MB. All the experiment data are generated randomly meeting some XML model requirements. The maximal number of the nodes in the data file is 200000. The span of valid time is $|VT| = VTs - Vte$. The probability of the same time period of the XML nodes in the same level, changes between 0.0 and 1.0 with an interval 0.1.

If the temporal path only involves with the parent-child node, then the path is called the basic path. If the temporal path expression L_t in querying *TQ* is the basic path, then *TQ* is called the basic temporal querying. General paths like *A*/*B* are composed of some basic path *C*/*D*, and general temporal XML querying can be composed of some basic querying without loss of generality. We only discuss the basic querying in this chapter. The basic querying can be divided into the following six types according to valid time profile:

① /*A*, ② *A*/*B*, ③ /*A* [*VT*(*A*)], ④ *A*[*VT*(*A*)]/*B*, ⑤ *A*/*B*[*VT*(*B*)], ⑥ *A*[*VT*(*A*)]/*B*[*VT*(*B*)]

"①" and "②" do not contain the temporal constraint and they can be treated as a particular case of the temporal querying. Both "④" and "⑤" can be treated as the particular case of "⑥", so we only talk about the absolute temporal path querying "③" and the relative temporal path querying "⑥".

10.5.2 Simulation and Evaluation

Simulations are divided into the case compared with DOM (massive data) and the case compared with DOM and *TXidm*(*D*) (choosing the original data in (Ye et al. 2007, section 4.2.4)).

10.5.2.1 Creating indexing schema

The relative simulated data are shown in Table 10.1, where
Tprob: probability of same temporal periods;
Lorder: number of linear order sets;
Peclass: number of equivalence classes of periods;
Sclass: number of semantic label classes with the same periods;
Dfile: number of bits of the data files;
Ifile: number of bits of the index files unit KB;
Tcons: time for constructing index, unit: ms.

Table 10.1 Data for building index

Tprob	Lorder	Peclass	Sclass	Dfile(KB)	Ifile(KB)	Tcons(ms)
0.0	112434	199914	200000	8490	59221	12109
0.1	107930	195350	199199	8500	58436	12453
0.2	99176	183159	197307	8501	56413	11610
0.3	92666	170594	193071	8526	54161	8781
0.4	83779	154124	186241	8497	51369	8219
0.5	67554	126297	162038	8500	45699	8671
0.6	59380	111907	166594	8509	43803	8937
0.7	51490	96675	164313	8497	41183	6719
0.8	33252	63446	126290	8499	34077	8750
0.9	20628	39708	109764	8509	29501	14390
1.0	283	525	730	8424	17619	31797

According to this table, the space size of indexing file is shown in Fig. 10.8 and 10.9.

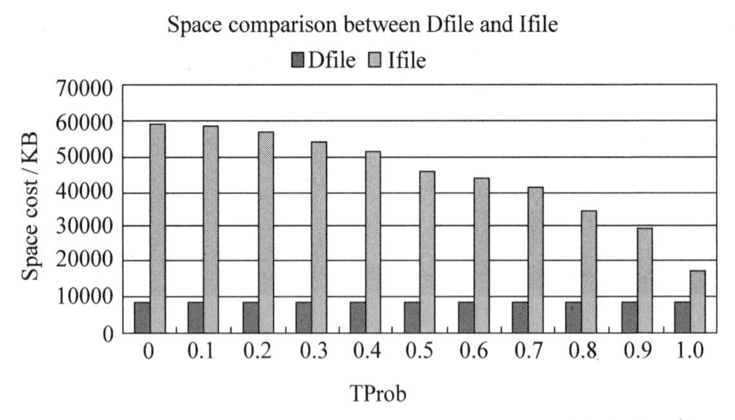

Figure 10.8 Comparison on the space between data and indexing file

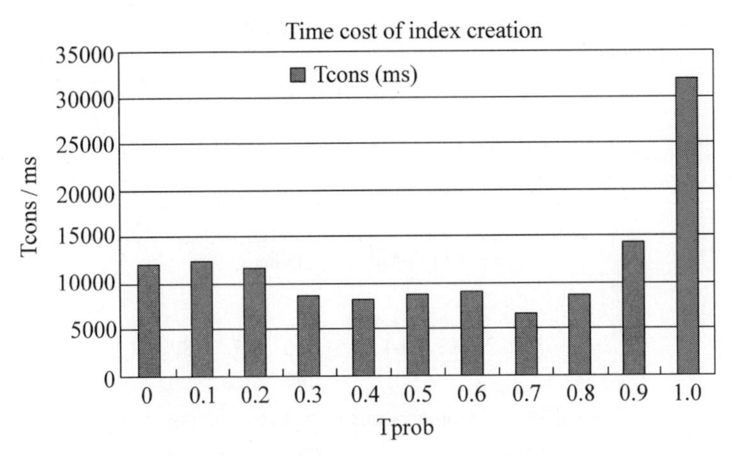

Figure 10.9 Time consumed in building index

The relation between XML data and index document is shown in Fig. 10.10.

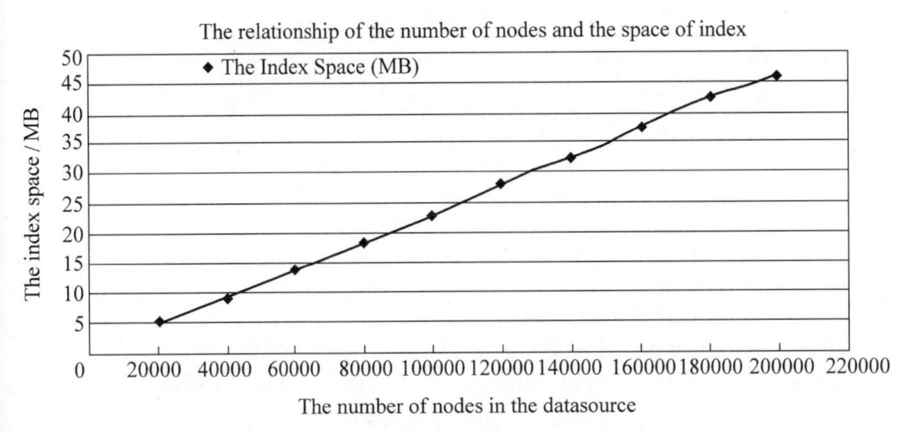

Figure 10.10 Relation between XML data and index document

From Fig. 10.10, we can see that the index document increases approximately linearly with the XML data.

10.5.2.2 Temporal XML querying

The comparison between the DOM and Tind is divided into two cases: based on the probability of the same temporal periods and the variety of the time span in the periods.

(1) The comparison between index querying and DOM querying

As shown above, for the sake of discussion, we only show the querying test outcome of Q_3 and Q_6. *TXind_Qi* denotes the *TXind*(*D*)-based querying, *Dom_Qi* denotes the DOM-based querying. The time consumed is shown in Fig. 10.11 and Fig. 10.12 ($i = 3,6$).

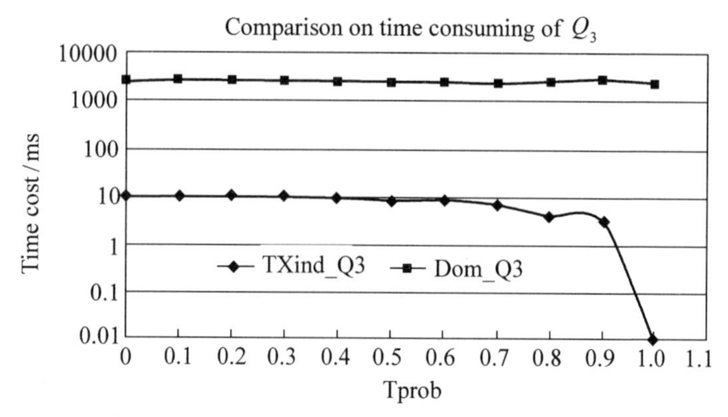

Figure 10.11 Comparisons on time consuming to Q_3

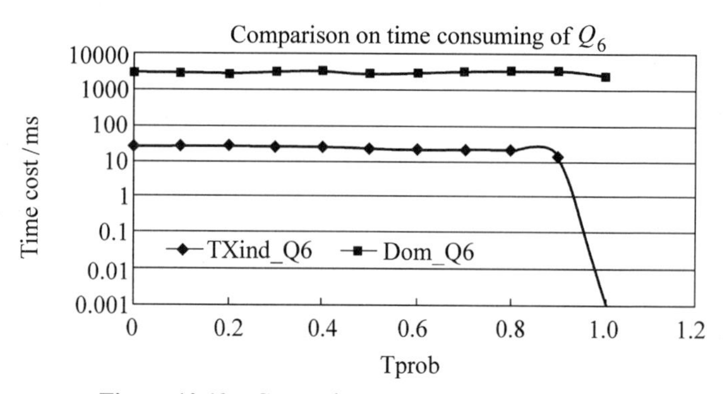

Figure 10.12 Comparison on time consuming to Q_6

(2) Case in variety of time span

Let the number of nodes be 200000. The probability of the same periods is 0.5, the span of querying periods change in 0~10000. The comparison of time consuming is shown in Fig. 10.13 and Fig.10.14.

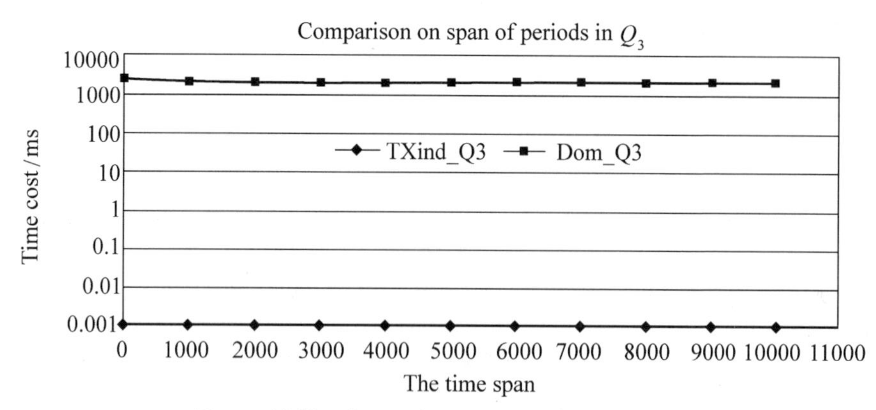

Figure 10.13 Comparison on span of periods in Q_3

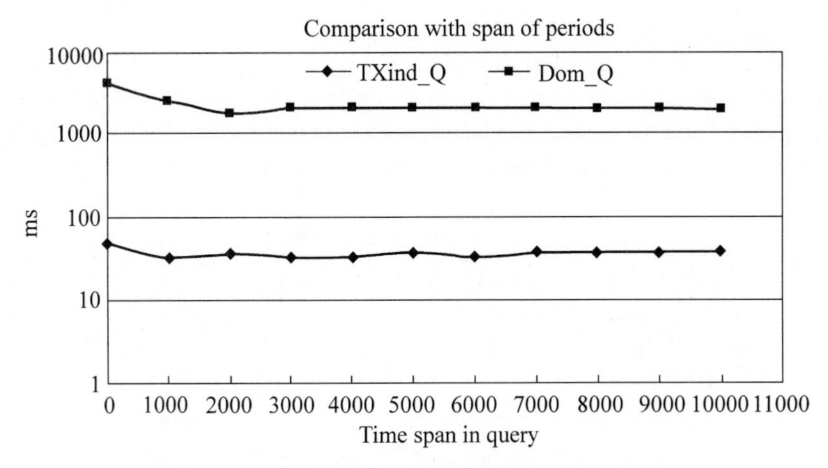

Figure 10.14 Comparison on span of periods in Q_6

10.5.2.3 Comparison between Tind and Txidm

In this section, *TXind*(*D*) will be compared with the temporal connection based XML index model *Txidm*(*D*). We can see that the performance of *TXind*(*D*) is superior to that of *Txidm*(*D*), in dealing with massive data. The following comparison uses the same data amount as (Ye et al. 2007). In, *DOM*, *Txidm* and *Tind* represent DOM tree-based traverse, temporal connection relation and linear order-based querying, respectively. The experiment environment is the same, and the maximum number of the temporal XML is 12000.

(1) Value-querying Test

The number of nodes that needs a visit in the same querying process is shown in Fig. 10.15.

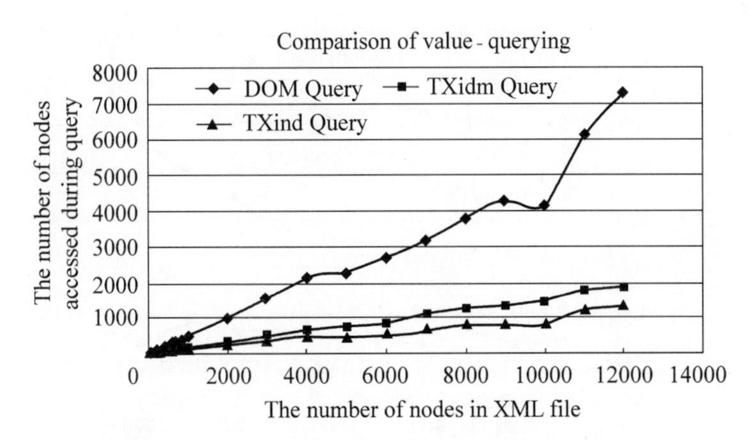

Figure 10.15 Visited nodes in the same querying process

(2) Path querying based Test

In (Ye et al. 2007), the querying of the absolute path and the relative path with temporal constraints are carried out in two steps. The first step begins with the deepest nodes and the second step begins with search of the interior nodes. The total time cost is the sum of these two parts. The comparison is shown in Fig. 10.16.

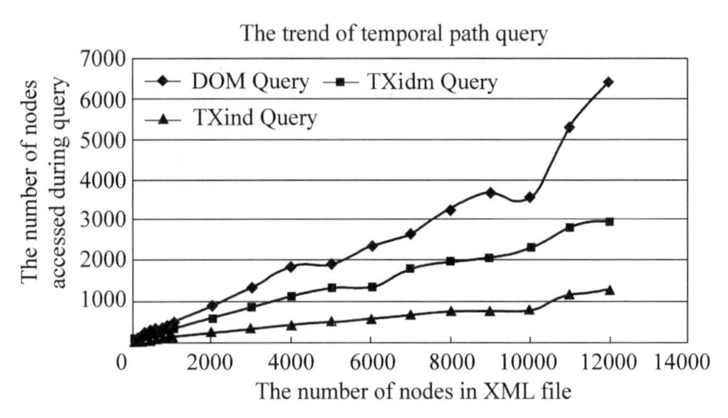

Figure 10.16 Comparison in path-querying

(3) Comparison on time spans and querying efficiency

The querying efficiency of *TXidm* may be decreased with the increase in the temporal period's span. In the simulation, let the time span of the leaf nodes change from 100 to 500, the corresponding interval is 50. The time span of the interior nodes changes from 200 to 1200 and the interval is 100. The total number of the nodes is 3004, creating 10 querying data for every time span and querying 100 times for every querying datum. The average number of the nodes being accessed in such querying is used to compare. The comparison is shown in Fig. 10.17.

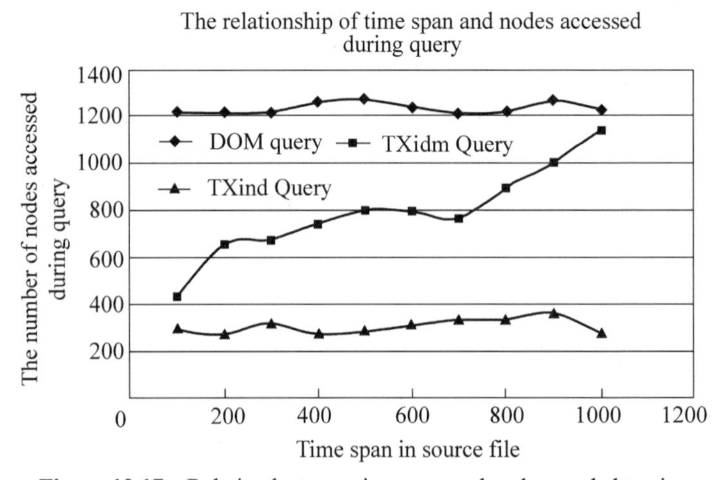

Figure 10.17 Relation between time span and nodes needed to view

References

[1] Alberto O. Mendelzon, Flavio Rizzolo, Alejandro Vaisman A (**2004**) Indexing Temporal XML Documents. Proceedings of the 30th VLDB Conference. 2004, 30: 216 – 227

[2] Gao. C., Snodgrass. R (**2003**) Temporal slicing in the evaluation of XML queries. Proceedings of the 29th International VLDB Conference. 2003: 632 – 643

[3] Goldman R, Widom J (**1997**) Dataguides: enabling query formulation and optimization in semistructured database. In: VLDB pp 436 – 445

[4] Flavio R, Alejandro A. V. (**2008**) Temporal XML: modeling, indexing, and query processing. The VLDB Journal. 2008.8, 17(5): 1179 – 1212

[5] Milo T, Suciu D (1999) Index structures for path expressions. In: Proceedings of the 7th International Conference on Database Theory, pp 277 – 255

[6] Zou QH, Liu SR, Chu WW (**2004**) Ctree: a compact tree for indexing xml data. In: WIDM'04, November 12 – 13, 2004, Washington, DC, USA

[7] Capitani DS (**2002**) An authorization model for temporal XML documents. In: Proceedings of SAC'02, pp 1088 – 1093

[8] Amagasa T, Yoshikawa M, Uemura S (**2000**) A temporal data model for XML documents. In: Proceedings of DEXA Conference, pp 334 – 344

[9] Dyreson CE, Bolen MH, Jensen CS (**1999**) Capturing and querying multiple aspects of semistructured data. In: Proceedings of the 25th VLDB Conference, pp 290 – 301

[10] Wang F, Zaniolo C (**2003**) Temporal queries in XML document archives and web warehouses. In: Proceedings of 10th International Symposium on Temporal Representation and Reasoning, Carirns, Australia, pp 47 – 55

[11] Wang F, Zaniolo C (**2004**) XBiT: an XML-based bitemporal data model. In: Proceedings of the 23rd International Conference on Conceptual Modeling, Shanghai, China, pp 810 – 824

[12] Wang F, Zhou X, Zaniolo C (**2005a**) Efficient XML-based techniques for archiving, querying and publishing the histories of relational databases. In: Time Center Technical Report http://www.cs.auc.dk/TimeCenter

[13] Wang F, Zhou X, Zaniolo C (**2005b**) Temporal XML? SQL strikes back! In: Proceedings of the 12th International Symposium of Temporal Representation and Reasoning (Time'05), Burlington, USA, pp 47 – 55

[14] Mendelzon AO, Rizzolo F, Vaisman A (**2004**) Indexing temporal XML documents. In: Proceedings of the 30th VLDB Conference, Toronto, Canada, pp 216 – 227

[15] Vaisman A, Mendelzon AO, Molinari E, Tome P (**2007**) Temporal XML: data model, query language and implementation. http://www.cs.toronto.edu/~avaisman/papers.html

[16] Campo M, Vaisman A (**2006**) Consistency of temporal XML documents. XSym 2006, LNCS 4156, pp 31 – 45

[17] Ye XP, Chen KY, Tang Y (**2007**) Techenology on temporal XML indexing (in Chinese with English abstract). Chinese Journal of Computers 30(7): 1074 – 1085

Part IV Temporal Database Management Systems

- Implementation of Temporal Database Management Systems
- Improvement and Extension to ATSQL2
- Design and Implementation of TempDB

11 Implementation of Temporal Database Management Systems

Hai Liu[1,2], Yong Tang[1,2+], Xiaoping Ye[1,2], and Huan Guo[2]

[1] Computer School, South China Normal University, Guangzhou 510631, P.R. China
[2] Department of Computer Science, Sun Yat-sen University, Guangzhou 512075, P.R. China

Abstract Dealing with temporal information has become a key technique in many database systems and information systems. We are in great need of a temporal database management system (TDBMS) to handle temporal data efficiently. This chapter gives a brief introduction to TDBMS and introduces two temporal data processing prototype systems—TimeDB and TempDB. The design principles, installation and user interfaces are presented. Several examples are given to demonstrate the basic functions of TDBMS.

Keywords *TDBMS, TimeDB, TempDB*

11.1 Introduction

There are already abundant temporal database theories including several temporal data models and query languages, which cover various characteristics of temporal data and the basic problems under consideration. Among these theories, there are several representative temporal query languages, such as TempSQL (Yau and Chat 1991), TQuel (Snodgrass 1987), TSQL2 (Snodgrass 1995), and ATSQL2 (Böhlen et al. 1995). ATSQL2 is a temporal extension for SQL and a representative language that integrates the spirits of TSQL2, Chronolog (Böhlen 1994; Böhlen and Marti 1994), and Bitemporal ChronoSQL (Pulfer 1995). However, the implementation techniques of temporal databases are relatively lagging behind the theoretical achievements. Prior to the design and implementation of a temporal data processing prototype system, we should first know what a temporal data management system (TDBMS) is and what functions it should include. In general, a TDBMS is a software system that helps users to store, access and maintain temporal data in a uniform way, that is, to manage time-varying data. TDBMS

+ Corresponding author: issty@mail.sysu.edu.cn

should provide:

(1) A temporal data definition language
(2) A temporal data manipulation language
(3) A temporal query language
(4) Temporal constraints (such as temporal referential integrity)

TimeDB (Time Consult; Steiner 1998) and TempDB (TempDB) are two of the famous temporal data processing prototype systems. Both of them are based on ATSQL2. The organization of this chapter is as follows: Section 11.2 introduces the functions of TimeDB, Section 11.3 covers the operations of TempDB.

11.2 TimeDB

TimeDB, which is developed by Andreas Steiner, is considered as a successful temporal database prototype system. TimeDB is a temporal relational database system based on the query language ATSQL2 and it runs as a front-end of the commercial relational database management system (e.g. Oracle). As a relatively mature TDBMS, TimeDB realizes the basic temporal management functions: temporal query, temporal update, temporal view and temporal integrity constraints. In TimeDB, ATSQL2 statements (queries, updates and assertions) are compiled into (sequences of) SQL-92 statements that are executed by the background database. This approach guarantees high portability between different platforms and different DBMS back-ends. TimeDB provides original temporal statements. TimeDB excels due to its seamless integration of time into database, and its upward compatibility and temporal upward compatibility, whilst it supports query language, data manipulation language, data definition language and assertions/constraints. Version 1 of TimeDB was written using SICStus Prolog and was ported to run with SWI Prolog. Version 2 of TimeDB is based on Java, uses JDBC, has a set of APIs and offers more functionality. Full version of TimeDB 2.0 is a commercial version, but TimeDB 2.0 Beta 4 is available for free download for research and study purposes. In the following sections, we will introduce TimeDB based on TimeDB 2.0 Beta 4.

11.2.1 Installation

1. Software requirements

Before running TimeDB, the following software is needed:

- Java Runtime Environment 1.1 (or newer)
- A DBMS, e.g., Oracle (Version 8), Sybase (Version 11.5)
- A JDBC driver for the DBMS

You will need the login and password for the database you will use, the JDBC driver name and the URL to connect to your database. In addition to the above, if you plan to develop applications that access temporal data via TimeDB, you also need a Java development kit (JDK).

2. Setup TimeDB 2.0 Beta 4

In the windows command console, enter the directory: 〈TimeDB_Home〉\class (suppose 〈TimeDB_Home〉 = C:\TimeDB2.0 B4. The driver's directory is C:\JDBCDriver\Oracle8i). Enter the following commands: Java-classpath C: \TimeDB2.0 B4\classes; C:\JDBCDriver\Oracle8i\calsses12.zip.

Then the following user interface of TimeDB2.0 will appear, as shown in Fig. 11.1.

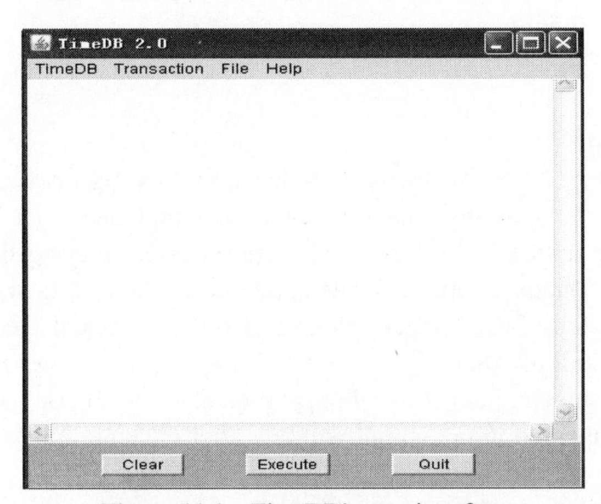

Figure 11.1 TimeDB's user interface

Click TimeDB|Preferences, then a Preferences dialog will appear, as shown in Fig. 11.2.

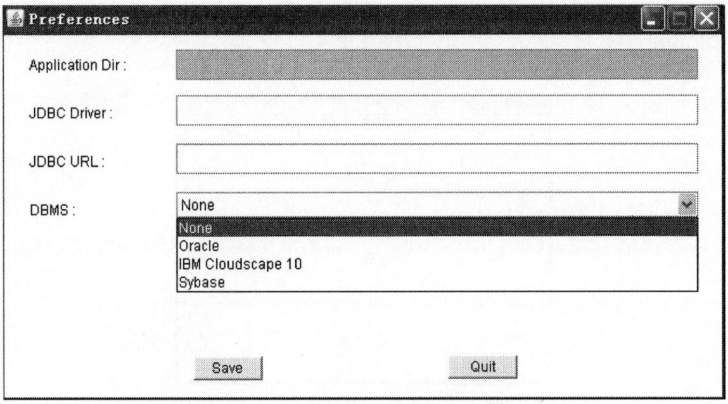

Figure 11.2 Attributes setting interface

Click the drop-down button under DBMS text box, then select the background DBMS. After the selection of the DBMS, the corresponding driver and its path will be added in the JDBC driver and JDBC URL text box by the system.

11.2.2 TimeDB 2.0 Beta 4's User Interface

This section gives a short introduction to TimeDB's user interface (UI).

1. User interface

TimeDB's UI is shown in Fig. 11.1, which can be divided into the following parts: the Title Bar, the Menu Bar, the Edit Area, and the three buttons with clear, execute and quit functions.

2. Menu bar

(1) TimeDB menu

As shown in Fig. 11.3, this menu includes the following items:

- Open DB: After clicking this item, a dialog that needs an input of user's password will appear. Click the "OK" button after entering the password. If you see that "Database opened" has appeared in the output window, then the database is successfully opened, else you need to check if the user name and password are still useable.
- Close DB: This menu item closes the specified database. If you see that "Database closed" has appeared in the output window after clicking this item, the database is successfully closed.
- Create DB: Before operating on the database, you need to initialize the database, that is, add auxiliary table to the existing database. Then you can make temporal data operations on this initialized database. If you see that "Database initialized" has appeared in the output window, then the database is successfully initialized.
- Clear DB: This menu item clears all the meta tables (auxiliary data table) in the database. If you see that "Database cleared" has appeared in the output window, then the database is successfully cleared.

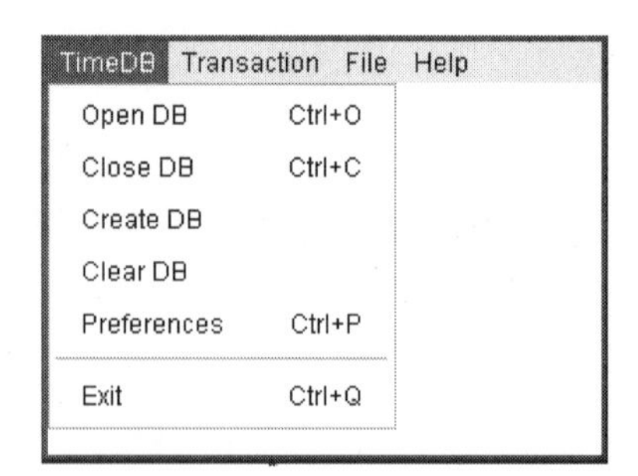

Figure 11.3 TimeDB menu

- Preferences: Set application program's path, JDBC's driver, JDBC URL and DBMS.
- Exit: This menu item exits the software.

(2) Transaction menu

As shown in Fig. 11.4, this menu includes the following items:

- Rollback: This operation makes the database recover to the last submitted state and the output window will show "Rollback done".
- Commit: Once executing modification operation on database, you need to use the "Commit" manipulation to commit the corresponding operation. Otherwise, after the rollback operation, the current modification will be lost. After the commit operation, the output window will show "Commit done".

  Note: Commercial databases are normally set as "Auto Commit" by default.
- AutoCommit: This operation automatically commits changes. This option is set to be selected by default. If you click it under unselected status, the output window will show "Auto Commit set on".

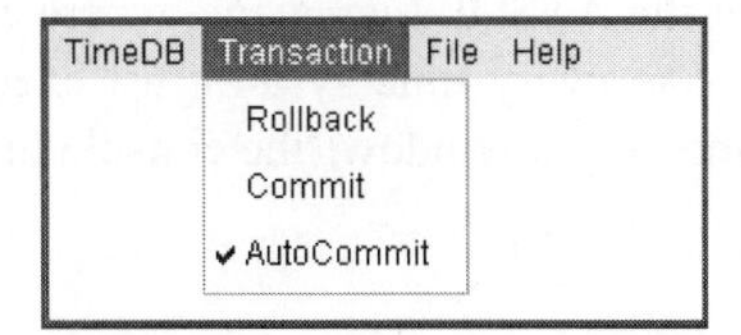

Figure 11.4 Transaction menu

(3) File menu

As shown in Fig. 11.5, this menu includes the following items:

- Execute File: A file selection dialog will pop up after selecting it. Once you select the file that will be executed, the execution results will be shown in the output window.
- Test File: A file selection dialog will pop up after selecting it. Once you select the test file, the execution results will be stored in a file with the same name, with ".out" as its suffix. The system will also compare the execution results with a selected reference file with suffix ".ref" in the database. If the content in both files are identical, system will point out that they have no difference. If the result file does not exist, system will report an open file error.

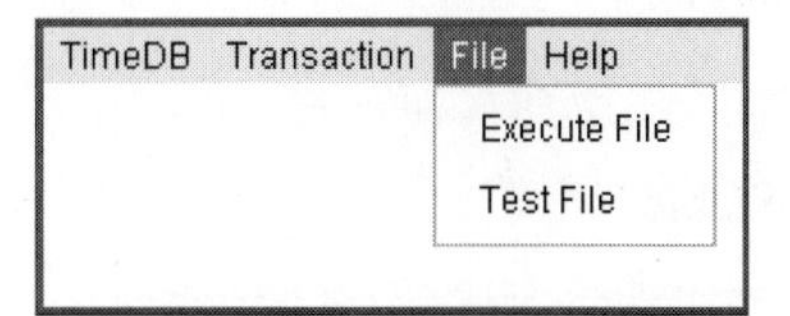

Figure 11.5 File menu

(4) Help item

As shown in Fig. 11.6, this menu includes the following items:

- About: This menu item opens a prompt box that displays TimeDB's authors, version number and the introduction to the version, etc.
- Grammar: This menu item opens a read-only text box, which records the ATSQL2 grammar used in TimeDB.
- Menu: This menu item opens a read-only text box, which provides the introduction about function of every button in menu.

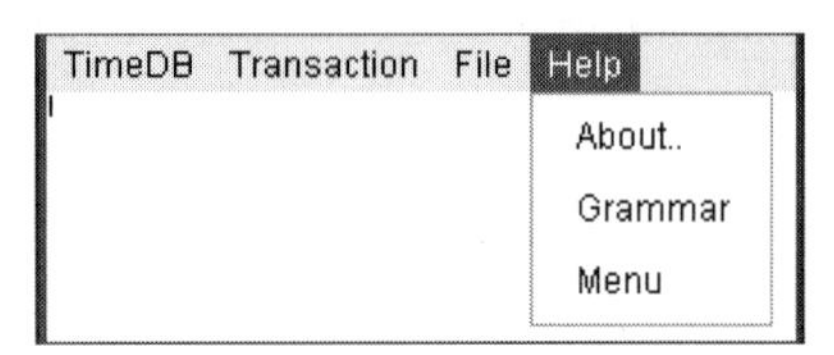

Figure 11.6 Help menu

3. Edit area

This area is used to input the ATSQL2 language, where every operation segment should be ended by ";", and every time system just executes one segment. The result will be shown in the output window (the console that appears after running the ". bat" file).

4. Buttons

- Clear button: Clears all the contents in edit area.
- Execute button: Executes the first program segment in the edit area.
- Quit button: Exits TempDB.

11.2.3 Examples

TimeDB is a front-end software of commercial DBMS. When users operate on non-temporal data, TimeDB is transparent to the users. However, when referring to temporal data, TimeDB's operation is easier than commercial DBMS. This is because commercial DBMS normally supports only TSQL language, but TimeDB supports ATSQL2. In the following section, we will prove the difference by an example based on Table 11.1.

1. Use Oracle 8i

(1) Create a database table

```
CREATE TABLE Employee (EmpID integer,
                       Name varchar(30),
                       Departmant varchar(30),
                       Salary integer,
                       ValidTimeStart data,
                       ValidTimeEnd data);
```

Table 11.1 Employee

EmpID	Name	Department	Salary	ValidTimeStart	ValidTimeEnd
10	Jemmy	Research	2000	1991	1995
10	Jemmy	Sales	3000	1995	1998
10	Jemmy	Sales	6000	1998	2002
11	Paul	Research	4000	1993	2000
12	George	Research	4000	2001	2002

(2) Insert data

```
INSERT INTO Employee VALUES (10,'Jemmy', 'Reaearch', 2000, '1991-01-01',
'1995-01-01');
```

(3) Query

Get all the people's work department and corresponding salary in 1998.

```
SELECT Name, Department, Salary
FROM Employee
WHERE ValidTimeStart<='1998-01-01' AND ValidTimeEnd >'1999-01-01'
```

The query results are:

```
NAME    DEPARTMENT      Salary
----------      ------------
Jemmy   Sales            6000
Paul    Research         4000
```

2. Use TimeDB

(1) Create a database table

```
CREATE TABLE Employee (EmpID integer,
                    Name varchar(30),
                    Departmant varchar(30),
                    Salary integer) AS VALIDTIME;
```

(2) Insert data

```
VALIDTIME PERIOD [1991-1995) INSERT INTO Employee VALUES (10,'Jemmy',
'Research', 2000);
```

(3) Query

Get all the people's work department and corresponding salary between the year 1998 and 1999.

```
VALIDTIME PERIOD [1998-1999)
SELECT Name, Department, Salary
FROM Employee
```

The query results are:

```
Validtime     NAME   DEPARTMENT    Salary
--------------------   ------------------
[1998-1999]  Jemmy   Sales         6000
[1998-1999]  Paul    Research      4000
```

By this example, you can see that TimeDB's processing for data with temporal information is easier than traditional commercial DBMS.

11.3 TempDB

The temporal process prototype **TempDB** is designed and implemented by the temporal development group of Co-soft R&D Center of Sun Yat-Sen University. The time process system is similar to that of TimeDB. TimeDB achieves success to a certain extent and supports a good reference for subsequent development of temporal database system. However, there are also limitations in these systems, mainly due to the following aspects: low performance, efficiency, compatibility and transportability, and the restriction for multi users' parallel usage. These shortages urge researchers to do further work. To meet the requirements of temporal database's development and with reference to the design principle of TimeDB, we designed and implemented an optimized temporal data process component, TempDB. This contains more matured functions. TempDB adopted the middleware architecture and the representative and improved ATSQL2 language. TempDB realized the following functions: lexical and syntax analysis, semantic processing, temporal DDL, temporal DML, temporal Query, results packing, partial temporal completeness constrains and temporal index MAP21 (Nascimento 1996) with the functions of *Now* semantic bound to improve the query performance. Version 2.01 of TempDB is based on Mysql, written in Java. It provides many programming APIs. Users can interact with TempDB using these APIs when building their own applications. TempDB can be downloaded free at the following URL: http://www.scholat.com/team.html?teamId=36.

In the following sections, we will give a brief introduction to TempDB.

11.3.1 Installation

1. System requirements

The following requirements must be met before installing TempDB.

- 32 bit Windows Operating System, including NT, 2000, XP or 2003
- Java Runtime Environment—JDK1.5 or above
- DBMS systems, currently MySQL5.0 or above is recommended
- TCP/IP Protocol

2. TempDB's installation

- Double click installation program to activate the set up program. The interface is shown in Fig. 11.7.

Figure 11.7 Welcome interface

- After self-extraction, the "welcome" interface appears. Click "Next" to continue. The procedure is complete.

11.3.2 TempDB's User Interface

The layout of TempDB UI is shown in Fig. 11.8, which can be divided into the following parts: Title Bar, Menu Bar, Toolbar, ATSQL Command Editor, Query Result Display, Tables Display and System Info. Display Area.

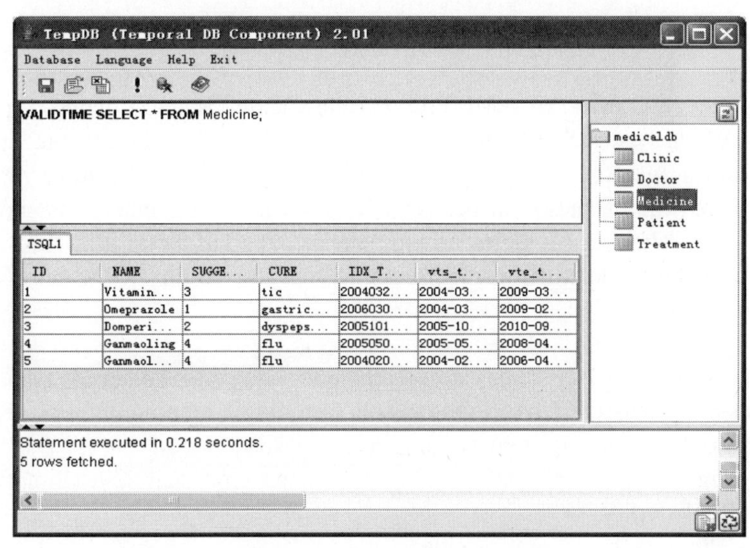

Figure 11.8 User interface

1. Menu Bar

Menu Bar includes menu items Database, Language, Help and Exit.

- Database -> Connection Configuration: Prompts the connection configuration dialog, which can configure the connection to backend DBMS, as shown in Fig. 11.9. The information in this dialog includes:
  - Server IP: the server's address where DBMS is installed, and the port number. The format is "Server IP: Port Number".
  - DB Name: the database name to connect to in the backend DBMS, or the schema name.
  - User Name: user name that is used to establish connection to backend DBMS.
  - Password: password of corresponding user name, displayed as mask in the dialog.

Note: The privilege set of user names in the backend DBMS will restrict the functionality of TempDB. Ensure that the user account has enough privileges to perform related operations.

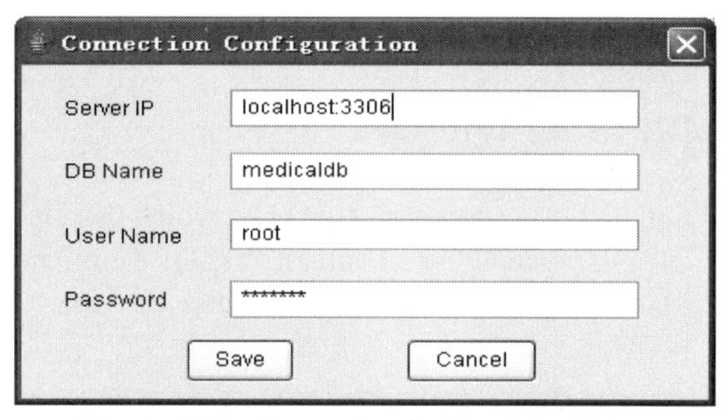

Figure 11.9 Connection configuration dialog

- Database -> Create Database: Prompts the Create Database dialog. Input the name of database (schema) into textbox, click "Create", and a console interface and a message box will be prompted. The system requires that after you have completed the execution in the console, click the "OK" button, as shown in Fig. 11.10. The user is required to input the password of root user in the console. Press any button to exit console program when you are finished. Press "OK" to close dialog.

  Note:
  - For MySQL DBMS, root password is required to create database (schema).
  - Do not press "OK" when the console program is running. Otherwise, error will occur in the program running in the console.
- Language: Two languages are supported at present: "Chinese (Simplified)" and "English (US)".
- Help: Prompts the "About" dialog.
- Exit: Exits the program.

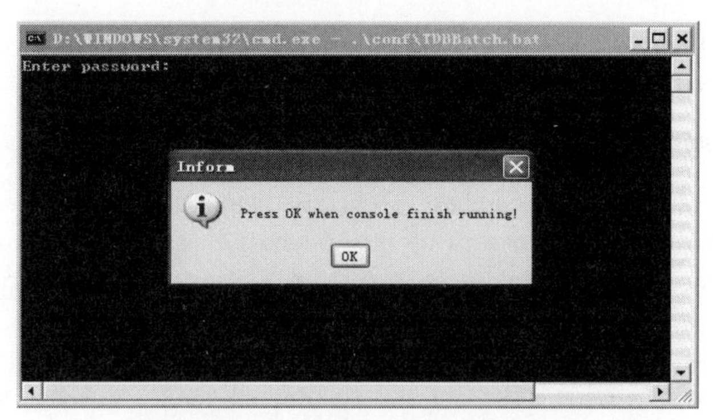

Figure 11.10 Create database (schema)

2. Toolbar

As shown in Fig. 11.11, buttons on toolbar are as follows:

- Save File: Saves the content in ATSQL Command Editor to files.
- Open File: Opens a text file and loads the content into ATSQL Command Editor.
- Close File: Closes the currently opened file.
- Execute: Executes all statements in ATSQL Command Editor.
- Cancel Execution: Stops statements that are currently being executed.
- Help: Opens the User Guide.

Figure 11.11 Toolbar

3. ATSQL Command Editor

ATSQL statements can be edited in ATSQL Command Editor. More than one statement can be input in this editor. Semicolons are used to separate each statement. Keywords will be highlighted in the editor, as shown in Fig. 11.12.

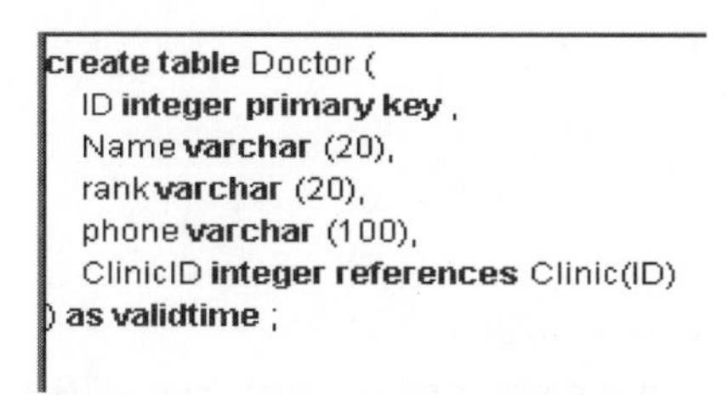

Figure 11.12 ATSQL command editor

4. Query Result Display

The results of the query statements will be displayed in this section. If there is more than one statement in the ATSQL Command Editor, more than one tab will be displayed in this section after execution, in the corresponding order of statements in ATSQL Command Editor.

5. Tables Display

This section displays the database (schema) name that TempDB currently is connecting to and all the names of recognized tables. Double clicking a table name can generate query statement in ATSQL Command Editor. Triple clicking a table name will generate a statement and will execute it as well.

Note: There is a refresh button on the upper right corner on this section. After executing a table creation statement, this section will not be refreshed immediately. Click the refresh button to refresh the Tables Display.

6. System Info. Display

This section shows the status information of statement execution. The information includes generated standard SQL statements, internal execution time, execution time, number of rows fetched, and other information.

There are two buttons at the bottom right corner of this section. They are used to clear information of this section and force garbage collection.

11.3.3 Examples

1. Temporal DDL with TempDB

With the TempDB, it is convenient for users to create tables, either temporal or non-temporal ones.

(1) Creating temporal tables

The temporal tables are created in the form of adding two columns, which are used to store valid time in the backend databases. These two columns are unified with the user-specified logic prime key to set the joint prime key.

```
CREATE TABLE customer (
  ID VARCHAR (100) PRIMARY KEY,
  Name VARCHAR (100),
  Address VARCHAR (200),
  Is_Vip INTEGER,
  ServedBy VARCHAR (100)
) AS VALIDTIME;
```

(2) Creating non-temporal tables

The creating of the non-temporal tables with TempDB follows the method of traditional DBMS.

```
CREATE TABLE department (
   DepNo varchar(100) PRIMARY KEY,
   DepName varchar(100),
   Location varchar(100)
);
```

2. Temporal DML with TempDB

Users can conveniently insert data and delete data to and from database by using TempDB. Under the temporal semantics, inserting and deleting operations are given richer connotation than the same operations under traditional non-temporal semantics. Depending on the valid time during which the operations are referenced, either the complex operations such as tuple's splitting or tuple's merging will appear. These operations are judged by the Allen's 13 Temporal Interval Relationships (Allen 1983). In this section, we will only give two examples on the two operations mentioned above.

(1) Inserting data with merging operation

After the following insert operation for Table 11.2, the resulting table is shown in Table 11.3.

```
VALIDTIME PERIOD [DATE "2009-01-01" - DATE "2010-01-01")
INSERT INTO employee VALUES ('0018', 'chenzicong', 10000, '088');
```

Note that the inserted tuple has the same attribute values with one existing tuple except the valid time, thus merging operation is needed.

Table 11.2 Temporal data before inserting

ID	NAME	SALARY	DEPID	vts_timeDB	vte_timeDB
001	zhang jin	2000	0001	1999-09-02 0...	2002-02-28 0...
001	zhang jin	2500	0001	2002-02-28 0...	2004-02-28 0...
001	zhang jin	3000	0001	2005-01-01 0...	2006-01-01 0...
001	zhang jin	3000	0001	2009-01-01 0...	*Now*
0018	chenzicong	10000	088	2006-01-01 0...	2009-01-01 0...
002	wang junwei	1900	0001	1998-02-15 0...	2001-01-30 0...
002	wang junwei	2200	0001	2001-01-30 0...	2004-05-30 0...

Table 11.3 Temporal data after inserting

ID	NAME	SALARY	DEPID	vts_timeDB	vte_timeDB
001	zhang jin	2000	0001	1999-09-02 0...	2002-02-28 0...
001	zhang jin	2500	0001	2002-02-28 0...	2004-02-28 0...
001	zhang jin	3000	0001	2005-01-01 0...	2006-01-01 0...
001	zhang jin	3000	0001	2009-01-01 0...	*Now*
0018	chenzicong	10000	088	2006-01-01 0...	2010-01-01 0...
002	wang junwei	1900	0001	1998-02-15 0...	2001-01-30 0...
002	wang junwei	2200	0001	2001-01-30 0...	2004-05-30 0...

(2) Deleting data with splitting operation

After the following deleting operation on the relation as listed in Table 11.3, the resulting table is shown in Table 11.4.

```
VALIDTIME PERIOD [DATE "2007-01-01" - DATE "2008-01-01")
DELETE FROM employee
WHERE ID='0018';
```

Table 11.4 Temporal data after deleting

ID	NAME	SALARY	DEPID	vts_timeDB	vte_timeDB
001	zhang jin	2000	0001	1999-09-02 0...	2002-02-28 0...
001	zhang jin	2500	0001	2002-02-28 0...	2004-02-28 0...
001	zhang jin	3000	0001	2009-01-01 0...	*Now*
0018	chenzicong	10000	088	2006-01-01 0...	2007-01-01 0...
0018	chenzicong	10000	088	2008-01-01 0...	2010-01-01 0...
002	wang junwei	1900	0001	1998-02-15 0...	2001-01-30 0...
002	wang junwei	2200	0001	2001-01-30 0...	2004-05-30 0...

This example shows the splitting operation that is caused by temporal deleting statement. The former tuple whose ID is 0018, with valid period [2006-01-01 00:00:00, 2010-01-01 00:00:00], has been split into two new tuples with the same attribute values as the split tuple except the valid time. This is because of the operation of deleting the entire 0018 employee's information between "2007-01-01 00:00:00" and "2008-01-01".

3. Examples of TempDB's temporal query

TempDB supports snapshot query, sequenced query and non-sequenced query. We will only give a special query example: query of historical changes in non-sequenced query.

Under the non-sequenced semantics, the valid time columns are treated as common user-defined columns, so special treatment to the two time columns can be avoided, e.g. temporal table connections using the intersection of valid time periods. Based on this observation, user can perform some special queries, such as historical changes with the help of non-sequenced semantics, while the same query under sequenced semantics will return an unexpected result because of additional restrictions.

The query result of the following query statement for Table 11.5 is shown in Table 11.6.

```
NONSEQUENCED VALIDTIME
SELECT emp.ID, emp.Name, emp.Salary AS BeforePromoteSalary, emp2.Salary
AS AfterPromoteSalary, emp.DepID, END (VALIDTIME (emp)) AS PromoteTime
FROM employee emp, employee emp2
WHERE emp.ID = emp2.ID AND emp.Salary < emp2.Salary AND
      VALIDTIME (emp) MEETS VALIDTIME (emp2);
```

Table 11.5 Employee table

ID	Name	Salary	DepID	vts_timeDB	vte_timeDB
001	Zhang Jin	2000	0001	1999-09-02 00:00:00	2002-02-28 00:00:00
002	Wang Junwei	1900	0001	1998-02-15 00:00:00	2001-01-30 00:00:00
001	Zhang Jin	2500	0001	2002-02-28 00:00:00	2004-02-28 00:00:00
002	Wang Junwei	2200	0001	2001-01-01 00:00:00	2004-05-30 00:00:00
001	Zhang Jin	3000	0001	2005-01-01 00:00:00	9999-01-01 00:00:00
002	Wang Junwei	2600	0001	2004-05-30 00:00:00	9999-01-01 00:00:00

Table 11.6 Result of the historical non-sequenced query

ID	Name	BeforePromote Salary	AfterPromote Salary	DepID	PromoteTime
001	Zhang Jin	2000	2500	0001	2002-02-28 00:00:00.0
002	Wang Junwei	1900	2200	0001	2002-01-30 00:00:00.0
002	Wang Junwei	2200	2600	0001	2004-05-30 00:00:00.0

This example clearly indicates the traits and advantages of historical non-sequenced query. When using sequenced query, by removing the keyword "NONSEQUENCED" from the ATSQL2 statement, the outcome will be *null* because of temporal table's self-connection.

11.4 Comparing TimeDB with TempDB

TimeDB realizes only the basic functions of temporal data processing, leaving many shortages in this system. TempDB is the first project for a complete software component based on temporal data model in China, which brings many improvements on TimeDB. Comparing TimeDB with TempDB, there are the following similarities and differences as shown in Table 11.7.

Table 11.7 Comparing TempDB 2.01 with TimeDB 2.0 Beta 4

	TempDB	TimeDB
Valid time	Fully supported	Fully supported
Transaction time	Not supported	Not supported
Temporal integrity	Fully supported as a core function	Partially supported
Temporal merger	Supports run-time merging and merging when updating	Supports run-time merging
Temporal variable *Now*	Supported	Not supported

(Continued)

	TempDB	TimeDB
Architecture	Clear division of the function modules	Functions between every module are relatively confusing
Software interface	Has a unified graphical interface to manage database and statement detection on execution	Has a graphic input interface, but no graphic output interface for query results; no detection function for statement execution
Temporal language standardization	With definition of a strict subset of ATSQL2, software's process modules are clear and standardized	Under a fuzzy temporal language, software's process modules are complex
Query performance optimization	Performance has not been set as the core requirements. TempDB reduces and systematically manages the interaction with background database	Responsibility division with background database management system is fuzzy, resulting in inefficient processing

We can see that TempDB is superior to TimeDB in the integrity of temporal data, temporal variable *Now* supporting, architecture design, and temporal language's standardization compliance, etc.

TempDB also provides a set of programming APIs for users, which allows users to interact with TempDB, using these APIs when implementing their own applications. The specific APIs can be seen in Appendix.

References

[1] Allen JF (**1983**) Maintaining knowledge about temporal intervals. Communications of the ACM 26(11): 832 – 843

[2] Böhlen MH (**1994**) Managing temporal knowledge in deductive databases. PhD thesis, Department Informatik, ETH Z urich

[3] Böhlen MH, Jensen CS, Snodgrass RT (**1995**) Evaluating and enhancing the completeness of TSQL2. Technical Report TR 95-5, Computer Science Department, University of Arizona

[4] Böhlen MH, Marti R (**1994**) On the completeness of temporal database query languages. In: Proceedings of the First International Conference on Temporal Logic, pp 283 – 300

[5] Nascimento M (**1996**) Efficient indexing of temporal databases via B+-trees. Ph.D. dissertation, School of Engineering and Applied Science, Southern Methodist University

[6] Pulfer D (**1995**) Optimierung von temporalen Queries. Master's thesis, Institute for Information Systems, ETH Zurich

[7] Snodgrass RT (**1987**) The temporal query language TQuel. ACM Transactions on Database Systems (TODS), 12(2): 247 – 298

[8] Snodgrass RT (**1995**) The TSQL2 temporal query language. Kluwer Academic Publishers, Norwell

[9] Steiner A (**1998**) A generalisation approach to temporal data models and their implementations. Ph.D. dissertation, Switzerland: Federal Institute of Technology

[10] Time Consult. http://www.timeconsult.com/

[11] TempDB. http://www.cosoft.sysu.edu.cn/TempDB/index.asp

[12] Yau C, Chat GSW (**1991**) TempSQL—A language interface to a temporal relational model. Inf Sci Tech: 44 – 60

12 Improvement and Extension to ATSQL2

Huan Guo[1], Yong Tang[1,2+], Xiaoping Yang[1], and Xiaoping Ye[1,2]

[1] Department of Computer Science, Sun Yat-sen University,Guangzhou 512075,P.R. China
[2] Computer School, South China Normal University, Guangzhou 510631, P.R.China

Abstract As the main interface for the interaction between user and temporal database, temporal query language determines temporal database's functionality and its implementation details. Thus, the choice of query language is a decisive step for the design of a temporal database. Current theoretical results of temporal database include various types of temporal data models and query languages. Typical query languages include TempSQL, Tquel, TSQL2 and ATSQL2. The query language chosen by the processing component of the temporal data TempDB is the improved ATSQL2. To explain ATSQL2 language accurately, it is necessary to study its syntax and semantics. As ATSQL2 has its limitations, a detailed description of the improved ATSQL2 is proposed in this chapter.

Keywords *ATSQL2, TempDB, improvement, extension*

12.1 Introduction

Structured Query Language **SQL** (Date 1989, Melton and Simon 1993), which is a language based on relational algebra and relational calculus, was proposed in 1974. It is very popular in the computer industry and the computer users for its rich functionality, flexibility in involving methods and its concise grammar. Its functions include data query, manipulation, definition and control. It is a general and powerful standard language of Relational Database. In October1986, ANSI approved SQL to be the American standard of relational database language. In June 1987, ISO adopted it as the International Standard, which is also called "SQL86".

In 1994, Richard Snodgrass began to cooperate with ANSI and ISO SQL3 committee. He added temporal support to SQL3 (SQL 1993) standard, which is called **SQL/Temporal**. Its main idea was to add TSQL2 language standard to SQL3. In 1994, M. Bohlen and R. Marti proposed the concepts of temporal

+ Corresponding author: issty@mail.sysu.edu.cn

semi-completeness and temporal completeness (Böhlen and Marti 1994, Böhlen 1994). In 1995, the theory was applied to evaluate the completeness of TSQL2 (Snodgrass 1995), which led to the redesign of the TSQL2. The work was finished by Michael Bohlen, Christian Jensen and Richard Snodgrass, and they named the new language **ATSQL2** (Applied TSQL2) (Böhlen et al. 1995). In 1996, ATSQL2 was accepted by ANSI (Snodgrass et al. 1996a, 1996b).

ATSQL2 is the extension of SQL3, which defines the grammar and semantics of temporal data query statements, temporal data alter statements, temporal data definition statements, temporal constraint and assertion. As the expansion of SQL3, ATSQL2 must first consider the compatibility of the two languages. Böhlen et al. (1995) proposed the requirements of **upward compatibility**. It points out that SQL3 language must be the subset of ATSQL2, so that every SQL3 statement in ATSQL2 has the same meaning as in SQL3. On the other hand, we should take into account that when migrating non-temporal databases into Temporal DBMS, the implementation of the original SQL query and alter statements, which are executed on current valid data, must have the same semantics as their original ones. This property is called "**temporal upward compatibility**" (Snodgrass et al.1996a).

In ATSQL2, according to different semantic expressions, temporal statements are divided into two categories: **sequential statements** and **non-sequential statements.** The sequential statement has snapshot reducible semantics. So the executing results of sequential query statement is equivalent to the cumulated results obtained by executing the corresponding non-temporal query statement on a series of snapshot states of Temporal Database.

Sequential statements always use the assumption that comparison and other operations can only be executed when the corresponding data exists at the same time point. However, in the real world we normally need to compare data of different time points, for example, comparing different states of the database or querying the change history of database. To meet this requirement, ATSQL2 provides non-sequential statements, which use multiple states of the database to get the results.

The content of this chapter is organized as follows. Section 12.2 discusses the properties of ATSQL2, Section 12.3 proposes the interpretation of semantics in ATSQL2 and Section 12.4 presents the improvements of ATSQL2.

12.2 Study on ATSQL2

12.2.1 Requirements and Expatiation

The design requirements of ATSQL2 are as follows:

(1) upward compatibility;

(2) temporal upward compatibility;

(3) orthogonal treatment of valid time and transaction time;

(4) similar grammar to SQL3;

(5) support of the statements of sequential semantics and non-sequential semantics;

(6) capable of using the statements of sub-queries to replace the corresponding tables (including the exported tables that appear in the "from" clauses and the sub-queries that appear in the "where" clauses);

(7) support of the compare predicate for five time intervals and two functions that were proposed by (Allen 1983);

(8) implement time interval merging operation.

In order to achieve the above design requirements, ATSQL2 extends SQL in four aspects as follows:

(1) Adding marked strings before statements (valid time, non-sequential valid time, or no prefix) to distinguish four different semantics (sequential semantics, non-sequential semantics, upward compatibility, temporal upward compatibility).

(2) Support of exported tables, whose status is equivalent to "view". To implement the main query, a database needs to evaluate on the exported table. Then it is required to supply the main query with the obtained intermediate result.

(3) Support of the compared predicate for five time intervals and converting it into the one for the corresponding time points compare. The "begin ()" and "end ()" functions should be implemented also to get the beginning and ending time stamp of a time interval, respectively.

(4) Adding the suffix "period" to the statement in order to merge the time intervals of table.

ATSQL2 not only extends standard SQL statements, which are most commonly used, but also retains and extends some more complex language features. These features include the support of multi-level nested sub-queries, merging operation for query results and temporal constraints. This language was generally accepted by the public for its excellent grasp of standard SQL (compatibility and extension).

In short, ATSQL2 is derived from TSQL2, based on the extension of standard SQL language. As it is well designed, it was accepted by ANSI and it has established its important status in temporal database.

12.2.2 Properties of ATSQL2

Building on standard SQL, ATSQL2 is a representative temporal query language, which integrates TSQL2, ChronoLog (Böhlen et al. 1994, Böhlen 1994), and Bitemporal ChronoSQL (Pulfer 1995). It has the following characteristics:

(1) ATSQL2 extends standard SQL. It is compatible with SQL, so it can support traditional applications. It is also convenient for users who are familiar with SQL to use ATSQL2.

(2) ATSQL2 integrates many temporal query language features. It supports both the query language, which supports temporal integrity, and the description

of the temporal integrity constraints, which are the requirements of a complete temporal database system.

(3) ATSQL2 supports bitemporal query, i.e., it supports the description of valid time and transaction time. It also supports the orthogonalization of valid time and transaction time. As a result, the realization of valid time and transaction time can be achieved in stages.

(4) ATSQL2 has been well accepted by researchers in the related fields and has been recommended to be the language standard for SQL/Temporal. ANSI has accepted it as the standard for SQL/Temporal.

(5) The representative temporal database system TimeDB (Time Consult, Steiner 1998) uses ATSQL2 as its query language, which can provide implementation reference for the development of TempDB.

The characteristics mentioned above of ATSQL convinced us to choose it as the temporal query language for TempDB. It should be noted that the language finally adopted by TempDB is not standard ATSQL2 language, but a more strict and concise language based on ATSQL2. Its complete definition of BNF grammar is in the Appendix. Because it is adapted from ATSQL2 and keeps the core content, we continue to use the original name, ATSQL2.

1. Upward compatible query

Upward compatibility refers to the normal use of non-temporal SQL statements, non-temporal views and tables in the temporal DBMS, just as operating on a temporal DBMS for users. Upward compatibility ensures a non-temporal database and the operation logic of non-temporal database can remain in use for the benefits of the original developers.

2. Temporal upward compatible query

When migrating a non-temporal database to a temporal DBMS, the non-temporal table should have a temporal expansion (for example, the table will be upgraded to a valid timetable by adding the valid time). After the expansion, new tables will be able to record the valid time or other dimensions of time for the relevant data. In order to ensure the validity of the original query statements, operation's temporal upward compatibility must be provided. This means that the original query can be used as usual, because it has been expanded to be a valid timetable. The evaluation of the query is obtained under the current state of the database (*Now*). The expression of the formula is $q(db) = q(db_{Now})$.

3. Query with sequential semantics

A sequential semantic query consists of operators and referenced temporal tables, which are "snapshot reducible". This means that the relational algebra operators used in the query will explain the valid time fields or the transaction time fields or both. In the premise of sequential semantic, all traditional algebra operators adopt the following assumptions—only when the two records exist at the same time point, the corresponding operation can be executed. In other words, the data is “synchronized”. From the perspective of grammar, sequential semantic query adds prefix “valid”, “transaction”, “valid and transaction” or “transaction and valid” to the query to specify the time dimension involved in the query.

4. Non-sequential semantic query

Non-sequential query allows the comparison between different states of a database, thus we can obtain the state change information of the database. At this point, the time-dimensional field added by the system (for example, field “valid” in the temporal RDBMS) is equal to the user-defined time interval field. Non-sequential query operator does not operate on the time interval, so the relational algebra operators of the query statements led by “non-sequenced” have no temporal semantics. If you want to quote the field “valid” of the temporal table, you must use the form “Valid (tableName)”.

12.3 Interpretation of ATSQL2 Semantics

The BNF definition of ATSQL2 can be divided into three parts: data definition statements, manipulation statements, and query statements. We will give examples to explain the definition of these parts.

12.3.1 Data Definition Statement

The data definition statement of ATSQL2 includes the following functions: create table, create view, delete table and delete view.

1. Table creation

The following statements are part of the BNF definition statements for table creation:

```
ddlTable    ::= 'create' 'table' identifier (tableDef  | ddlQuery)
tableDef    ::= '(' colDefList ')' ['as' 'validtime']
ddlQuery    ::= ['(' colList ')'] 'as' query
colDefList  ::= colDef {',' (colDef | tableConstraint)}
colDef      ::= identifier dataType [columnConstraint]
```

Just like general SQL statements, ATSQL2 statements use the keywords “create table” to create a database table, “identifier” represents the identifier of table name, which has the same naming pattern as commercial databases. There are two ways to create a table—defining a list of fields or using a query to obtain the fields of the new table. Keyword “as validtime” indicates the creation of a temporal table, which supports valid time.

Example 12.1 Create a valid timetable with student ID and name as its fields:

```
create table t1(
       id integer primary key,
       name varchar(200)
       )as VALIDTIME;
```

2. View creation

The following statement is the BNF definition for view creation:

```
ddlView       ::='create' 'view' identifier ddlQuery
```

The definition shows that ATSQL2 can use query statements to create a view.

Example 12.2 Create a view for the table "employee", which contains all the information of workers who left before 1999-9-9:

```
Create view employee_v as Validtime
              select * from employee
              Where VALIDTIME(employee) before date "1999-9-9";
```

Note that VALIDTIME(R) represents the valid time interval of tuple R.

3. Table and view deletion

```
dropTable   ::='drop' 'table' identifier
dropView    ::='drop' 'view' identifier
```

The keywords "drop table" and "drop view" are used to delete tables and views. They have the same semantics as standard SQL language and even they need the corresponding operation privileges of the underlying database.

12.3.2 Data Manipulation Statement

ATSQL2 data manipulation statements include insertion and deletion functions, where insertion operation has two forms: value-based insertion and query-based insertion.

(1) Value-based insertion is defined as:

```
insertByValues       ::=['validtime' 'period' intervalExp] 'insert'
                       'into'identifier 'values'' ( 'valList') '
```

Compared to standard SQL statement, the above statement has time flag "['validtime' 'period' intervalExp]" as its prefix, which is used to represent the valid time interval of the inserted tuple.

Example 12.3 The information of table "staff members", before and after adding staff members' information into it are shown in Table 12.1 and Table 12.2.

Table 12.1 Before record insertion

Employee ID	Name	Department	Valid Time
112562	Wang Linlu	Information department	[2005-4-2, 2008-4-5]

Table 12.2 After record insertion

Employee ID	Name	Department	Valid Time
112562	Wang Linlu	Information department	[2005-4-2, 2008-4-5]
132202	Zhang Jinning	Information department	[2008-1-1, *Now*]

(2) As for query-based insertion, the table name is followed by a query statement, thus insertion operation is executed based on the query results.

(3) Deletion statement is similar to the above definition, which deletes tuples in a specified time range.

12.3.3 Data Query Statement

The query part of ATSQL2 language has a complex definition. This section will combine the definitions of selection, projection and connection in temporal relational algebra to introduce data query statements by stages.

1. Related concepts

- **Snapshot query**: This is a query on current state of the database. For non-temporal table (snapshot database), snapshot query is an upward compatible query method, which can query data of non-temporal database. For temporal table, snapshot query is executed on current state of a database and returns query results on current state.
- **Sequential query**: Sequential query is a temporal query, which adds the prefix keywords "validtime", "transaction", "validtime and transaction" or "transaction and validtime" to indicate the time dimension to which the query operator refers. The system uses Temporal Relational Algebra to process time stamp (the valid time, transaction time or a combination of the two) automatically.
- **Non-sequential query**: When executing non-sequential query, the system does not explain time stamp according to Temporal Relational Algebra. Instead, it just treats them as user-defined attributes. Non-sequential query can meet some special query requirements, for example, queries that refer to state changes in the database.
- **Temporal merging**: Temporal merging is a reconstruction operation for databases, which merges tuples with the same attribute values and adjacent or overlapped time intervals into one tuple. This process keeps the temporal state of the database unchanged.

2. TimeFlag of temporal query statement

```
query      ::= [ timeFlag ] queryExp
timeFlag  ::= [ 'nonsequenced' ] 'validtime' [ identifier | interval ]
```

Different from standard SQL statements, ATSQL2's temporal query begins with timeFlag, which includes the following keywords:

- nonsequenced: This flag means that all fields are treated as user-defined attributes, including valid time and transaction time fields.
- validtime: This flag means that this query is valid time query.
- identifier: This flag is the variable used to represent time intervals.
- interval: This flag is the variable used for marking time intervals.

The default query was snapshot query.

"timeFlag" has the following effects on query operation: Firstly, it specifies the query types (nonsequenced, validtime and snapshot). Secondly, it specifies the time range of the queries, that is, constraining the time interval of each SFW statement (including "Select", "From" and "Where" clauses).

Example 12.4 The following statement obtains tuples whose valid time is between 2001-2-1 and 2005-6-2:

```
VALIDTIME period [date "2001-2-1" - date "2005-6-2")
SELECT * FROM Table;
```

3. Relational algebra for temporal query

The BNF definition of ATSQL2 includes operations of union, intersection and difference:

```
queryExp    ::= queryTerm { ('union' | 'except') queryTerm}
queryTerm ::= queryFactor { 'intersect' queryFactor}
queryFactor ::= '(' query ')' [coal] | sfw
```

Union operation: Union operation of temporal relation means that if relations $r1$ and $r2$ satisfy the union conditions, we can merge the temporal elements in $r1$ and $r2$, including time intervals.

Example 12.5 Rrelation $r1$ and $r2$ are shown in Tables 12.3 and 12.4. Here, we get $r3 = r1 \cup r2$ by merging $r1$ and $r2$, as shown in Table 12.5.

Table 12.3 Relation $r1$

Employee ID	Department	Valid Time
10885632	market department	[1999, 2001]
	technical support department	[2001, 2003]

Table 12.4 Relation $r2$

Employee ID	Department	Valid Time
10885632	technical support department	[2002, 2006]
	Ministry of Personnel	[2006, 2007]

Table 12.5 Relation $r3 = r1 \cup r2$

Employee ID	Department	Valid Time
10885632	market department	[1999, 2001]
	technical support department	[2001, 2006]
	Ministry of Personnel	[2006, 2007]

Intersection operation: This operation of temporal relation means that if relations $r1$ and $r2$ meet the merging condition, we can take the tuples that exist in both $r1$ and $r2$ with the same time intervals as one tuple in the intersected relation, taking the overlapped time interval as its time interval.

Example 12.6 $r4 = r1 \cap r2$ is shown in Table 12.6.

Table 12.6 Relation $r4 = r1 \cap r2$

Employee ID	Department	Valid Time
10885632	technical support department	[2002, 2003]

Difference operation: This operation means that if relations $r1$ and $r2$ meet the merging condition, we can get the resulting tuples by taking tuples that exist in $r1$ but not in $r2$.

4. Temporal query

The main and most commonly used definitions for temporal query are as follows:

```
sfw    ::= 'select' selectItemList
                'from' tableRefList
                                    [ 'where' condExp ]
                                    [ 'group' 'by' groupByList ]
                                    [ 'having' condExp ]
selectItemList   ::=  '*' | selectItem { ',' selectItem }
tableRefList     ::=  tableRef { ',' tableRef }
tableRef         ::=  '(' query ')' [ coal ] alias [ '(' colList ')' ] |
                                      identifier [ coal ] [ alias ]
coal             ::=  '(' 'period' ')'
```

Similar to standard SQL statements, ATSQL2 also contains "Select", "From", "Where", "Group By" and "Having" clauses, where "Select" and "From" clauses are the necessary parts.

(1) "Select" clause

"Select" clause corresponds to the projection operation of relational algebra, which is the selection of the specified attributes from relation while ignoring the rest of the attributes. The keyword "Select" can be followed by "*" or a scalar expression list, where "*" has different meaning in different queries:

- When the snapshot query is implemented, all the user-defined fields of current state except the time stamps (valid time and transaction time) will be chosen.

- When the sequential query is implemented, all the user-defined fields of the specified time intervals except the time stamps will be chosen.
- When the non-sequential query is implemented, all the fields including time stamps will be chosen.

(2) "From" clause

"From" clause corresponds to the Cartesian production operations in relational algebra. For sequential query whose time stamp is valid time, temporal Cartesian product operations combine the two tuples. The time interval of the resulting tuple is the common part of the two original time intervals.

Example 12.7 Given two relations *R* and *S*, as shown in Tables 12.7 and 12.8, the query results of "VALIDTIME SELECT * FROM R, S WHERE R. Emp_ID = S. Emp_ID;" is shown in Table 12.9.

Table 12.7 Relation *R*

Emp_ID	Emp_Dept	The Valid Time Interval
0011	market department	[1994, 2004]
0022	Ministry of Personnel	[1998, 2007]

Table 12.8 Relation *S*

Emp_ID	Emp_Salary	The Valid Time Interval
0011	2000	[1996, 2008]
0022	3000	[1997, 2008]

Table 12.9 Relation $T = R \times S$

Emp_ID	Emp_Dept	Emp_Salary	The Valid Time Interval
0011	market department	2000	[1996, 2004]
0022	Ministry of Personnel	3000	[1998, 2007]

Similar to general SQL statements, "FROM" clause can be nested in an inner query and the query results' field can be renamed.

The keyword "(period)" following the table name or a sub-query means that the system should take temporal merging operations on the query results to eliminate redundant data.

Example 12.8 Given query results as shown in Table 12.10, we can see that the product whose Prod_ID is 1008 has the same price in time interval [1998 2004] as in [2004 2008]. As a result, on an output of the results, these two tuples can be merged into one tuple with larger time interval: $\langle$ 1008, 211, [1998 2008] $\rangle$.

The same operation can also be executed on the product whose Prod_ID is 1852. The results are shown in Table 12.11.

Table 12.10 Before merging

Prod_ID	Price	The Valid Time Interval
1008	102	[1996, 1998]
1008	211	[1998, 2004]
1008	211	[2004, 2008]
1852	1054	[1998, 2007]
1852	1054	[2007, *Now*]

Table 12.11 After merging

Prod_ID	Price	The Valid Time Interval
1008	102	[1996, 1998]
1008	211	[1998, 2008]
1852	1054	[1998, *Now*]

5. Other clause

The structures of "WHERE", "GROUP BY" and "HAVING" clauses are similar to the standard SQL statement, but the comparing operation and operators in "WHERE" clause have a new definition. The improved syntax definition of ATSQL2 that is used by temporal data-processing components TempDB, will be elaborated in the following sections.

12.4 Improved ATSQL2

The improvements of ATSQL2's grammar made by temporal data-processing component, TempDB include the following aspects: clearly regulating the semantic of corresponding operators, redefining scalar expression and clearly regulating the usage method of common operators and the temporal operators in conditional statements.

12.4.1 Clear Regulation to the Semantic Operator

Operators in ATSQL2 are defined as:

```
condOp ::=      '<' | '>' | '<=' | '>=' | '<>' | '=' |
                'precedes' | 'overlaps' | 'meets' | 'contains'
```

This definition only includes four temporal operators (Allen 1983) (precedes, overlaps, meets, contains). Other temporal relational operators (for example, begins) will reuse traditional comparison operators (<, >, <=, etc.), which extend the type scope of traditional operators. Thus, we must re-interpret traditional operators. However, ATSQL2 language does not give a detailed syntax specification, so the definition may lead to ambiguity in the operating process. The modification to operator definition by temporal data-processing components is as follows:

```
condOp      ::=    commonOp | timeOp
commonOp    ::=   '<' | '>' | '<=' | '>=' | '<>' | '='
timeOp      ::=   'before' | 'contains' | 'overlaps' | 'meets' |
                  'starts' | 'finishes' | 'equals'
```

This definition puts traditional comparing operators and temporal operators under the definition of non-terminals to show the different types of usage. commonOp is used for traditional value types' comparison. For example, in "Employee.Salary> 2000", ">" is an integer comparison operator. "timeOp" covers Allen's 13 kinds of time interval relations (Allen 1983) and five relations between time intervals and time points. These 13 kinds of time interval relations have the following equivalence relations:

```
Before (t1, t2)= After (t2, t1)
During (t1, t2)= Contains (t2, t1)
Overlaps (t1, t2)= OverlappedBy (t2, t1)
Meets (t1, t2)=MetBy (t2, t1)
Starts (t1, t2)= StartedBy (t2, t1)
Finishes (t1, t2)= FinishedBy (t2, t1)
```

Therefore, "timeOp" only defines 7 kinds of non-equivalent operators and the remaining 6 operators can obtain the corresponding operator by exchanging the operands.

These seven operators can express the relationship between time intervals, time interval and time point, and time points. For example, "Before (*t*1, *t*2)", if *t*1 is a time point, *t*2 is a time interval, it means that *t*1 happens before the starting point of *t*2. If *t*1 and *t*2 are all time points, it means that *t*1 happens before *t*2.

12.4.2 Re-Definition of Scalar Expression

In its original definition, scalar expression is defined as follows:

```
scalarExp    ::= term { ('+' | '-') term }
term         ::= factor { ('*' | '/') factor }
factor       ::= [ ('+' | '-') ] simpleFactor
simpleFactor  ::= colRef | const | '(' scalarExp ')' | 'abs' '(' scalarExp ')'
colRef       ::= identifier [ '.' identifier ]
const        ::= integer | float | ''' string ''' | interval | event | span
```

This definition is very rough, because the non-terminal operator "const" almost contains all the basic types supported by a temporal database, but not all of these types can take arithmetic operations and complementation. For example, the operation "interval + interval" has no practical meaning. As a result, in the temporal data-processing components, the scalar expression is redefined to regulate different data types and different results types for different operations, as shown in Table 12.12.

Table 12.12 Operations between improved data types

Data Type	Operation	Return Value	Explanation
Integer, float	Arithmetic Operation, complementation operation, absolute value computing	Integer, float value	Same as the conventional explanation
String type	" + " operation	String	Connection of the string
Between the time span	" + "and" – " operations	Time span	Time span can be accumulated, for example: 3days + 3days = 6days
Between the time point	" – " operation	Time span	Compute the distance between the two time points on the timeline
Time point and time span	" + " and " – " operations	Time point	Find the time point that has a certain time span from some time point. " + " operation means the return point is after the time point involved in the operation, while " – " operation means before the time point involved in the operation

At the same time, various types of attribute values in database tables can also participate in computing of the scalar expression. In order to ensure that these attributes can take part in the operation of the scalar expression, we must make the definition of attribute operation compatible with the above description. For example, if the type of the attribute is a time point, operation " – " can be carried out between time points and can take the " + " or " – " operations with time span. As a result, "colRef '-' event, colRef {('+' | '–') span}" can be added to the BNF definition, where "colRef" represents the referenced fields of the database table.

The revised ATSQL2 syntax is presented in the Appendix.

12.4.3 Clearly Regulate the Usage of Common Operators and Temporal Operators in Conditional Statements

Non-terminal operator "simpleCondFactor" defines a conditional expression for comparison. This expression can return a Boolean value "true" or "false" after the interpretation and execution The expression defined by simpleCondFactor in the conditional expression of “Where” clause is connected by the operator “and”, “or” and “not” to constitute the selective condition.

The original definition of “simpleCondFactor” is listed below:

```
simpleCondFactor ::= '(' condExp ')'      |
                     'exists' '(' query ')'      |
                     scalarExp condOp scalarExp  |
                     scalarExp condOp ('all' | 'any' | 'some') '(' query ')'|
                     scalarExp [ 'not' ] 'between' scalarExp 'and' scalarExp|
                     scalarExp [ 'not' ] 'in' '(' query ')'      |
```

The detailed definition of “simpleCondFactor” in temporal data-processing components TempDB is presented in the Appendix.

In the improved definition, it is clearly regulated that we should use general operator (commonOp) to calculate the common expression (constScalarExp), the temporal operator (timeOp) to calculate temporal expression (tempScalarExp), and the expression “between ··· and” is only applicable to the relational comparison between general expression and time point (eventTerm).

References

[1] Allen JF (**1983**) Maintaining knowledge about temporal intervals. Communications of the ACM 26(11): 832 – 843

[2] Böhlen MH (**1994**) Managing temporal knowledge in deductive databases. Ph.D. dissertation, Switzerland: Departement Informatik, ETH Zurich

[3] Böhlen MH, Jensen CS, Snodgrass RT (**1995**) Evaluating and enhancing the completeness of TSQL2. Technical Report TR 95-5, Computer Science Department, University of Arizona

[4] Böhlen MH, Jensen CS, Snodgrass RT (**1995**) Evaluating the completeness of TSQL2. In: Clifford J, Tuzhilin A (eds) Recent Advances in Temporal Databases, pp 153 – 172

[5] Böhlen MH, Marti R (**1994**) On the completeness of temporal database query languages. In: Proceedings of the First International Conference on Temporal Logic, pp 283 – 300

[6] Date CJ (**1989**) A guide to the SQL standard. Addison-Wesley Publishing Company

[7] Melton J, Simon AR (**1993**) Understanding the new SQL: a complete guide. Morgan Kaufmann Publishers

[8] Pulfer D (**1995**) Optimierung von temporalen Queries. Master D dissertation, Switzerland: Institute for Information Systems, ETH Zurich

[9] (**1993**) SQL: American National Standards Institute. ANSI X3H2-93-091/YOK-003, ISO-ANSI (WorkingDraft) Database Language SQL3

[10] Snodgrass RT (**1995**) The TSQL2 temporal query language. Kluwer Academic Publishers, Norwell, MA

[11] Snodgrass RT, Böhlen MH, Jensen CS, Steiner A (**1996a**) Adding valid time to SQL/Temporal. SQL/Temporal Change Proposal, ANSI X3H2-96-501r2, ISO/IEC JTC1/SC21/WG3 DBL MAD-146r2

[12] Snodgrass RT, Böhlen MH, Jensen CS, Steiner A (**1996b**) Adding transaction time to SQL/Temporal. Change Proposal. ANSI X3H2-96-502r2, ISO/IEC JTC1/SC21/WG3 DBL-MAD-147r2

[13] Steiner A (**1998**) A generalisation approach to temporal data models and their implementations. Ph.D. dissertation, Switzerland: Federal Institute of Technology

[14] Time Consult. http://www.timeconsult.com/

13 Design and Implementation of TempDB

Yong Tang[1,2], Huan Guo[2], and Xiaoping Ye[1,2]

[1] Computer School, South China Normal University, Guangzhou 510631, P.R. China
[2] Department of Computer Science, Sun Yat-sen University, Guangzhou 512075, P.R. China

Abstract Generally speaking, implementation techniques of temporal database management systems are lagging behind the corresponding theoretical achievements. In this chapter, the design and implementation details of a temporal data processing prototype system, TempDB, are presented. TempDB improves the grammar of ATSQL2 and corresponding temporal integrity constraints. The temporal index with *Now* binding function is introduced to improve the query performance. TempDB combines and applies relevant theories and development methods of temporal databases. TempDB also accumulates relevant development experience for future implementation of temporal database.

Keywords *TempDB, ATSQL2, temporal integrity, temporal index*

13.1 Introduction

Dealing with temporal information has become a key technique in many databases and information systems, such as e-commerce, e-government, data warehouse, and decision support system, where temporal data processing plays an important role. Compared to the booming theoretical achievements, the implementation techniques of temporal database lag the theoretical achievement. Currently, there are very few real temporal data management systems. TimeDB (Time Consult; Steiner 1998) is considered as a successful temporal database prototype system. As a front-end component of commercial DBMS, TimeDB implements the basic query functions on temporal data. TimeDB shows the feasibility of the development of a temporal database system and brings many theories into practice. A relatively new prototype TANGO (Slivinskas et al. 2001), for the first time, proposed the development of a temporal database system based on the middleware architecture.

issty@mail.sysu.edu.cn, mcsyxp@mail.sysu.edu.cn

To meet the requirements of temporal database development by referencing TimeDB's design principles, we designed and implemented an optimized and more mature system of temporal data process, TempDB. TempDB adopts the middleware architecture and uses the improved representative ATSQL2 (Snodgrass et al. 1996) language. TempDB implements the functions, lexical and syntax analysis, semantic processing, temporal DDL, temporal DML and temporal Query, result packing, partial temporal completeness constrains and temporal index MAP21 (Nascimento 1996) with semantic bounding of *Now* to improve the query performance.

The organization of this chapter is as follows. Section 13.2 introduces the framework of TempDB, Section 13.3 gives a discussion on the implementation, Section 13.4 proposes the temporal integration constraints, and Section 13.5 investigates the optimization of queries.

13.2 Framework of TempDB

13.2.1 Middleware Architecture

According to the extension of temporal support and the difficulty for software development, temporal data model-based software component development schemes can be classified into four types: using DATE type to support temporal manipulation, realizing abstract time type, extending non-temporal DBMS layers to construct the middleware for temporal database and establishing a temporal DBMS from scratch. On an analysis of the advantages and disadvantages of these four schemes, we found that the first scheme cannot provide sufficient temporal support or cannot realize the integrated temporal data model. The second scheme uses the idea of ODBMS as a reference, thus it is incompatible with traditional RDBMS-oriented applications. The fourth scheme is ideal in theory but impractical, as a great amount of labor and other resources are needed to develop a complete temporal DBMS, and producers of DBMS will not introduce processing functions for temporal information *Now*, when both technology and theory of temporal DBMS are immature. However, the third scheme is a good trade off, which extends traditional DBMS in a layered manner to construct a middleware platform of temporal DBMS. It can also make use of the theories of temporal DBMS and accumulates beneficial experience to the design of a complete temporal DBMS.

Trop and others (Al-Kateb et al. 2005; Torp et al. 1998) argued that under current research of TDBMS, extending traditional DBMS in layered way to establish temporal DBMS middleware is the best way both in the short term and in the long term. TempDB thus chooses the middleware architecture as its architecture based on the third schema.

13.2.2 Platform of Implementation

Another decision to make during the realization process is the choice of the backend database. As the realization schema of the temporal software component to be developed is based on traditional RDBMS, the choice of background DBMS has its fair influence on TempDB's implementation.

For example, when using Oracle as the basic platform, the whole workload of the middleware platform will be reduced by means of the powerful interfaces provided by Oracle (for example, *TGREATEST*, *LEAST*) and other operations can help more when operating on time points and time intervals. Compared to Oracle, MySql provides less support, but it makes the deployment simple and fast.

In our opinion, when it comes to semantic processing, the middleware should avoid using platform related functions provided by backend DBMS with its best effort, which could keep the converted language compatible with the SQL standard and guarantee the portability of the component platform. The overall investigations lead to the decision that TempDB component should be based on MySql.

13.2.3 Architecture of TempDB

The architecture of TempDB is showed in Fig. 13.1.

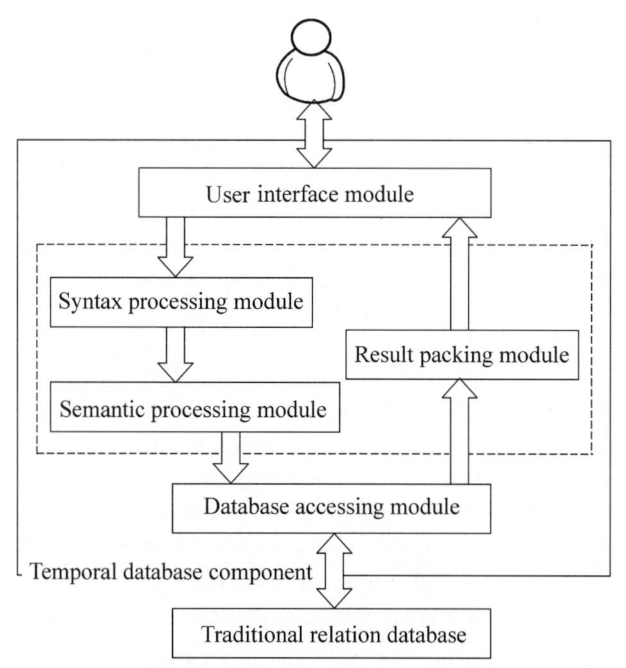

Figure 13.1 Architecture of TempDB

Figure 13.1 shows that TempDB can be divided into five modules—the user interface module, the syntax processing module, the semantic processing module, the database accessing module and the result packing module. Arrows in Fig. 13.1 denotes the primary dataflow directions, but do not cover all possible directions for simplicity purpose. For example, semantic processing module may get its meta-information from a background RDBMS via the database access module.

(1) The **user interface module** is responsible for receiving the input ATSQL2 statements and sending them to the syntax processing module, receiving data from the result packing module and showing them to the user as well.

(2) After receiving ATSQL2 statements from the user interface, the **syntax processing module** will execute lexical and syntax analysis of the statements according to the improved BNF definition of ATSQL2 (please refer to the Appendix for more details of the BNF definition of ATSQL2.). Then, an ATSQL2 statement will be converted into a syntax tree and sent to the semantic processing module. If some abnormal situations occur during the lexical and syntax analysis process, it will inform the users and cancel successive processes.

(3) The **semantic processing module** will conduct some semantic analysis on the syntax tree according to the temporal data model, convert the corresponding ATSQL2 statements of Temp DDL, Temp DML and Temp Query into the equivalent standard SQL statements supported by the backend RDBMS. To meet user's requirements of temporal data processing, it also provides query and its optimization operations, and temporal integrated constraints control.

The semantic processing phase is divided into two phases.

The first phase is the conversion of the syntax tree. Original syntax tree records the whole information of ATSQL2 statements including non-standard SQL parts, which needs to be preprocessed for the final conversion from ATSQL2 to the standard SQL. The statements of the target language (standard SQL) can also be considered as a syntax tree, so the semantic processing module performs the conversion of different syntax trees. With the syntax tree of the target language, it would be easier to get the statements in the target language. In essence, this conversion process converts temporal data model based operations to the corresponding traditional data model based operations.

Example 13.1 In Fig. 13.2, temporal operator *Overlaps* about two time intervals t_1 and t_2. *Overlaps* is converted into a set of standard SQL constraints on the starting and ending time points of corresponding time intervals.

The second phase is to generate standard SQL statement recursively. Based on the syntax tree constructed during the first phase, we can get corresponding standard SQL statements by traversing them into the converted syntax tree.

Example 13.2 Given a fragment of an ATSQL2 statement: "t_1 OVERLAP t_2", the corresponding standard SQL fragments after the syntax analysis and semantic conversion are (obtained by inorder traversing the syntax tree as shown in Fig. 13.2):

$$t_1.\text{start} < t_2.\text{start AND } t_1.\text{end} > t_2.\text{start AND } t_1.\text{end} < t_2.\text{end}$$

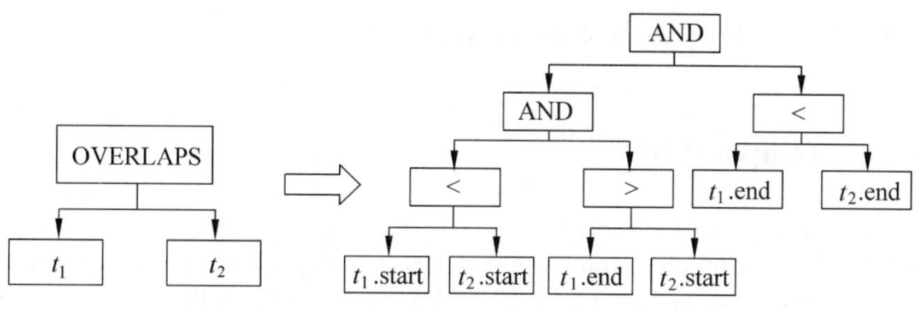

Figure 13.2 Transformation of overlaps' parse tree

Up to now, TempDB accomplishes the conversion from temporal query language to standard SQL.

(4) The **database accessing module** receives the standard SQL statements from the semantic processing module, executes the corresponding operations on the backend database, and then sends the results to the result packing module.

(5) After the above processes, we can get the corresponding result. However, they cannot be displayed on user interface directly. The result should be processed by the **result packing module** to convert it to different expression forms of different data models. To be specific, temporal data model is based on non-1NF data structure, while relational data model follows 1NF data structure strictly. Therefore, TempDB must hide the low-lever 1NF data structure and show temporal data model to users. That is, the result packing module should pack the temporal information recorded by many data columns in the background DBMS into the one with only one data column. Additionally, variable *Now* is denoted by a concrete data value "9999-01-01 00:00:00" in the backend DBMS. If the results contain "9999-01-01 00:00:00", it must be changed into *Now*. Thus, it is necessary to convert the expression form of different data models by this module.

The syntax and semantic processing modules are two core modules of TempDB, which are the key processes of the temporal model based on traditional RDBMS. These two core modules are designed into two separate components based on the following considerations. On one hand, though ATSQL2 has many advantages of temporal query language, it has not become the ultimate standard and may be changed in the future. On the other hand, temporal information processing technology lacks practical experience and it is not maturely developed. The processing methods may be changed. Thus, the middleware architecture was adopted to decouple syntax analyzing (language processing) from semantic processing (temporal processing) to accommodate the potential changes in the future. Classic TimeDB system architecture did not consider these factors. This is the main improvement of our TempDB.

As the core modules of TempDB, the main tasks of syntax and semantic processing modules are to convert the syntax tree of the input ATSQL statements into the standard SQL statements.

13.3 Implementation of TempDB

13.3.1 Temporal DDL

The conversion algorithm of temporal DDL statements is a relatively easy part in temporal conversion algorithms of TempDB. Corresponding to the construction of DDL statements, the conversion algorithm mainly deals with four kinds of sub statements, which are *CREATE TABLE*, *CREATE VIEW*, *DROP TABLE* and *DROP VIEW*, where *DROP TABLE* and *DROP VIEW* only need very small amounts of processing related to time. On the other hand, the work for *CREATE VIEW* is composed by *CREATE TABLE* and *DATA QUERY*, so the work for temporal DDL mainly focuses on *CREATE TABLE*, which is also the main discussion topic of this section.

Based on the improved ATSQL grammar, there are mainly two ways to create tables: one by listing the table name, type and the composition, constraint, name and type of table's column directly (this method corresponds to "tableDef" of ATSQL's BNF definition. In the following part, we will call this way the "tableDef" method). The other one creates a new table through the query results of a given query (this method corresponds to the "ddlQuery" of ATSQL's BNF definition). The treatment for "ddlQuery" method mainly focuses on the temporal query part, so it will not be discussed in this section. In the following, we will discuss the algorithm of the "tableDef" method.

In TempDB, the basic idea for syntax tree conversion is to use top-down analysis method, and combine it with some necessary parent nodes' backtracking. This can convert the sub syntax tree, which describes the temporal semantics, into one of standard SQL language structure. The conversion method of the sentence, which creates a table by "tableDef" method, follows the same idea.

The main conversion parts for "tableDef" method include the conversion of temporal table's marking "AS VALIDTIME", primary key and foreign key, the processing for user-defined constraint. Because TempDB only supports temporal views and the creation of temporal indexes at present, so the processing for other functions are ignored. According to the scanned keywords, TempDB creates the corresponding temporal table, view and index objects, respectively. If an error is found in the analytic process, TempDB will throw an exception with error message to the users. After scanning, once the table to be created is found to be a temporal table, the conversion for columns and the whole table's constraints are executed. The main processing in this step is to adjust referenced primary keys and foreign keys, which allows the created table to store temporal data and express temporal constraints. The typical operations include adding two columns to the table that will be created to express the starting and ending points of the valid time period, setting the two new added columns as the primary keys, rechecking and converting the referenced relation of foreign keys. Note that because the constraints involved in table creating process always spans multi-columns, TempDB puts all the column's constraints as the table's constraints.

13.3.2 Temporal DML

In the improved ATSQL, which is used by TempDB, temporal DML includes two sub-statements, *INSERT INTO* and *DELETE FROM.* In this section, we will introduce the conversion algorithm of DML in detail.

Temporal DML statements can be divided into three categories, which are temporal insertion, temporal deletion, and temporal update statements. With temporal semantic, the insertion and deletion operations for data tables have richer semantics than that in traditional databases under non-temporal semantics. The main reason is that temporal insertion and deletion should be executed under valid time's constraint, which will induce complicated operations of tuples such as merging and splitting operations. To be specific, every tuple in a temporal data table corresponds to a valid time interval, and the temporal insert statements and the temporal delete statements contain valid time constraints on the tuples to be inserted and deleted. Thus, once temporal inserting is executed, it may invoke the merging operation if the information of the inserted tuple fills up the missing information of one existing object, and the splitting operation if a deleting operation deletes a tuple's partial information. Whether these operations happen or not can be decided based on Allen's famous 13 time interval relationships (Allen 1983). Because of the complex overlapping relationships between valid time intervals, the specific processing flows are different between temporal insertion and deletion operations. At the same time, TempDB also needs to convert ATSQL2-based temporal update sentences to SQL sentences, which must confirm to the SQL standard and the semantics of temporal relation data model. In order to guarantee the execution efficiency, optimization is also needed for the DML's algorithm.

13.3.2.1 Conversion algorithm for temporal inserting statement

The main processes for temporal insertion statement are identifying the tuples that have the same information as the tuples to be inserted, except the valid time intervals, and checking whether this insertion will cause merging operation because of the overlap between the time intervals. The algorithm for temporal insertion is as follows:

(1) Get the temporal relation *Table* specified by the ATSQL insertion sentence.

(2) Get the inserted time interval *ITime* (*ITstart*, *ITend*) specified by the ATSQL inserting sentence.

(3) Decide whether to insert the whole attributes or partial attributes. If former, go to Step (7), else go to Step (4).

(4) Get the partial attributes and their values, which need to be inserted.

(5) Extract all the attributes of the selected table.

(6) Transform the inserting sentence that inserts partial attributes into the one inserting the whole attributes by setting the attributes, which are not to be inserted, with null value.

(7) Combine all the attributes of *Table* and their values into one query condition (denoted as *SelectWhere*).

(8) Get all the tuples (denoted as *CandidateCoalesce*) that satisfy condition *SelectWhere* by querying *Table*.

(9) If there does not exist any tuple in *CandidateCoalesce*, go to Step (10). Otherwise, for every tuple in *CandidateCoalesce*:

(9.1) Get tuple's valid time *ETime*(*ETstart*,*ETend*).

(9.2) Merge *ITime* with *ETime*, the resulting time interval is recorded in *ITime*.

(10) Delete all the tuples whose valid time interval *VT* (*VTsatrt*,*VTend*) satisfies the following condition in the selected temporal relation table:

SelectWhere and VTsatrt >=ITstart and VTend <=ITend.

13.3.2.2 Conversion algorithm for temporal delete statement

The main processing for temporal deletion statement is: identify all the tuples that satisfy delete condition and check whether this operation will cause splitting operation based on the time periods' overlapping between the selected tuples' valid time and the time interval in the delete condition. Optimization method is introduced for judging time period's relationships to avoid traversing the whole table. The algorithm for temporal deletion is:

(1) Get the temporal relation *Table* specified by the ATSQL deletion sentence.

(2) Get temporal delete condition (*Dwhere*, allowed to be empty) specified by the ATSQL delete sentence.

(3) Get the delete time interval *DTime*(*DTstart*,*DTend*) specified by the ATSQL delete sentence.

(4) Get all the tuples (denoted as *TSet*) of the selected table that satisfies the delete condition *DWhere* and where the ending time is greater than or equal to *DTstart*.

(5) Sort all the tuples in *TSet* by the valid time *ETstart* and *ETend* in ascending order.

(5.1) Get current tuple's next tuple. If the current tuple does not exist, set the *TSet*'s first tuple as the next tuple.

(5.2) Get tuple's valid time period *ETime*[*ETstart*,*ETend*].

(5.3) Out of Allen's 13 kinds of relationships, determine the one that satisfies *DTime* and *ETime*.

(5.4) According to different relationships, perform the corresponding transformation process, which converts the ATSQL delete statement into the corresponding SQL statement with the same temporal delete semantic.

(5.5) Execute the transformed SQL sentence.

13.3.2.3 Conversion algorithm for temporal update statements

An ATSQL temporal update statement corresponds to a number of SQL statements. The concrete contents of these SQL statements are unpredictable. They relate to the specified time interval of the update statement, the update condition and the set of time intervals of the tuples, which satisfy *WHERE* condition in the temporal relation table. Thus, the temporal update algorithm mainly studies how to convert an ATSQL temporal update statement into corresponding SQL statements, which confirm to SQL standards and semantic of temporal relation data model. The algorithm for temporal deletion is:

(1) Get the temporal relation *Table* specified by the ATSQL update sentence.

(2) Get temporal update condition (*Uwhere*, allowed to be empty) specified by the ATSQL update statement.

(3) Get the update time interval *UTime*(*UTstart*,*UTend*) specified by the ATSQL update sentence.

(4) Get all the tuples to be updated (denote as *TSet*) that satisfy the update condition *Uwhere*, and where the ending time is greater than or equal to *UTstart*.

(5) Sort all the tuples in *TSet* by the valid time *ETstart* and *ETend* in ascending order:

(5.1) Get current tuple's next tuple. If the current tuple does not exist, set the *TSet*'s first one as the next tuple.

(5.2) Get tuple's valid time period *ETime*[*ETstart*, *ETend*].

(5.3) Determine which one of Allen's 13 types of relationships that *UTime* and *ETime* satisfy.

(5.4) According to different relationships, perform the corresponding transformation process, which converts the ATSQL delete statement into corresponding SQL statements with the same temporal delete semantic.

(5.5) Execute the transformed SQL statement.

13.3.3 Temporal Query

Query statements are the SQL statements with the richest and most flexible syntax and semantics of the standard SQL, which is the same as that of ATSQL. The conversion of the Query statement is also more complicated. In this section, we will emphasize on the conversion process of the query statements.

As ATSQL is the natural expansion of standard SQL, in essence, they are similar. It is known from the BNF syntax definition of ATSQL2 that the query statement of ATSQL2 is similar to that of standard SQL in syntax. The main body is composed of the classic key words *SELECT*, *FROM*, *WHERE*, *GROUP BY*, *HAVING*. The semantics of these key words are consistent with that of standard SQL. The main differences are as follows.

Firstly, the flag "timeFlag" used to express the time interval related to the current query appears in the front part of the query statement, which is the biggest difference between ATSQL's query sub-language and that of standard SQL. ATSQL2 provides two different query patterns: sequenced query (default query pattern) and non-sequenced query, where the former considers time interval as a whole, while the latter permits users to treat the starting and ending points of time interval separately, and query them as two ordinary columns.

Secondly, the operators used in specifying query conditions are extended. As temporal query language allows users to operate on time interval and other new time types directly, it must introduce new operators. For example, temporal operation "timeOp" (such as *BEFORE*, *CONTAINS*, *OVERLAPS*) may appear in the expression of the *WHERE* statement. Based on Allen's 13 kinds of relationships between time intervals, these operators specify the relations between the time intervals, reflecting the language characteristics of temporal query.

Thirdly, traditional query operations are implicitly extended under the temporal semantics. For example, multi-table join is needed under multi-table queries. Under this circumstance, TempDB will accomplish this operation implicitly in the background based on temporal Cartesian product, temporal connection and other operations in temporal data model theory, which can guarantee the accuracy of the queries under temporal semantics.

13.4 Processing Mechanism of Temporal Integrity Constraints

TempDB is a temporal middleware for RDBMS. Its data model is temporal relational data model. To ensure temporal data's correctness, it is necessary to introduce temporal integrity constraint mechanism. Otherwise, when temporal insert, update and delete operations are executed on the data, they will produce a large amount of garbage data in basic data table. The results may violate the semantics of reality modeling, which may make the temporal middleware lose its basic function of being transparent to users while correctly dealing with temporal data and being able to reflect the real-world semantics.

The following sections are organized as follows. We will first investigate the properties of temporal integrity constraint by extending relation database's referential integrity into the temporal relation data model. Then study the processing mechanism when corresponding operations violate the integrity constraints. We also realize parts of processing mechanism of temporal referential integrity and temporal update's operation.

The following sections mainly present the definition of temporal referential integrity and the processing mechanism that is adopted to maintain the correctness of data in temporal database when update operations violate temporal referential integrity.

13.4.1 Basic Concepts

Temporal integrity is used to ensure the correctness of the data stored in temporal database, where correctness means that the stored data must accord to the semantics of real world. Temporal integrity constraints are also divided into three classes: temporal entity integrity, temporal referential integrity and user-defined integrity. Supports for temporal entity integrity and temporal referential integrity are compulsory. Temporal integrity contains one data and the corresponding time, where the time is data's valid time. In this section, we mainly introduce concepts about temporal referential integrity.

Definition 13.1 Let us assume that A is an attribute or an array of attributes of temporal relation R, but it is not the primary key. If A corresponds to the **primary key** (Pk, for short) of temporal relation S, then A is called as **Foreign Key Attribute** of R. The time interval of attribute A is called **Foreign Key Time Interval** (Ft_i, for short), temporal relation R is called **Referencing Temporal Relation**, temporal relation S is called **Referenced Temporal Relation**.

According to Definition 13.1, in temporal relation data model there exists not only the reference of attribute value between the referencing temporal relation and the referenced temporal relation, but there also exists the reference of time interval of corresponding attributes. Therefore, we define the **Foreign Key** of temporal relation data model as follows.

Definition 13.2 (Foreign key of temporal relation model) In temporal relation data model, the foreign key of the referencing relation R is defined as: $\{Fk_1, \cdots, Fk_n\}$, where $Fk_i(1 \leqslant i \leqslant n)$ is represented as (Fa_i, Ft_i). Fa_i corresponds to one foreign key attribute of the referencing relation R and Ft_i is the corresponding time interval. When $i = 1$, it means that there is only one foreign key in R.

Definition 13.3 (Principles of temporal referential integrity) Assuming $\{Fk_i = (Fa_i, Ft_i)|(1 \leqslant i \leqslant n)\}$ is the foreign key of temporal relation R, Fa_i corresponds to the primary key of temporal relation S, then every $Fk_i(1 \leqslant i \leqslant n)$ corresponding to one tuple of R must satisfy the following conditions:

(1) The value of Fa_i is *null* or equals to the Pk of one tuple of S.

(2) If the value of Fa_i is not *null*, Ft_i must be contained by the valid time of Pk (denoted as Pvt).

From definition 13.3, we can see that to reflect correct semantics of real world modeling in temporal data model, temporal referential integrity must consider the corresponding relationship between the time interval Ft_i of every foreign key of referencing temporal relation R and the corresponding Pvt in referenced temporal relation S.

13.4.2 Temporal Insertion

In temporal insertion operation, only the insertion of a new tuple into referential temporal table will violate integrity constrains. Thus, we only consider the situation of inserting a tuple into referencing temporal table.

1. Temporal constrained insertion

When inserting the tuple T into a referencing temporal relation table, if all $Fa_i(1 \leqslant i \leqslant n)$ of T is not *null*, the corresponding inserting operation will be executed only if the following condition is fulfilled:

There exists corresponding tuples in the referenced temporal relation, whose Pk is the same as the value of corresponding Fa_i in the foreign key set $\{Fk_i = (Fa_i, Ft_i) | 1 \leqslant i \leqslant n\}$ of T, and the corresponding valid time Ft_i is contained by corresponding Pkt of referenced temporal relation. Otherwise, this operation will be executed.

2. Temporal recursive insertion

When inserting the tuple T into a referencing temporal relation table, if the value of Fa_i or Ft_i of some element in the foreign key set $\{Fk_i = (Fa_i, Ft_i) | 1 \leqslant i \leqslant n\}$ of T does not meet temporal referential integrity constrains, the corresponding tuple can be inserted into its referenced temporal table, whose Pk equals to Fa_i and corresponding Pkt equals to Ft_i, until each element of $\{Fk_i | 1 \leqslant i \leqslant n\}$ meets temporal referential integrity constrains. Then you can insert tuple T.

13.4.3 Temporal Deletion

In temporal relation data model, tuple's deletion of referencing temporal table will not violate integrity constrains, so we only need to consider the situation of deleting tuple in the referenced temporal table S.

1. Temporal constrained deletion

The corresponding deleting operation will be executed, only if the following conditions are fulfilled:

(1) There does not exist such a tuple in the corresponding referencing temporal table, where the value of Fa_i in its foreign key equals to the Pk of the tuple to be deleted.

(2) There exists such a Pk but the time interval of the tuple to be deleted is not relevant to the corresponding Ft_i in the referencing temporal relation table.

2. Temporal deletion by setting *null* value

Delete the specified tuple of referenced temporal relation table and set the foreign key of corresponding tuples in the referencing temporal relation to be *null* during the time interval of the deleted tuple.

3. Temporal cascaded deletion

Delete the specified tuple of the referenced temporal relation table and delete the tuples in the corresponding referencing temporal relation, such that all the values $\{Fa_i | i_1 \leqslant i \leqslant i_n\}$ of its foreign keys are the same as the *Pk* of the deleted tuple and corresponding $\{Ft_i | i_1 \leqslant i \leqslant i_n\}$ are contained by the time interval of the deleted tuple. If this referencing temporal relation is also the referenced temporal relation of another relation, then this deleting operation will continue in the cascaded form.

13.4.4 Temporal Modification

The basic purpose of the temporal modifying mechanism is to control temporal modifying operations by making sure that foreign keys and the corresponding time interval will not be modified to the value that does not exist in the referenced temporal table. We can consider this problem from the following aspects: when referencing temporal table is modified and when referenced temporal table is modified.

(1) If the modified relation is a referencing temporal relation table, we must check the referenced temporal relation to see whether there are such tuples where *Pkts* are contained by the time interval of referencing temporal relation that needs to be modified, and to see whether there are such tuples where *Pks* equal to the foreign key of the modified tuple. The processing principles of these two situations are as follows:

① Temporal refused modification

The modifying operation will be executed, only if the value of Fa_i of any tuple's foreign key in referenced temporal relation is not the same as the primary key to be modified, or such tuples exist but the corresponding Ft_i is irrelevant to the specified time interval. Otherwise, the operation will be refused. That is, if the referencing temporal table has the reference of the tuple to be modified in the referenced temporal table, this tuple will not be modified. This reference contains double references to the primary key and the corresponding time interval of the tuple to be modified.

② Temporal modification by setting *null* value

Modify the specified tuple of the referenced relation and set the foreign keys of the corresponding tuples in the referencing temporal relation with *null* value during the time interval of the modified tuple.

③ Temporal cascaded modifying

Modify the specified tuple of the referenced relation and get all the tuples of the referencing temporal relation whose foreign keys equal to the *Pk* of the modified tuple. Then modify these tuples with the new value during the time interval that is contained by the valid time of the modified tuple. That is, when the primary key is modified, all the referenced temporal tables corresponding to this primary key need to be modified.

(2) When modifying a primary key of referenced temporal relation, there are two situations to be considered: ① *Pk* of the temporal relation table is not allowed to be modified. In this situation, it will not violate the semantics of temporal referential integrity constrains when modifying the referenced relation's tuple. ② *Pk* of temporal relation table is allowed to be modified. When modifying the tuple in the referenced relation, we must check the referencing temporal relation to see whether there exists such tuples whose foreign keys equal to the modified *Pk* of the referenced relation, and the corresponding time interval is contained by the corresponding *Pkt*. The processing principles of these two situations are as follows.

① Temporal constrained modification:

The modifying operation will be executed, only if:

- There exists such tuples in the referenced temporal relations, whose *Pk* equals to the new value of the foreign key Fa_i of the tuple to be modified in the referenced temporal relations;
- The new valid time Ft_i is contained by the valid time interval of the foreign key in the referenced temporal relations.

To be specific, the original tuples cannot be modified, if the new value of the modified foreign key does not exist in the referenced temporal table, or it exists in the referenced temporal table but the new valid time interval is not contained by the corresponding *Pkt* in the referenced temporal table.

② Modification for recursive temporal insertion

When one tuple in the referenced temporal relations is modified and if the value of Fa_i or Ft_i of some elements in the foreign key set $\{Fk_i|1 \leqslant i \leqslant n\}$ of the modified tuple does not meet the requirements of the referenced temporal integrity constraint, the corresponding tuples should be inserted into the referenced temporal table, whose *Pk* and *Pkt* equal to the new foreign keys and valid time interval of the modified tuple in the referenced temporal relations respectively, until each element in $\{Fk_i|1 \leqslant i \leqslant n\}$ meets the requirements of the referenced temporal integrity constraint. We can then modify the original tuple in the referenced temporal relations with the new value.

③ The processing principles for modifying temporal table, which is both a referencing and a referenced temporal table are as follows: If the temporal relation table to be modified is not only a referencing temporal table but also a referenced one, there are two situations to be considered:

- if the primary key is not allowed to be modified, we only need to consider the modification of the reference relational table;
- if the primary key can be modified, this temporal relation can have six possible combinations referenced to the process principles mentioned in ① and ②.

In TempDB, we implement four temporal referential integrity-processing mechanisms, which are the temporal constrained insertion, the temporal constrained deletion, the temporal deletion by setting *null* value and the temporal constrained modification.

13.5 Optimization of Performance

In order to improve query efficiency, the conventional relation database management systems mainly use indexes. In the current database management systems, B-tree (Bayer and Mccreight 1972) and B^+-tree (Comer 1979) are widely used. In temporal database, we can also use temporal index technology to improve query efficiency. During the process of TempDB's design and development, we have done a preliminary study and implemented temporal indexing and querying performance optimization.

13.5.1 Temporal Indexes and MAP21

Document (Bliujute et al. 2000) proposed the 4-R indexing methods, which index bitemporal data efficiently by using R*-Tree on the bottom layer. Based on R* tree (Beckmann et al. 1990), document (Šaltenis and Jensen 2002) firstly proposed an efficient and general indexing method for temporal data, which is called R^{ST}-tree. This method not only indexes spatial data, but also indexes temporal data and also supports temporal variable *Now*. Document (Nascimento 1996) proposed the representative MAP21 algorithm by summarizing various indexing techniques and constructed the MAP21 index tree model to index data whose valid time is a closed interval and then improved the MAP21 tree to process the valid time data with *Now*. TempDB is based on the conventional relational database, while MAP21 temporal index method can use B^+-tree implemented by the backend database management systems directly. Thus, TempDB adopts MAP21 temporal index method.

MAP21 method is used to index the time interval with integer type starting point and the ending point. In the meanwhile, it can also be extended to open intervals (that is, the ending point of interval is temporal variable *Now*). Its main idea is to map the time interval expressed by the starting point and the ending point into one point value, that is, to map two-dimensional data into one-dimensional data. Then the mapped one-dimensional data can be indexed by the standard B^+-tree. The indexing and querying methods were described in (Nascimento 1996).

13.5.2 Binding on *Now*

In TempDB, we introduce the variable *Now*. The document (Creem 2005) compared several expressed methods of *Now*, including NULL, MIN/MAX time stamp, and the point-expressed method. However, whichever method is adopted, we need to specially deal with *Now* and replace it with the current time.

There are two strategies for *Now* binding, timely updating and real-time updating. The timely updating method updates all temporal index fields in a temporal table when the system time arrives at a specified time, which will take up some computer resources. It is suitable for the databases with coarse time granularity and high occurrence rate of *Now* in time stamp. The real-time updating method updates and queries temporal index fields of the relevant data table before executing query operation, which reduces the query efficiency and concurrency, and is suitable for data tables with fine time granularity and low occurrence rate of *Now* in time stamp. The shortcoming of this method is that it needs real-time updating or timely updating of query-related data tables, which brings extra querying cost and concurrent efficiency.

A better method is to use "OR" sub-expression to add more query conditions to separate the close interval query and the open interval query in different logical clauses. When generating "*Where*" clause, Temporal index fields are generated by MAP21-B method, but they do not update the open interval mapping values before execution. TempDB also adopts this method.

13.5.3 MAP21-B

Based on the seven temporal operators' characteristics in the improved ATSQL2 grammar and the processing method of *Now* binding, we improve the MAP21 method to make it work with the function of binding *Now*, called bounded MAP21 (MAP21-B, for short) method.

In the MAP21-B method, *Now* is bound to the current time value by *Bind*(.) function. The binding operation is as follows:

(1) if $V_e = Now$, then $Bind(V_e) =$ the current time value:

(2) Else $Bind\ (V_e) = V_e$.

In the following discussion, we use T to represent time point, V to express time interval $V = [V_s, V_e]$. Here, V_s and V_e represent the starting and ending time of time interval V, respectively, with a value domain {the time range that a computer can express $\cup$ *Now*}. *Now* is expressed by the maximum system time. That is, *Now* = "9999-01-01 00:00:00" in background database. The domain of *Bind*(.) is the time range that a computer can express. Detailed instruction about mapping function *F*(.) is presented in (Nascimento 1996). Time point T can be seen as a time interval with length 0.

Different query conditions for different temporal operators under the MAP21-B method will be discussed, where *Now* represents current statement's execution time, $\Delta = \max_k\{V_e^k - V_s^k, \forall k\}$ is the max span of indexed time interval.

1. BEFORE operator

(1) Given a time point T, to find all the time interval V that satisfy *Before* (T, V). Only the time intervals whose mapping values satisfy: $F(V) > T \cdot 10^{\alpha} + T$ need to be searched.

(2) Given a time point T, to find all the time intervals V that satisfies *Before* (V, V^j). Only the time intervals whose mapping values satisfy: $F(V) < T \cdot 10^{\alpha} + T$, and one with $V_e = Now$ and $Now < T$ need to be searched.

(3) Given interval V^j, to find all the time intervals V that satisfy *Before*(V, V^j). Only the time intervals whose mapping values satisfy: $F(V) < V_s^j \cdot 10^{\alpha} + V_s^j$ and the one with $V_e = Now$ and $Now < V_s^j$ need to be searched.

(4) Given interval V^i, to find all the time intervals V that satisfy *Before* (V^i, V), if $V_e^i \neq Now$. Only the time intervals whose mapping values satisfy $F(V) > V_e^i \cdot 10^{\alpha} + V_e^i$, and the one with $V_e^i = Now$ and $Vs > Now$ need to be searched.

2. CONTAINS operator

(1) Given a time point T, to find all the time intervals V that contain T. Only the time intervals whose mapping values satisfy: $(T - \Delta) \cdot 10^{\alpha} + T \leqslant F(V) \leqslant T - \Delta \cdot 10^{\alpha} + T + \Delta$ and the one with $V_e = Now$ and $V_s \leqslant T \leqslant Now$ need to be searched.

(2) Given a time interval V^j, to find all the time intervals V that contain V^j. only the time intervals whose mapping values satisfy: $(V_e^j - \Delta) \cdot 10^{\alpha} + V_e^j \leqslant F(V) \leqslant V_s^j \cdot 10^{\alpha} + V_s^j + \Delta$ and the one with $V_e = Now$ and $V_s^i \leqslant V_s$ need to be searched.

(3) Given a time interval V^i, to find all the time intervals V, which are contained by V^i. Only the time intervals whose mapping values satisfy: $V_s^i \cdot 10^{\alpha} + V_s^i \leqslant F(V) \leqslant V_e^i \cdot 10^{\alpha} + V_e^i$ and the one with $V_e = Now$ and $V_s^i \leqslant V_s$ and $Now \leqslant V_e^i$ need to be searched.

3. OVERLAPS operator

(1) Given a time interval V^j, to find all the time intervals V that satisfies *Overlaps*(V, V^j). Only the time intervals whose mapping values satisfy: $(V_s^j - \Delta) \cdot 10^{\alpha} + V_s^j \leqslant F(V) \leqslant V_s^j \cdot 10^{\alpha} + V_s^j + \Delta$ and the one with $V_e = Now$ and $V_s \leqslant V_s^j \leqslant Now$ need to be searched.

(2) Given interval V^i, to find all the time intervals V that satisfies *Overlaps* (V^i, V). Only the time intervals whose mapping values satisfy: $V_s^i \cdot 10^{\alpha} + V_s^i \leqslant F(V) \leqslant V_e^i \cdot 10^{\alpha} + V_e^i + \Delta$ and the one with $V_e = Now$ and $V_s^i \leqslant V_s \leqslant V_e^i$ need to be searched.

4. MEETS operator

Given an interval V^j, to find all the time intervals V that satisfy *Meets*(V, V^j). Only the time intervals whose mapping values satisfy: $(V_s^j - \Delta) \cdot 10^{\alpha} + V_s^j \leqslant F(V) \leqslant V_s^j \cdot 10^{\alpha} + V_s^j$, or $V_e^j \cdot 10^{\alpha} + V_e^j \leqslant F(V) \leqslant V_e^j \cdot 10^{\alpha} + V_e^j + \Delta$, and the one with $V_e = Now$ and $V_s^j = Now$ or $V_e^j = V_s$ need to be searched.

5. STARTS operator

(1) Given a time point T, to find all the time intervals V that satisfies *Starts*(T,V). Only the time intervals whose mapping values satisfy: $T \cdot 10^{\alpha} + T \leqslant F(V) \leqslant T \cdot 10^{\alpha} + T + \Delta$, and the one with $V_e = Now$ and $T = V_s$ need to be searched.

(2) Given a time interval V^j, to find all the time intervals V that satisfies *Starts*(V, V^j). Only the time intervals whose mapping values satisfy: $V_s^j \cdot 10^\alpha + V_s^j \leqslant F(V) \leqslant V_s^j \cdot 10^\alpha + V_e^j$, and the one with $V_e = Now$ and $V_s = V_s^j$ and $Now < V_e^j$ need to be searched.

(3) Given a time interval V^i, to find all the time intervals V that satisfies *Starts*(V^i, V). Only the time intervals whose mapping values satisfy: $V_s^i \cdot 10^\alpha + V_e^i \leqslant F(V) \leqslant V_s^i \cdot 10^\alpha + V_s^i + \Delta$, and the one with $V_e = Now$ and $V_s^i = V_s$ and $V_e^i < Now$ need to be searched.

6. FINISHES operator

(1) Given a time point T, to find all the time intervals V that satisfy *Finishes*(T,V). Only the time intervals whose mapping values satisfy: $(T - \Delta) \cdot 10^\alpha + T \leqslant F(V) \leqslant T \cdot 10^\alpha + T$ and the one with $V_e = Now$ and $T = Now$ need to be searched.

(2) Given a time interval V^j, to find all the time intervals V that satisfy *Finishes*(V, V^j). Only the time intervals whose mapping values satisfy: $V_s^j \cdot 10^\alpha + V_e^j < F(V) < V_e^j \cdot 10^\alpha + V_e^j$, and the one with $V_e = Now$ and $V_e^j = Now$ and $V_s > V_s^j$ need to be searched.

(3) Given a time interval V^i, to find all the time intervals V that satisfy *Finishes*(V^i, V). Only the time intervals whose mapping values satisfy: $(V_s^i - \Delta) \cdot 10^\alpha + V_e^i \leqslant F(V) < V_s^i \cdot 10^\alpha + V_e^i$, and the one with $V_e = Now$ and $V_e^i = Now$ and $V_s < V_s^j$ need to be searched.

7. EQUALS operator

Given a time interval V^i, to find all the time intervals V that equals to V^i. Only the time intervals whose mapping values satisfy: $F(V) = F(V^i) = V_s^i \cdot 10^\alpha + V_e^i$, and the one with $V_e = Now$, $V_s^i = V_s$ and $V_e^i = Now$ need to be searched.

References

[1] Al-Kateb M, Mansour E, El-Sharkawi ME (**2005**) CME: a temporal relational model for efficient coalescing. In: Proceedings of the 12th International Symposium on Temporal Representation and Reasoning (TIME'05), pp 83 – 90

[2] Allen JF (**1983**) Maintaining knowledge about temporal intervals. Communications of the ACM, 26(11): 832 – 843

[3] Bayer R, Mccreight E (**1972**) Organization and maintenance of large ordered indexes. Acta Informatica 1(3): 173 – 189

[4] Beckmann N, Kriegel H-P, Schneider R, Seeger B (**1990**) The R*-tree: an efficient and robust access method for points and rectangles. In: Proceedings of the 1990 ACM SIGMOD International Conference on Management of Data, pp 322 – 331

[5] Bliujute R, Jensen CS, Saltenis S, Slivinskas G (**2000**) Light-weight indexing of general bitemporal data. In: Proceedings 12th International Conference on Scientific and Statistical Database Management, Berlin, Germany
[6] Comer D (**1979**) The ubiquitous b-tree. ACM Computing Surveys 11(2): 121 – 137
[7] Creem KN (**2005**) A comparison of approaches to modeling NOW in bitemporal databases. 21st Computer Science Seminar. Hartford. USA
[8] Nascimento M (**1996**) Efficient indexing of temporal databases via B+-trees. Ph.D. dissertation. USA: School of Engineering and Applied Science Southern Methodist University
[9] Šaltenis S, Jensen CS (**2002**) Indexing of now-relative spatio-bitemporal data. The VLDB Journal—The International Journal on Very Large Data Bases, 11(1): 1 – 16
[10] Slivinskas G, Jensen CS, Snodgrass RT (**2001**) Adaptable query optimization and evaluation in temporal middleware. ACM SIGMOD 2001. Santa Barbara, California, USA
[11] Snodgrass RT, Böhlen MH, Jensen CS, Steiner A (**1996**) Adding valid time to SQL/Temporal. ANSI X3H2-96-501r2,ISO/IEC JTC1/SC21/WG3 DBL MAD-146r2
[12] Steiner A (**1998**) A generalisation approach to temporal data models and their implementations. Ph.D. dissertation, Switzerland: Federal Institute of Technology
[13] Time Consult. http://www.timeconsult.com/
[14] Torp K, Jensen CS, Snodgress RT (**1998**) Stratum approaches to temporal DBMS implementation. In: Proceedings of the 1998 International Symposium on Database Engineering & Applications

Part V Temporal Application and Case Study

- Research on Temporal Extended Role Hierarchy
- Temporal Workflow Modeling and Its Application
- Temporal Knowledge Representation and Reasoning
- Temporal Application Modes and Case Study

14 Research on Temporal Extended Role Hierarchy

Wei Dao[1,2], Yong Tang[1,2+], Jianguo Li[2], and Hanjiang Lai[2]

[1] Computer School, South China Normal University, Guangzhou 510631, P.R. China
[2] Department of Computer Science, Sun Yat-sen University, Guangzhou 512075, P.R. China

Abstract Role hierarchy is one component of the RBAC model. It can reduce the workload of permission assignment. Nowadays, most research on RBAC cannot deal with temporal constraints. The main purpose of this chapter is to study the effect of temporal constraints acting on role hierarchy. We first propose an extended model to solve the inheritance granularity. Based on it, we add temporal constraints and propose a temporal extended role hierarchy model. We analyze the effect of time constraint on the transferring of different permission in different inheritance modes and the character of temporal extended role hierarchy model. Finally, we analyze the space and time efficiency of the model.

Keywords *RBAC, role hierarchy, inheritance, temporal constraints, restricted special permission*

14.1 Introduction

Role hierarchy is an important component of **RBAC** models. Senior roles get the junior roles' permission by inheriting or activating. A properly designed role hierarchy brings efficient specification and management of access control structures for a system (Crampton 2003; Kahtani and Sandhu, 2003). The inheritance of permission can be viewed as a kind of authorization. One simple way is complete inheritance. However, in reality the inheritance relation is not always complete. Some permissions of junior role should not be inherited by senior role (Chaari et al. 2004). For example, in a task every member has some private data. These data can only be accessed by the owner and the access authority for these data cannot be inherited by the senior role. This is **private permission**. Another kind of permission can be inherited by senior role only in some restricted scope. The

+ Corresponding author: issty@mail.sysu.edu.cn

restricted special feature can be area or time. This kind of permission is called **restricted special permission** (McDaniel 2003; Ahn and Sandhu 2000). The value of special feature is dynamic.

This chapter discusses the effect of time constraints on inheritance relationship, presents a **temporal role hierarchy** model based on **time constraints** and inheritance granularity. The chapter is organized as follows. Section 14.2 introduces related work, Section 14.3 proposes an extended role inheritance relationship, Section 14.4 proposes a temporal extended role hierarchy model and analyzes the efficiency of time and space for the model, and Section 14.5 discusses future research on temporal role hierarchy.

14.2 Related Work

Role hierarchy is a partial ordering relation on role aggregate. It is defined as follows.

Definition 14.1 (Partial ordering) The relation "$>$" of aggregate S is called partial ordering. The aggregate with partial ordering is called poset, noted as $(S, \geqslant)$. For each a, b, $c \in S$, the following conditions hold:

(1) Reflexivity: $i \geqslant i$;

(2) Anti-symmetry: $i \geqslant j, j \geqslant i \rightarrow i = j$;

(3) Transmit: $i \geqslant j, j \geqslant k \rightarrow i \geqslant k$.

Role inheritance is a standard partial ordering relation (Moffett 1998; Suroop and Joshi 2005; Fereaiolo et al. 2001). As we described above, in real-life the inheritance relation is not always complete. Some permission of junior role is not required to be inherited by senior role.

References (Kahtani and Sandhu 2003; Yu and Zhang 2004) propose the concept of private role. It adds a **private role** r' to take the private permissions of a junior role. All the private permissions are put into r'. As Fig. 14.1 shows, in order to keep some private permission of T_1, T_2, P_1, S_1 not to be inherited, we add private roles T_1', T_2', P_1', S_1'. Thus, the number of the roles in the system will increase rapidly and the role inheritance relation will be much more complex. All this reduces the management advantages of the RBAC model.

Aiming at the incomplete inheritance in role hierarchy, we extend the role inheritance relationship by introducing *Public Permission*, *Private Permission* and *Restricted Special Permission* to enable multiple inheritance granularities. Based on them, we add time constraint to form an integrated temporal role hierarchy model.

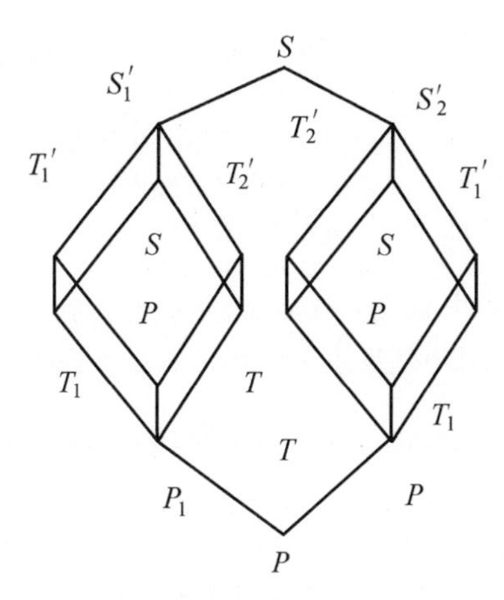

Figure 14.1 Role hierarchies of adding private roles

14.3 Extended Role Hierarchy

A role's permission can be divided into three categories:

- **Public permission:** Permission aggregation that can be inherited by senior role;
- **Private permission:** Permission aggregation that cannot be inherited by senior role;
- **Restricted special permission:** Whether this kind of permission can be inherited is not definite. It lies on the inheritance method and feature value scope.

According to the difference of the three kinds of permission, role inheritance mode can extend to *normal inheritance*, *special privatizing inheritance*, *special publicizing inheritance.*

In case of normal inheritance, the transferring way of restricted special permission from junior role to senior role will not change. Other two inheritance modes will change the transferring way.

(1) NMI (normal inheritance): define a binary relation of R and R. $I_{\text{NMI}} \subseteq R \times R$, $(r_1, r_2) \in I_{\text{NMI}}$ express role r_2 normal inherit r_1, shown as $r_1 \rightarrow r_2$. If $r_1 \rightarrow r_2$ and $\forall p \in P_{\text{RSPP}}(r_1)$, then $p \in P_{\text{RSPP}}(r_2)$.

(2) SPVI (special privatizing inheritance): define a binary relation of R and R. $I_{\text{SPVI}} \subseteq R \times R$, $(r_1, r_2) \in I_{\text{SPVI}}$ express role r_2 privatizing inherit role r_1, shown as $r_1 \rightarrow {}^{-}r_2$. If $r_1 \rightarrow {}^{-}r_2$ and $\forall p \in P_{\text{RSPP}}(r_1)$, then $p \in P_{\text{PRP}}(r_2)$.

(3) SPUI (special publicizing inheritance): define a binary relation of R and R. $I_{\text{SPUI}} \subseteq R \times R$, $(r_1, r_2) \in I_{\text{SPUI}}$ express role r_2 publicizing inherit role r_1, shown as $r_1 \rightarrow {}^{+}r_2$. If $r_1 \rightarrow {}^{+}r_2$ and $\forall p \in P_{\text{RSPP}}(r_1)$, then $p \in P_{\text{PUP}}(r_2)$.

Different permissions obey different inherit rules:

① Private permission cannot be inherited. Let $r_1, r_2 \in R$. If $r_1 \rightarrow r_2$ or $r_1 \rightarrow {}^{-}r_2$ or $r_1 \rightarrow {}^{+}r_2$ and if $\forall p \in P_{\text{PRP}}(r_1)$, then $p \notin P(r_2)$. The transfer depth of private permission $n = 0$.

② Public permission will be inherited and the inherited permission will be still public permission. Let r_1, $r_2 \in R$. If $r_1 \to r_2$ or $r_1 \to^{-} r_2$ or $r_1 \to^{+} r_2$ and if $\forall p \in P_{PBP}(r_1)$, then $p \in P_{PBP}(r_2)$. The transfer depth of public permission $n = \infty$.

③ As far as the inheritance of *restricted special permission* is concerned, we should first analyze the type of special permission. According to the value scope, we can judge whether the inheritance relation exists. Judging special values is a dynamic process and the time constraint is a key factor. Always the transfer depth of restricted special permission is $0 < n < \infty$, in NMI n--, in SPVI $n = 0$, in SPUI $n = \infty$.

Table 14.1 describes the transferring relation of different permissions in different inheritance modes.

Table 14.1 Extended role inheritance rule

Permission type	Inherit type		
	NMI	SPVI	SPUI
Public permission	Public permission	Public permission	Public permission
Restricted special permission	Restricted special permission	Private permission	Public permission
Private permission	—	—	—

In extended role hierarchy, the combination of role permissions is shown as Fig. 14.2.

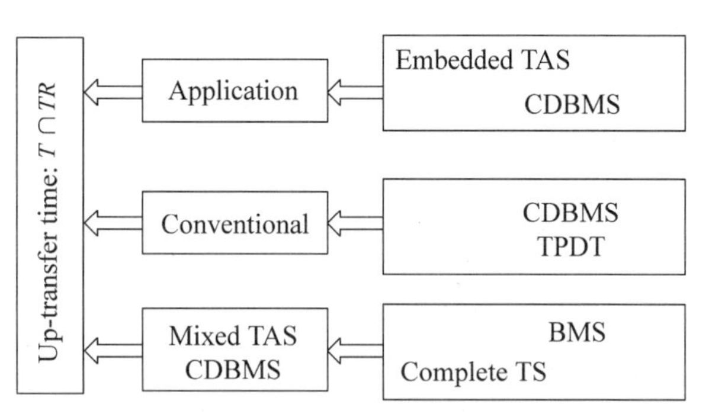

Figure 14.2 Role permission of extended role hierarchy

Aiming at Fig. 14.1, private roles S_1', S_2', T_1', T_2', T_3', T_4', P_1', P_2' can be instituted by the restricted special permission of roles S_1, S_2, T_1, T_2, T_3, T_4, P_1, P_2. The new role hierarchy relation is shown in Fig. 14.3.

From Fig. 14.3, we can see that the number of roles change from 14 to 10. In the new role hierarchy, not only the number of roles reduces, but also the structure keeps legible. At the same time, the extended hierarchy can simulate real-life situations much better.

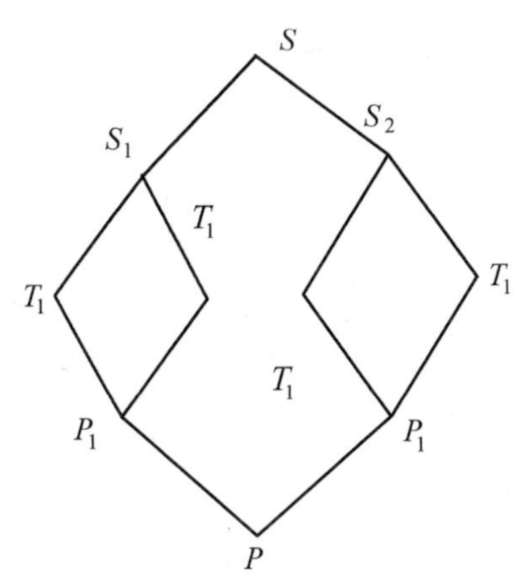

Figure 14.3 Role hierarchy of extended inheritance mode

14.4 Temporal Role Hierarchy

14.4.1 Time Constraint on the Inheritance of Restricted Special Permission

In this section, we propose an extended inheritance mechanism by introducing private permission, public permission and restricted special permission. Role hierarchy on this mechanism can overcome the redundancy caused by private role.

In this section, we discuss the effect of time constraint on the inheritance relationship of restricted special permission, that is, restricted special means time constraint.

The transferring of restricted special permission can be expressed as follows:

$$Inherit_{rs} = (R_1, R_2, P_{rs}(R_2), TR, S)$$

Here:

- R_1, R_2 are two roles, and R_2 is the senior role of R_1.
- P_{rs} is the restricted special permission of R_2.
- TR is the time constraint of special feature.
- S is the inheritance mode, such as NMI, SPVI, SPUI.
- $\forall P_{rs},(R_2 \geqslant_t R_1) \wedge \text{canbe\_acquired}(P_{rs}, R_1, t) \wedge t \in TR \rightarrow \text{canbe\_acquired}(P_{rs}, R_2, t)$.

Whether the restricted special permission can up-transfer to the senior role is estimated by function execute ($Inherit_{rs}$, t), where $Inherit_{rs}$ is a triple-group of restricted inheritance, t is a time point. The main function of execute($Inherit_{rs}$, t) is to judge the value of $t \in TR$. If the value is true, then restricted special permission can up-transfer, otherwise it cannot at time point t.

In different inheritance modes, the transferring path of restricted special permission is different. For all inheritance modes, the time constraint on restricted special permission should up-transfer. The transferring is shown in Fig. 14.4.

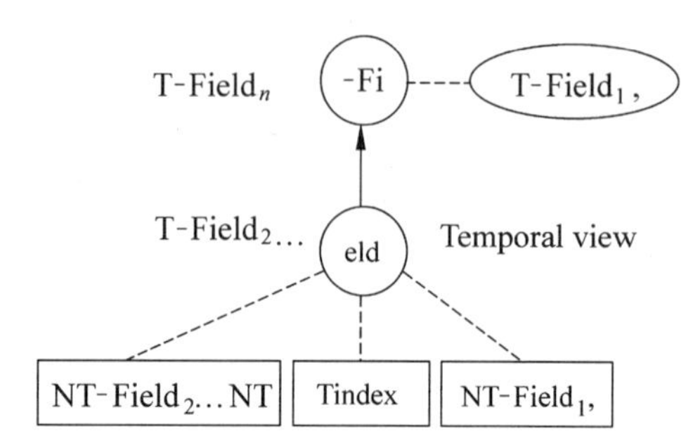

Figure 14.4 Up-transferring of time constraint on $Permission_{RS}$

In NMI, the restricted special permission of a junior role and time constraint on it will up-transfer to its senior role and then it becomes $Permission_{RS}$ of the senior one. Senior role can add its own special feature on it to combine a new $Permission_{RS.}$

$$PATH = P_{RS}(R_1) \rightarrow P_{RS}(R_2) \rightarrow P_{RS}(R_3) \rightarrow \cdots \rightarrow P_{RS}(R_n)$$

$$TR = TR_1 \cap TR_2 \cap TR_3 \cap \cdots \cap TR_n$$

(We do not consider time τ for the sake of simplicity.)

Transfer depth $n = n - 1$. When $n = 0$, the transfer stops.

In SPVI, the restricted special permission and time constraint of a junior role up-transfers to senior role just one time and then becomes $Permission_{PRI}$ of the senior role. It cannot be inherited anymore. Senior role can add its own special feature on it to combine a new $Permission_{RS.}$ The transfer depth $n = 0$.

In SPUI, the restricted special permission and time constraint of a junior role up transfers to senior role just one time and then becomes $Permission_{PUB}$ of the senior role. The transfer depth $n = \infty$. *Now*, time constraint not only acts on permission, but also keeps pace with the role state.

Algorithm 14-1 Compute permissions of a role in temporal extended role hierarchy

Algorithm name: $RP(P, R, RH, r, t)$

Input: permission aggregation P, role aggregate R, role hierarchy RH, specific r, time point t

Output: permission aggregation of the specific r

Step 1: Build role net-graph $RG = (R, E)$ for role hierarchy RH, $E = \{(r_i, r_j) | r_i \geqslant_t r_j\}$. This is a directed acyclic graph.

Step 2: Use Breadth-First-Search. Only consider the output direction of Directed Arc, look at Role r as start node, compute all the nodes that r can reach, then put them into stack. There are r nodes of role hierarchy domain in the stack.

Step 3: Set role hierarchy level count $lv=0$.

Step 4: When STACK is not null, execute Step 5 – Step 13.

Step 5: Pop-up a node r_i from STACK, lv++.

Step 6: Compute the direct permission aggregation $RP(r_i)$ of r_i.

Step 7: If (lv>1), then for each direct junior role r_j of r_i, $(r_i, r_j) \in RH$, execute Step 8 – Step 12.

Step 8: Compute the permission aggregation that r_i directly inherits from r_j, $Temp = \{p \mid p \in p_{pub}(r_j) \cup (p_{rs}(r_j) \cap t \in TR)\}$.

Step 9: $P_{\text{direct}}(r_i) = P_{\text{direct}}(r_i) \cup Temp$.

Step 10: If (lv>2), compute the permission aggregate that r_i indirectly inherits from r_j. For each direct junior role r_k for r_j, $(r_j, r_k) \in RH$, execute Step 11 – Step 12.

Step 11: $Temp = \{p \mid p\ t \in p_{pub}(r_K) \cup ((p_{rs}(r_k) \cap t \in TR) \cap ((Inherit.S = \text{"NMI"}) \cup (Inherit.S = \text{"SPUI"})) \cap n>0)\}$.

Step 12: $P_{\text{indirect}}(r_i) = P_{\text{indirect}}(r_i) \cup Temp$.

Step 13: Return all the permission aggregates $P(r_i) = \{P_{\text{direct}}(r_i) \cup P_{\text{indirect}}(r_i)\}$ of r_i.

Step 14: Return $P(r)$.

From the algorithm, we can see the time-consumption for computing a role's permission aggregation is $O(|R|*(|R|+|RH|))$.

14.4.2 Temporal Inheritance Character

In order to solve private role, we divide permission into public permission, restricted special permission and private permission. Because of the division, the inheritance of senior role from junior role is incomplete. On one side, senior role can inherit all or a part of permissions of the junior role. On the other side, in the up-transferring process, the junior role's permission is contracted according to the condition. The situation does not exist where some permissions of junior role is not inherited by the direct senior role but by the higher senior role. This phenomenon is constraint inheritance. Similarly, in the up-transferring process, the time constraint is contracted.

Character 14.1 Time constraint on $Permission_{\text{RS}}$ is transferable and contractible.

Proof Permission aggregation $P(r)$, which a role gets by inheriting, can be divided into two parts: propagatable $P(r)$ and unpropagatable $P(r)$. The former consists of $Permission_{\text{PUB}}$ and $Permission_{\text{RS}}$ inherited by NMI and SPUI. The latter consists of $Permission_{\text{PRI}}$ and $Permission_{\text{RS}}$ inherited by SPVI. The time constraint is transferring with the propagatable permission. For two roles $r_i \geqslant_t r_j$, P_{ij} is the permission of r_i inherited from r_j, TR is the time constraint on P_{ij}, P'_{ij} is part of P_{ij} that can up-transfer from role r_i, TR' is the time constraint on P'_{ij}, τ is the time extent that r_i is enable. So, $TR' = TR \cap \tau$, $TR' \subseteq TR$. Therefore, time constraint appending on the propagatable permission is transferable and contractible.

Character 14.2 The restricted inheritance relationship of temporal role hierarchy is transferable and contractible.

Proof Permission aggregation $P(r)$ that a role gets by inheriting can be divided into two parts: propagatable $P(r)$ and unpropagatable $P(r)$. The permission that senior role inherits from junior role can only be propagatable $P(r)$ of junior role. For two roles $r_i \geqslant_t r_j$, P_{ij} is the permission of r_i inherited from r_j, TR is the time constraint on P_{ij}, P'_{ij} is part of P_{ij} that can up-transfer from role r_i, TR' is the time constraint on P'_{ij}, $P'_{ij} \subseteq P_{ij}$, $TR' \subseteq TR$.

Role r_i, r_j, r_k have inheritance relationship: $r_i \geqslant_t r_j$, $r_j \geqslant_t r_k$, from $r_i \geqslant_j r_k$, P_{jk} is the permission of r_j inherited from r_k, $t \in TR_{jk}$, only $Permission_{\mathrm{PUB}}$ and $Permission_{\mathrm{RS}}$ inherited through NMI can be inherited by r_i, that is, P'_{jk}, $t \in TR'_{jk}$, permission can up-transfer from r_i is P''_{jk}, $P''_{jk} \subseteq P'_{jk} \subseteq P_{jk}$, $TR''_{jk} \subseteq TR'_{jk} \subseteq TR_{jk}$. Therefore, propagatable permission and time constraint in temporal role hierarchy is transferable and contractible.

14.4.3 Space and Time Efficiency Analysis

First, we analyze space efficiency of temporal extended role hierarchy. The temporal extended role hierarchy adds definition of permission type to each of the permissions (public, private and restricted special), adds inheritance mode to inheritance relationship (NMI, SPUI and SPVI), reduces private role. We do a comparative analysis. For traditional role hierarchy with private role, the number of private roles is npr, the space for the role is $srt = sr$(role space) + st(time space), the space for inheritance relationship is $spt = sp$(inheritance relationship space) + st(time space), then the total space for private role is $s = npr*(srt + spt)$. For the temporal extended role hierarchy in this chapter, the added space is to reserve permission type and inheritance mode. The worst situation is that all the permissions are private permissions, then the space $S' = |RH| + |P|$. Express the space for inheritance relationship and permission space (we have considered the time factor). Compare s' and S. When $npr > (|RH| + |P|/(sr + sp)$, the temporal extended role hierarchy can economize space. For enterprises, such as Guangzhou Telecom Company, IBSS is an important business support system. It contains over 120 businesses, nearly 12 substations and 10 centers to use it. The system contains about 550 permissions. They are assigned to over 100 station roles. Considering private role and department role, the total roles are over 400. Table 14.2 shows the extended role hierarchy. It only needs 120 roles to finish the same function.

Table 14.2 Role comparison of two models

model	Role type				
	Leader role	Business role	Private role	restricted role in department	public role in department
private role model	10	90	140	150	20
extended inheritance model	10	90			20

According to this example, we execute an exercise to analyze the space efficiency. When the private role attached is 140, the space of extended inheritance model reduces than that of private role model. Result is illustrated in Fig. 14.5.

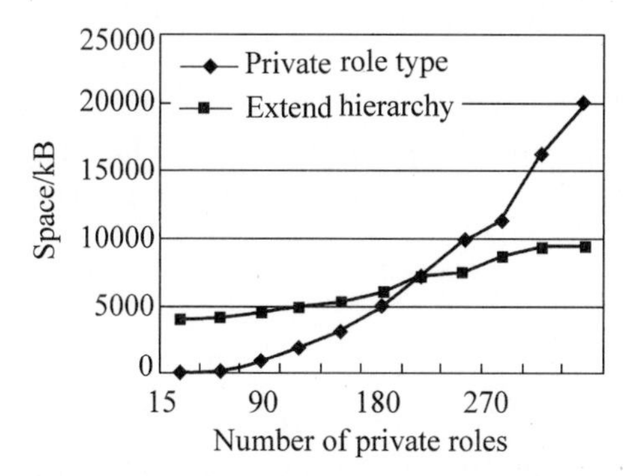

Figure 14.5 Space comparisons of the two models

For temporal extended role hierarchy, the authorization to a user consists of 5 processes, their time costs are: permission assignment to role T_{PR}, the building of role hierarchy T_{RH}, user assignment to role T_{UR}, computation of the direct and indirect authority $T_{\mathrm{PRAccount}}$, appending and judging for time constraint $T_{\mathrm{tconstraint}}$. From these processes, let us analyze the time cost of private role model and extended hierarchy model.

- T_{PR}: No matter which model, they all need to assign permission to role. Since the total number of permissions is same, so the time cost of permission assignment is equal for two models $T_{\mathrm{PR}} = T'_{\mathrm{PR}}$.
- T_{RH}: Since in extended role hierarchy we use *permission*$_{\mathrm{PRI}}$, *permission*$_{\mathrm{RS}}$ to substitute the private role, so the total number of roles reduces. Then number of inheritance relationship reduces, that is, $|RH'| < |RH|$, so $T'_{\mathrm{RH}} < T_{\mathrm{RH}}$.
- T_{UR}: Since inducting private permission, the number of roles reduces. Face to the same user, user-role assignment relationship reduces, so the time cost reduces. $T'_{\mathrm{UR}} < T_{\mathrm{UR}}$.
- $T_{\mathrm{PRAccount}}$: In terms of calculating the direct and indirect permission of roles, as mentioned in Algorithm 14.1, the time cost is decided by $O(|R| + |RH|)$. Since $|R'|<|R|$ and $|RH'|<|RH|$, we can get $T'_{\mathrm{PRAccount}} < T_{\mathrm{PRAccount}}$.

We discuss the temporal extended role hierarchy, so we should analyze the time constraint in inheritance relationship.

- $T_{\mathrm{tconstraint}}$: In temporal role hierarchy, time constraint acts on inheritance relationship, role state and restricted special permission. Since $|RH'|<|RH|$, the time cost of time constraint on extended model is less than that of private model. For time constraint on role state, in private model, in virtue of private role, the number of roles is enlarged. The complexity of how to compute the time constraint's influence on role state is related to $|R|$, so the same time cost of time constraint on extended model is less than that of private one. For time constraint on $permission_{RS}$, $permission_{RS}$ is specific for extended model, $permission_{RS} + permission_{PRI} + permission_{PUB}$ = private role + common role. The number of $permission_{RS}$ is less than that of private roles, so the time cost of time constraint on extended model is less than that for private one. Above all, $T'_{tconstraint} < T_{tconstraint}$.

The time cost of temporal extended role hierarchy $T' = T'_{PR} + T'_{RH} + T'_{UR} + T'_{PRAccount} + T'_{tconstraint}$. The time cost of private role one is $T = T_{PR} + T_{RH} + T_{UR} + T_{PRAccount} + T_{tconstraint}$, $T' < T$. So the time efficiency of temporal extended role hierarchy is better.

References

[1] Ahn GJ, Sandhu R (**2000**) Role-based authorization constraints specification. ACM Transactions on Information and System Security 3(4): 207 – 226

[2] Chaari S, Biennier F, Amar CB, Favrel J (**2004**) An authorization and access control model for workflow. In: Proceedings of the lst International Workshop on Computer Supported Activity Coordination, CSAC, Porto, Portugal, pp 21 – 30

[3] Crampton J (**2003**) On permissions, inheritance and role hierarchies. In: Proceedings of l0th ACM Conference on Computer and Communication Security, Washington, pp 85 – 92

[4] Fereaiolo D, Sandhu R, Gavrila S, Kuhn DR, Chandramouli R (**2001**) Proposed NIST standard for role-based access control. ACM Transactions on Information and System Security 4(3): 224 – 274

[5] Kahtani MAA, Sandhu R (**2003**) Induced role hierarchies with attribute-based RBAC. In: Proceedings of the 8th ACM symposium on Access Control Models and Technologies, Como Italy, pp 142 – 148

[6] McDaniel P (**2003**) On context in authorization policy. In: Proceedings of the 8th ACM Symposium on Access Control Models and Technologies

[7] Moffett JD (**1998**) Control principle and role hierarchies. In: Proceedings of the 3rd ACM Workshop on Role-based Access Control, Farifax,VA, USA, pp 63 – 69

[8] Suroop MC, Joshi JBD (**2005**) Towards administration of a hybrid role hierarchy. In: Proceedings of Information on Reuse and Integration Conference, pp 500 – 505

[9] Yu W, Zhang Q (**2004**) The research on the private permission of access control based on RBAC. Computer Application and Research 21(4): 50 – 51

15 Temporal Workflow Modeling and Its Application

Yong Tang[1,2], Guohua Chen[1], Yan Pan[1], and Yang Yu[1]

[1] Department of Computer Science, Sun Yat-sen University, Guangzhou 512075, P.R. China
[2] Computer School, South China Normal University, Guangzhou 510631, P.R. China

Abstract Time-information management in workflow has been recognized as one of the most significant tasks in workflow management. The uncertainties in time and time-related constraints in workflow models should be taken into consideration. This chapter first presents the meta-model of Temporal Workflow by extending and modifying the WfMC's Basic Process Definition Meta-model. The temporal attributes of elements and their relations in the model are analyzed in detail. Based on the introduction of time constraints on elements in Fuzzy-timing Petri Nets, this chapter proposes a new workflow model named Fuzzy Temporal Workflow Nets (FTWF-nets). The calculation of temporal elements in FTWF-nets is given. Subsequently, time modeling and time possibility analysis of temporal phenomena in FTWF-Nets are investigated. Finally, an example is given to illustrate how to use these methods.

Keywords *workflow, FTN, time modeling, time possibility analysis*

15.1 Introduction

In recent years, time management in **workflow** has been one of the most active research areas in both academic and industrial communities. There are many kinds of time constraints in business processes. For instance, a work task or a whole process should be finished in a limited duration. Enterprises may suffer great losses when these time constraints are violated. For example, in a customer claim handling process, claims that are not being handled in time may depress the customers. An enterprise should pay penalty when a commercial contract has not finished execution on time. Thus, it is of great importance to manage the time information in workflows effectively and to avoid the violation of time constraints.

The research on time management in workflows mainly focuses on time planning of workflow execution, estimation of activity duration, avoidance of

issty@mail.sysu.edu.cn

violation of time constraints on activities or processes and exception handling of time constraint violation (Li and Fan 2002a). The key for an effective approach of time management in workflows is an effective time modeling and analysis method. **Petri Nets** have attracted researchers' attention because of their mathematical basis and powerful descriptive ability. Some workflow models based on extended Petri nets have been proposed to describe time information in workflows (Ling and Schmidt 2000; Du et al. 2003; Li and Fan 2002b; Li et al. 2002; Li and Fan 2004). In these models, the time information is supposed to be certain. However, in the real world, there are many uncertainties in time information in workflows and the time information is hard to describe precisely. (Murata 1996; Zhou and Murata 1999) put forward **Fuzzy-Timing Petri Nets** (FTN) to describe time uncertainty in Petri nets. This chapter first presents the **meta-model** of **Temporal Workflow** by extending and modifying the WfMC's Basic Process Definition Meta-model. The chapter, subsequently, analyzes in detail the temporal attributes of elements and their relations in the model. Then a workflow model named Fuzzy Temporal Workflow Nets (FTWF-Nets) based on the extension of FTN is proposed. The temporal behavior analysis in workflows using FTWF-Nets is discussed.

Performance analysis of workflow plays an important role in implementing successful workflow management. Conventionally, Petri nets have been widely used for performance analysis of business processes (Li and Fan 2002; Murata 1996; Zhou and Murata 1999; Ling and Schmidt 2000; Du et al. 2003; Li et al. 2002; Li and Fan 2004). However, most of the existing research works are based on the assumption that the execution time of each activity (transition) is exponentially distributed, which is hard to be satisfied in practical environment. In many cases, especially in the early stages of the development of a workflow system, workflow designers may have known the execution time of activities from previous execution or estimation by experience. They may have required an evaluation method based on time performance to estimate the average turnaround time of workflow processes.

In this chapter, we extend the proposed FTWF-nets for time performance evaluation in workflow models. We present an algorithm to decompose an extended FTWF-net model into a set of subnets without alternative control structures. Then we estimate the turnaround time of each subnet and finally compute the average turnaround time of the workflow models.

15.2 Related Work

Petri Nets based workflow technologies have become a hot research topic in recent years. At the same time, researchers have paid increasing attention to time information management in workflow. Many Petri Nets based time related workflow models have been proposed (Ling and Schmidt 2000; Du et al. 2003; Li et al. 2002;

Li and Fan 2004).

(van der Aalst 1998) has put forward Workflow Nets (WF-Nets), which are used in modeling and analysis of workflow control structures. He has also defined in WF-Nets "soundness", an important evaluating criterion of workflow control structures.

(Ling and Schmidt 2000) has proposed Time Workflow Nets (TWF-Nets), the extended WF-Nets. In TWF-Nets, each transition is attached with an interval and the execution duration of the transition must be within that interval. They also defined "time safety" in TWF-Nets.

Based on TWF-Nets, (Du et al. 2003) has proposed Extended Time Workflow Nets (XTWF-Nets). An XTWF-Net consists of several TWF-nets based on specific rules and it can describe concurrent time constraints in workflows. XTWF-Nets also introduced time-zoom related time mapping functions, which can be used to describe time constraints on workflows of different time zooms.

Based on time constraint requirements in real world workflow processes, (Li and Fan 2002b, 2004) have proposed Time Constraint Workflow Nets (TCWF-Nets). TCWF-Nets combine the analysis methods in Time Constraint Petri Nets and TWF-Nets. It can describe the arrival time of a workflow instance, activity enabled time, allowable interval of activity firing and activity execution duration. Li has also proposed a method to analyze the time constraints in workflows.

Nonetheless, in the models above, time information is certain and precise. They all lack the ability to describe time uncertainty. However, in the real world, because of the dynamic characteristics of resources and activities in workflows, much of the time information is uncertain and cannot be described precisely.

On the other hand, (Murata 1996; Zhou and Murata 1999) have proposed Fuzzy-Time Petri Nets (FTN). FTN are a kind of colored Petri Nets and introduced four fuzzy time functions, which are fuzzy time stamp function, fuzzy enabled time function, fuzzy occurrence time function, and fuzzy delay function. FTN can describe and analyze the fuzzy temporal behavior in workflows. However, it cannot describe the time constraints on transitions and resources. (Zhou et al. 2000) has proposed Extended FTN (EFTN), which introduced the valid interval constraints on transitions. Unfortunately, EFTN still ignores the valid interval constraints on resources, which means the life cycle duration of resources. To the best of our knowledge, there is little research on valid interval constraints on resources in workflow, except for a brief discussion in (Yu et al. 2004).

15.3 A Modified Workflow Meta-Model and Temporal Attributes

A meta-model defines a language for expressing a model, which is used to define and construct the rules of a semantic model. The workflow meta-model is used to describe the elements, their relations and the attributes of these elements and relations in a workflow system. The first step of Temporal Workflow research is to endue these elements and relations with temporal attributes.

WfMC presented a Basic Process Definition Meta-model (Hollingsworth 1995). The elements that should be defined in meta-model depend on the requirements of the application domain. We present a meta-model, including a Build-time Meta-model and a Run-time Meta-model. For many applications, the conceptions defined in this meta-model are helpful.

15.3.1 Build-Time Meta-Model

Build-Time Meta-Model consists of four sub-models: organization meta-model, information meta-model, application meta-model, and process meta-model.

1. Organization meta-model

As shown in Fig. 15.1, organization meta-model describes the resource-relevant conceptions and their relations in workflow. Activities are executed by resources. Resources can be divided into organizational units according to the structure of the organization, or roles according to the functional characteristics.

Considering that resources are often allocated to activities by the intersection of organizational units and roles, a conception "UnitRole" is introduced. An organizational unit consists of several UnitRoles, e.g., market department consists of leaders, supporting engineers, salespersons. At the same time, these leaders belong to a subset of department leader, and supporting engineers belong to a subset of engineer, and so on. A user can be either a person or an agent (Chiu et al. 1999).

One user can entrust another user with his UnitRoles. "RR" includes "Belong" relation and "Peer" relation (Yu et al. 2004).

Putting the organization meta-model into time dimension, we can see that an organization has its created time and valid time. For example, a company is founded on a particular day and its business license is valid for 10 years. A user is employed on a certain day and the valid time is defined in the labor contract. A person is appointed a role from a particular day, and the valid time is 3 years. A user will have a holiday for 3 days from the next day. He entrusts his roles to another user. The "entrust" relation has its start time (the next day) and valid time (3 days), and so on.

2. Information meta-model

The data perspective of workflow deals with production data and control data (Chiu et al. 1999). Production data are information objects (e.g., documents, forms). They are handled by the applications. Control data are data introduced solely for

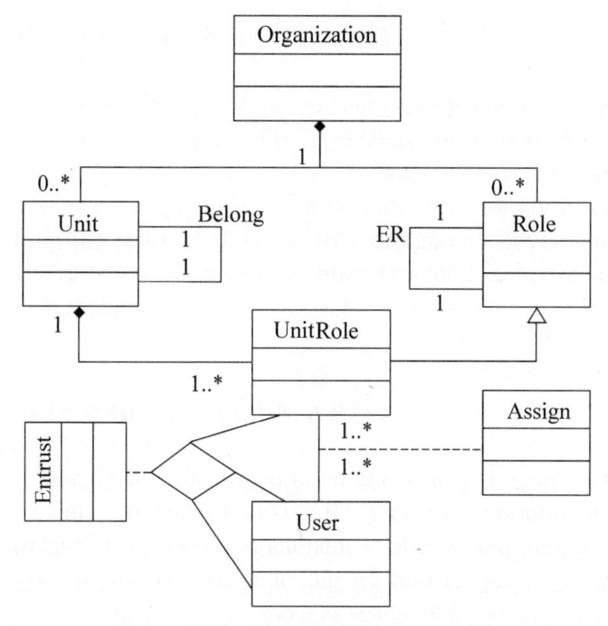

Figure 15.1 Organization meta-model

workflow management purposes. They are produced and consumed by workflows, such as the input and output parameters of activities, variables introduced for routing purposes, the index or control parameters for production data (e.g., identifier, priority, deadline). As shown in Fig. 15.2, information meta-model describes the conceptions and their relations about control data.

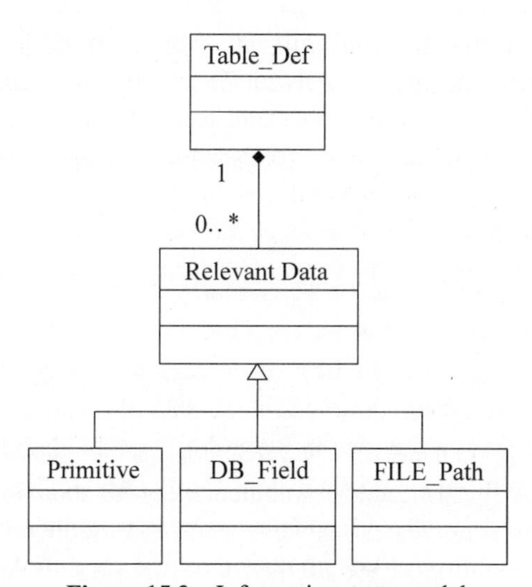

Figure 15.2 Information meta-model

In this meta-model, relevant data can be a primitive type (e.g., parameters, variables), or a DB_Field type that links to a field of database table, or a FILE_Path type that includes the file path. Table_Def is a structured abstract of a group of relevant data, which always expresses a meaning.

The elements in this meta-model also have their own temporal attributes, e.g., in a Claim Handling System, a claim table should be created when receiving a customer's claim, and the table should be handled within 3 days. This means the claim table has a created time and a valid period of 3 days. If the claim table is not handled within its valid period, an exception-handling process should be started.

3. Application meta-model

As shown in Fig. 15.3, the application meta-model describes the conceptions and their relations about invoked applications. An invoked application can be a common table-handling application, a conventional application or a URL to a web service.

According to the rules specified in ISO 9000, as a tool, an invoked application must be checked with its validity and veracity periodically. Therefore, an application has its own valid period.

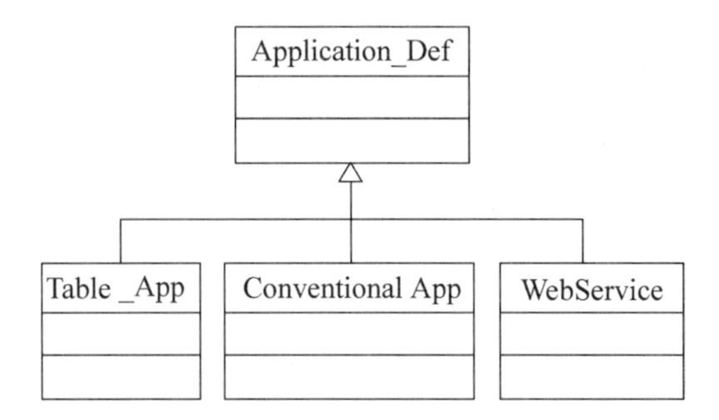

Figure 15.3 Application meta-model

4. Process meta-model

As shown in Fig. 15.4, relevant conceptions are defined in process meta-model to specify the activities that need to be executed along with the order in which they should be executed (i.e., the routing or control flow). In order to support the structured process definition (Tian and Li 2003), the conception sub-process is introduced. Profiting from the idea in literature (Tian et al. 2004), the description of the structures (split, join) and their constraints (AND, OR, XOR) is separated from the description of activity and a conception connector is introduced to express it. In fact, a connector can be considered as an activity with some special functions. The executing order of activities is determined by "Transition-Rule".

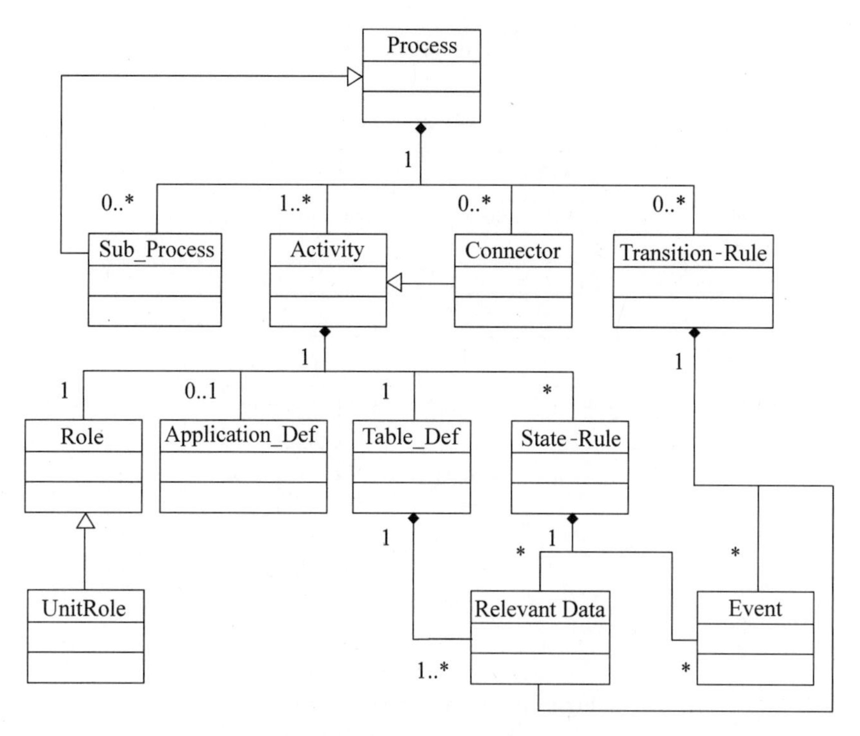

Figure 15.4 Process meta-model

One activity can be described as, one role operates on a table with an invoked application during which the state of activity changes according to a set of stated rules. The changes in activity states and the transitions among the activities are all inspired by events. Transition rules and state rules consist of events and relevant data.

Each version of a process definition has its created time. Once a new version is created, new instances will be generated from it. An activity definition has its created time and valid period also. To meet the needs of workflow management, a process or an activity is always assigned an earliest finish time and a latest finish time (Murata 1996).

15.3.2 Run-Time Meta-Model

As shown in Fig. 15.5, run-time meta-model describes the conceptions and their relations in workflow run-time period. In this meta-model, process instances, activity instances, applications, and tables are all instantiated from the relevant conceptions in the build-time meta-model. Here, roles have been mapped into concrete users.

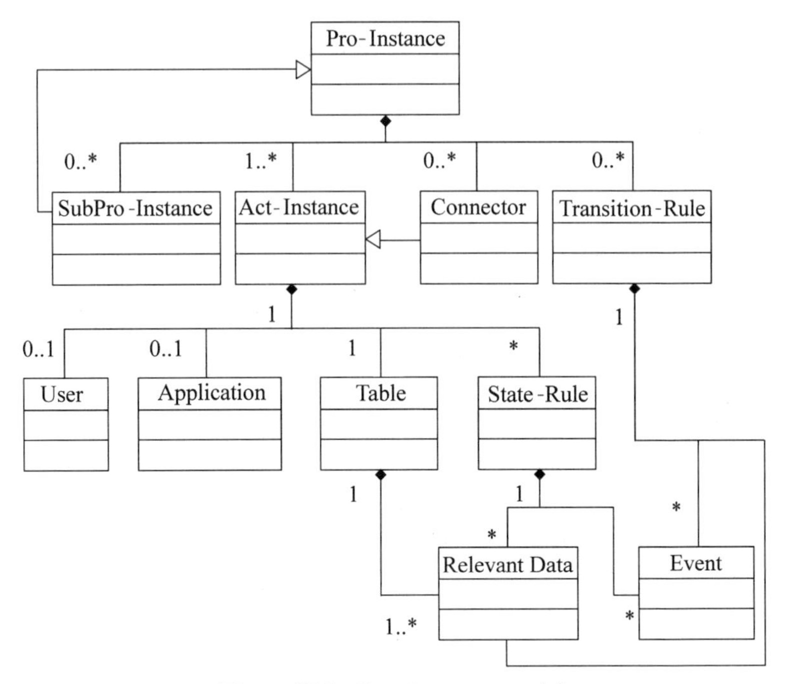

Figure 15.5 Run-time meta-model

Every process instance or activity instance has its own created time and lifetime. In the build time, the temporal attributes of processes and activities are always defined in relative time. In run time, the temporal attributes of process instances and activity instances are always defined in absolute time.

15.3.3 A Formal Model of Temporal Workflow

From the research on workflow meta-model, we get the primary elements and relations and we have analyzed their temporal attributes. Based on these results, we can describe the temporal workflow as a formal model. Here, we only formalize the primary entities.

Definition 15.1 A workflow process $WP = \langle Pid, Pn, Ver, TS, AS, F, tp, Tp \rangle$, where *Pid* is the identifier of the process template, *Pn* is the process name, *Ver* is the version number, *TS* is the set of transition rules, *AS* is the set consisting of activities, connectors and sub-processes, *F* is the set of mapping rules from *AS* to *TS*, *tp* is the creation time of *WP*, *Tp* is the valid period of *WP*.

From Definition 15.1, we can see that there is a good mapping from the process definition to Petri Net. For example, an activity can be mapped to a TRANSITION, and a set of transition rules can be mapped to a PLACE.

Definition 15.2 An activity $WA = \langle Aid, An, Ver, Pid, Rid, Tid, APid, Lim, ta, Ta, State, SRule \rangle$, where *Aid* is the identifier of *WA*, *An* is the activity name, *Ver* is the version number, *Pid* is the identifier of the process that *WA* belongs to, *Rid* is the identifier of the role, *Tid* is the identifier of the table instance, *APid* is the identifier of the application, *Lim* is the limit for activities to access relevant data, *ta* is the creation time of *WA*, *Ta* is the valid period of *WA*, *State* is a set of activities states, *SRule* is a set of stated rules.

Definition 15.3 A process instance $Ip = \langle Ipid, Pid, TS, AS, F, PS, tp, Tp \rangle$, where *Ipid* is the identifier of the process instance, *Pid* is the identifier of the process that *Ip* belongs to, *TS* is the set of transition rules, *AS* is the set consisting of instances of activities, connectors and sub-processes, *F* is the set of mapping rules from *AS* to *TS*, *PS* is the current state of *Ip*, *tp* is the creation time of *Ip*, *Tp* is the valid period of *Ip*.

Definition 15.4 An activity instance $Ia = \langle Iaid, Ipid, Aid, Uid, Tid, APid, ta, Ta, SRule, s \rangle$, where *Iaid* is the identifier of the activity instance, *Ipid* is the identifier of process instance that *Ia* belongs to, *Aid* is the identifier of activity that *Ia* belongs to, *Uid* is the identifier of the user, *Tid* is the identifier of the table instance, *APid* is the identifier of the application, *ta* is the creation time of *Ia*, *Ta* is the valid period of *Ia*, *SRule* is a set of stated rules, $s \in State$ is the current state of *Ia*.

Definition 15.5 A role $WR = \langle Rid, Rn, C, A, tr, Tr \rangle$, where *Rid* is the identifier of *WR*, *Rn* is the name of *WR*, *C* is the set of *WR's* abilities, *A* is the set of limit rules for *WR* to access relevant data, *tr* is the created time of *WR*, *Tr* is the valid period of *WR*.

Definition 15.6 A table $WT = \langle Tid, Tn, Ver, DataList, Frame, td, Td \rangle$, where *Tid* is the identifier of *WT*, *Tn* is the name of *WT*, *Ver* is the version number, *DataList* is a list of workflow-relevant data, *Frame* is the definition of table frame, *td* is the creation time of *WT*, *Td* is the valid period of *WT*.

15.4 Fuzzy Temporal Workflow Nets (FTWF-Nets)

15.4.1 Fuzzy Time Point

Definition 15.7 Fuzzy time point is the possibility distribution of a function mapping from time scale Γ to real interval [0,1], which restricts the more or less possible value of a time point. Let π_a denote the possibility distribution function attached to a time point *a*, then $\forall \tau \in \Gamma, \pi_a(\tau)$ denotes the numerical estimate of the possibility that *a* is precisely τ. Let fuzzy set *A* be the possible range of time point *a* and μ_A denote the membership function of *A*, then we have $\forall \tau \in \Gamma$, $\pi_a(\tau) = \mu_A(\tau)$. In this chapter, fuzzy time point is denoted by trapezoid possible distribution, which must be normal and convex. Thus, a fuzzy time point can be represented by $h(\pi_1,\pi_2,\pi_3,\pi_4)$. An example is shown in Fig. 15.6.

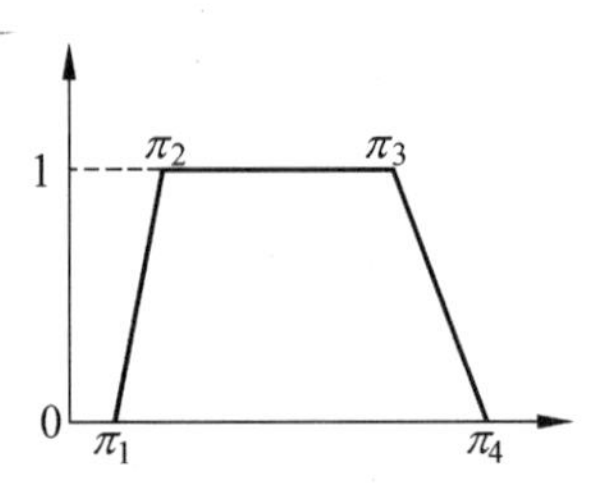

Figure 15.6 Trapezoid function of a fuzzy time point

15.4.2 Formal Definition for FTWF-Nets

Definition 15.8 A Fuzzy Temporal Workflow Net (FTWF-Nets) has 9 tuples $(P,T,A,FT,FE,FO,D,RVT,TVT)$:

- P is a set of places, T is a set of transitions and $P\cap T=\varnothing$, $P\cup T\neq\varnothing$; A is a set of arcs, which is a subset of $(P\times T)\cup(T\times P)$. The triple (P,T,F) forms a Workflow Net (van der Aalst 1998).
- FT denotes a set of fuzzy time stamps, which is attached to tokens (resources). Let $\pi(\tau)$ denote the fuzzy time stamp function, which describes the possibility distribution of a token's arrival time on a place.
- FE denotes a set of fuzzy enabled time of transitions. Let $e_t(\tau)$ be the fuzzy enabled time function of transition t. $e_t(\tau)$ describes the possibility distribution of t being enabled at time point τ.
- FO denotes a set of fuzzy occurrence time of transitions. Let $o_t(\tau)$ be the fuzzy occurrence time function of transition t, which describes the possibility distribution of t firing at time point τ.
- D denotes a set of fuzzy delays of transitions. Let $d_t(\tau)$ be the fuzzy delay function of transition t, which describes the possibility distribution of the duration from the firing time of t to the time when t outputs tokens to its output places.
- RVT: $TOKEN\rightarrow Q^+\times(Q^+\cup+\infty)$ denotes a set of valid intervals constraints on tokens (resources). Each of them can be represented by a relative interval $[a,b]$ (a and b are both time points, $0\leqslant a<b$). Let τ denote the fuzzy time stamp of a token. Then the valid interval constraint $[a,b]$ on the token means that the token is available only between $a+\tau$ and $b+\tau$.
- TVT: $T\rightarrow Q^+\times(Q^+\cup+\infty)$ denotes a set of valid internal constraints on transitions (activities). Each of them can be represented by a relative interval $[c, d)$ (c and d are both time points, $0\leqslant c<d$). Let τ denote the fuzzy enabled time of a transition. Then the valid internal constraint $[c, d)$ on t means that the transition can only fire between $(c+\tau)$ and $(d+\tau)$.

15.4.3 Time Related Calculation in FTWF-Nets

Definition 15.9 The fuzzy enabled time $e_t(\tau)$ of transition t denotes the possibility distribution of the latest arrival time of the input token of t. It can be computed by $e_t(\tau) = latest\{\pi_i(\tau) \oplus [a_i, b_i], i = 1, 2, \cdots, n\}$, where *latest* is the operator that computes the "latest-arrival/lowest possibility distribution" from n distributions (Murata 1996), and $\oplus$ is the extended addition(Dubios and Prade 1989). $\pi_i(\tau)$ is the fuzzy time stamp of the enabled token i arriving at the input place p_i of transition t and $[a_i, b_i]$ is the relative valid interval constraint on token i. Let $h_i(\pi_{i1}, \pi_{i2}, \pi_{i3}, \pi_{i4})$ be the trapezoid function of $\pi_i(\tau)$, then

$$\begin{aligned}\pi_i(\tau) \oplus [a_i, b_i] &= h_i(\pi_{i1}, \pi_{i2}, \pi_{i3}, \pi_{i4}) \oplus I(a_i, a_i, b_i, b_i)\\ &= \min\{h_i, I\}(\pi_{i1} + a_i, \pi_{i2} + a_i, \pi_{i3} + b_i, \pi_{i4} + b_i)\\ &= h_i(\pi_{i1} + a_i, \pi_{i2} + a_i, \pi_{i3} + b_i, \pi_{i4} + b_i)\end{aligned}$$

According to the approximate computation of latest operator in (Murata 1996), $e_t(\tau)$ can be computed by

$$\begin{aligned}e_t(\tau) &= latest\{\pi_i(\tau) \oplus [a_i, b_i], i = 1, 2, \cdots, n\}\\ &= latest\{h_i(\pi_{i1} + a_i, \pi_{i2} + a_i, \pi_{i3} + b_i, \pi_{i4} + b_i), i = 1, 2, \cdots, \mathrm{n}\}\\ &= \min\{h_i\}(\max\{\pi_{i1} + a_i\}, \max\{\pi_{i2} + a_i\}, \max\{\pi_{i3} + b_i\}, \max\{\pi_{i4} + b_i\}),\\ &\qquad i = 1, 2, \cdots, n\end{aligned}$$

Definition 15.10 The fuzzy occurrence time of transition t_k denotes the possibility distribution of the firing time point of t_k. Suppose there are n quasi-enabled transitions t_i, $i = 1, 2, \cdots, k, \cdots, n$. Let their fuzzy enabled times be $e_{ti}(\tau)$ and their valid internal constraints be $[c_i, d_i]$. Then the fuzzy occurrence time of t_k can be computed by $o_{tk}(\tau) = MIN\{e_{tk}(\tau) \oplus [c_k, d_k], earliest(e_{ti}(\tau) \oplus [c_i, d_i], i = 1, 2, \cdots, n)\}$, where *earliest* is the operator that selects the earliest enabled time of the quasi-enabled transitions. *MIN* is the intersection of distributions(Dubios and Prade 1989). Let $h_{ti}(\pi_{i1}, \pi_{i2}, \pi_{i3}, \pi_{i4})$ be the trapezoid function of $e_{ti}(\tau)$, $i = 1, 2, \cdots, n$, then

$$\begin{aligned}e_{ti}(\tau) \oplus [c_{ti}, d_{ti}] &= h_{ti}(\pi_{i1}, \pi_{i2}, \pi_{i3}, \pi_{i4}) \oplus I(c_i, c_i, d_i, d_i)\\ &= \min\{h_{ti}, I\}(\pi_{i1} + c_i, \pi_{i2} + c_i, \pi_{i3} + d_i, \pi_{i4} + d_i)\\ &= h_i(\pi_{i1} + c_i, \pi_{i2} + c_i, \pi_{i3} + d_i, \pi_{i4} + d_i)\end{aligned}$$

According to the approximate computation of earliest operator in (Murata 1996),

$$\begin{aligned}&earliest(e_{ti}(\tau) \oplus [c_i, d_i], i = 1, 2, \cdots, n)\\ &= earliest(h_{ti}(\pi_{ti1} + c_{ti}, \pi_{ti2} + c_{ti}, \pi_{ti3} + d_{ti}, \pi_{ti4} + d_{ti}), i = 1, 2, \cdots, n)\\ &= \max\{h_{ti}\}(\min\{\pi_{ti1} + c_{ti}\}, \min\{\pi_{ti2} + c_{ti}\}, \min\{\pi_{ti3} + d_{ti}\}, \min\{\pi_{ti4} + d_{ti}\})\end{aligned}$$

Thus $o_{tk}(\tau)$ can be computed by $o_{tk}(\tau) = MIN\{h_{tk}(\pi_{k1} + c_k, \pi_{k2} + c_k, \pi_{k3} + d_k, \pi_{k4} + d_k), \max\{h_{ti}\}(\min\{\pi_{ti1} + c_{ti}\}, \min\{\pi_{ti2} + c_{ti}\}, \min\{\pi_{ti3} + d_{ti}\}, \min\{\pi_{ti4} + d_{ti}\})\}$, $i = 1, 2, \cdots, n$.

Definition 15.11 Suppose *To* is an output token of transition *t*, $o_t(\tau)$ is the fuzzy occurrence time of *t* and $d_t(\tau)$ is the fuzzy delay of *t*, then the fuzzy time stamp of token *To* can be computed by $\pi_{To}(\tau) = o_t(\tau) \oplus d_t(\tau)$. Let $h_1(\pi_1, \pi_2, \pi_3, \pi_4)$ be the trapezoid function of $o_t(\tau)$ and $h_2(d_1,d_2,d_3,d_4)$ be the one of $d_t(\tau)$, then

$$\begin{aligned}\pi_{To}(\tau) &= h_1(\pi_1, \pi_2, \pi_3, \pi_4) \oplus h_2(d_1, d_2, d_3, d_4)\\ &= \min\{h_1, h_2\}(\pi_1 + d_1, \pi_2 + d_2, \pi_3 + d_3, \pi_4 + d_4)\end{aligned}$$

15.5 Time Modeling and Time Possibility Analysis

A good workflow model can not only model the logical control structures in workflows, but also describe and analyze time information and temporal behavior of resources and activities. Generally speaking, the typical temporal phenomena includes execution delays of activities, valid occurrence interval constraints on activities, valid interval constraints on resources (life cycle limits of resources), execution duration limits of processes, time distance between two activities, etc. FTWF-Nets have inherited the four fuzzy time functions in FTN. They have also introduced valid interval constraints on resources and activities. Therefore, FTWF-Nets can be used to describe fuzzy time information.

In FTWF-Nets, the fuzzy delay function $d_t(\tau)$ can describe an activity's execution delay. A resource's life cycle can be limited by the valid interval constraint $[a,b]$ on that resource. The valid interval in which an activity can fire can be limited by the valid interval constraint on the activity.

The execution duration of a process, one of the most important elements in workflow time management, means the executing time limit of a workflow instance between its start time and end time. In FTWF-Nets, we introduce a temporal logic operator Π to describe and analyze the time possibility of the execution duration of a process.

Definition 15.12 Let *a* and *b* be fuzzy time points, whose possibility distributions are $\pi_a(\tau)$ and $\pi_b(\tau)$, respectively. Let (C,A,B,D) and (G,E,F,H) be the trapezoid functions of $\pi_a(\tau)$ and $\pi_b(\tau)$, as shown in Fig. 15.7. Then the

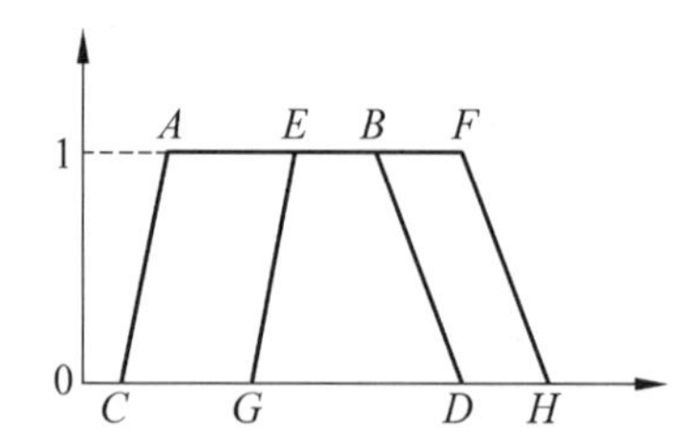

Figure 15.7 $a \leqslant b$ (*b* is a fuzzy time point)

possibility distribution of the temporal relation of b before a can be computed by

$$\Pi(a \leqslant b) = \frac{\text{Area}(\text{trapezoid}ADCB \cap \text{trapezoid}EFHG)}{\text{Area}(\text{trapezoid}EFHG)} = \frac{\text{Area}(\text{trapezoid}EBDG)}{\text{Area}(\text{trapezoid}EFHG)}$$

If b is a precise time point, then $\pi_b(\tau)$ can be represented by (t,t,t,t) as shown in Fig. 15.8. Thus,

$$\Pi(a \leqslant b) = \frac{\text{Area}(\text{trapezoid}AEFC)}{\text{Area}(\text{trapezoid}ABDC)}$$

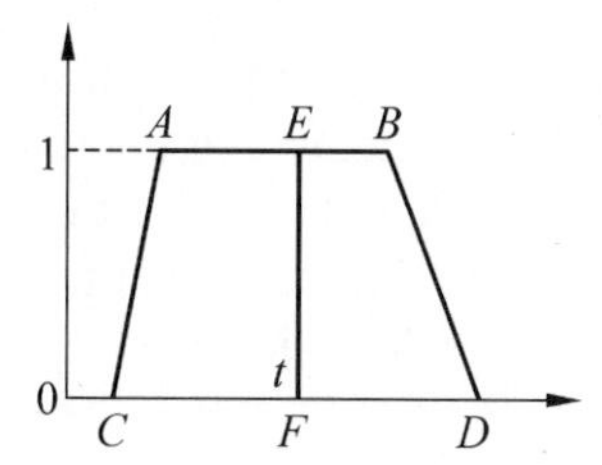

Figure 15.8 $a \leqslant b$ (b is a precise time point)

Suppose a workflow instance arrives at the initial place i of FTWF-Nets and the expected execution duration is $f(\tau)$. Then the expected fuzzy time stamp when the instance arrives at the end place o of the FTWF-Net should be $\pi_i(\tau) \oplus f(\tau)$. On the other hand, we can use the formulas in Section 15.3 to calculate the fuzzy time stamp $\pi_o(\tau)$ when the workflow instance arrives at the end place o according to every transition sequence from i to o. Thus, the time possibility Tian and Li 2003; Tian et al. 2004 that the workflow instance can finish before the execution duration constraint can be represented by $\Pi(\pi_o(\tau) \leqslant (\pi_i(\tau) \oplus f(\tau)))$.

The constraint of the time distance between two activities means that the time length between the activities should be less than a specific duration. In this chapter, the time distance is represented by the absolute value of the distance between the occurrence times of the two activities. Let A and B be two activities in a workflow instance and the expected duration between them be $f(\tau)$. Assume the occurrence times of A and B in that instance are $o_A(\tau)$ and $o_B(\tau)$, respectively, after calculation. Then the time possibility that the time distance between A and B is more than $f(\tau)$ can be represented by $\Pi((o_A(\tau) \oplus f(\tau)) \leqslant o_B(\tau))$, and the one that the time distance between A and B is less than $f(\tau)$ can be represented by $\Pi(o_B(\tau) \leqslant (o_A(\tau) \oplus f(\tau)))$.

15.6 An Illustration

In the development of several workflow-related applications, including ERP (Enterprise Resource Planning) systems and laboratory management systems, FTWF-nets have been used to model uncertain time information in work processes and to analyze their temporal behavior. The practice results show that FTWF-nets can help users to analyze the temporal behavior of workflows effectively.

The following example is selected from a laboratory management system. It illustrates the method to use FTWF-nets to analyze the time possibility of execution duration limit of a process.

Example 15.1 Suppose the valid percentage of some vitamin C, which is suspected to have been oxygenated, has to be determined in a chemical laboratory. The process is as follows and we assume that hour is used as the unit of time in this example.

Step 1: Before the determination procedure, the use of the lab should be applied. Assume the fuzzy delay of the "examine and approve" activity is 1(1,2,3,4). A 3-hour duration will be assigned when the application is examined and approved.

Step 2: Because of its easy oxygenation in the air, the vitamin should be dissolved in some acidic liquor before the determining experiment. The fuzzy delay of the dissolving step is 1(0.05,0.1,0.2,0.3). The vitamin C in the liquor can be kept stable for 0.5 hour.

Step 3: Potassium dichromate can react with potassium iodine to form iodine, which can be used in the following step. The fuzzy delay of this step is 1(0.1,0.2,0.3,0.5). Iodine is also easy to be oxygenated in the air. It can be kept stable for 0.4 hour.

Step 4: The iodine can be used to titrate the acidic liquor with vitamin C. The fuzzy delay of this step is 1(0.3,0.4,0.5,0.6).

It needs 0.1 hour for preparation before executing the experiments as specified in Steps 2, 3 and 4. We assume that each of these three steps must start within 0.05 hour after the preparation is over.

What is the time possibility that the process can finish within 7 hours?

FTWF-net representation of Example 15.4 is shown in Fig. 15.9.

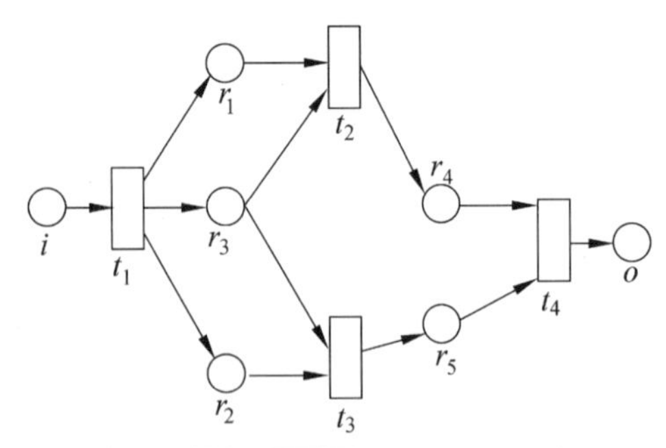

Figure 15.9 FTWF-net representation

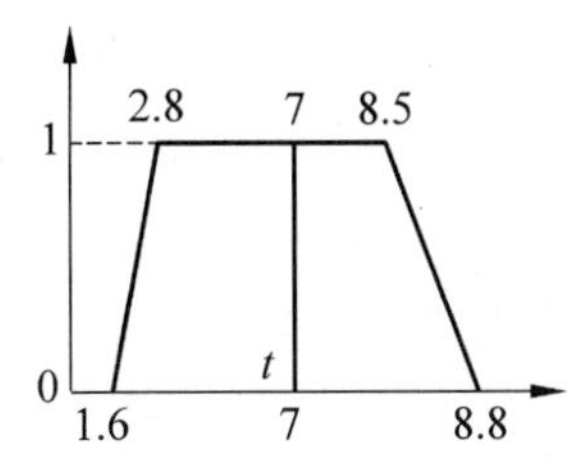

Figure 15.10 $\Pi(\pi_o(\tau) \leqslant \pi_f(\tau))$ of the determining process

The meaning of the elements in Fig. 15.9 are as follows:

i: initial place in the FTWF-Net

o: end place in the FTWF-Net

r_1: vitamin C, acidic liquor

r_2: potassium dichromate, potassium iodine

r_3: laboratory

r_4: the acidic liquor, which has dissolved vitamin C

r_5: iodine

t_1: examining and approving step of the use of the laboratory

t_2: the step when the vitamin has dissolved in the acidic liquor

t_3: the step when potassium dichromate reacts with potassium iodine and forms iodine

t_4: the step when the iodine is used to titrate the acidic liquor with vitamin C.

Let $\pi_i(\tau)$ be the fuzzy time stamp of a workflow instance at the initial place i and $\pi_i(\tau) = 1(0,0,0,0)$. There is no valid interval constraint on the token, so we can set the valid interval to be [0,0]. Thus the fuzzy enabled time of t_1 can be computed by $e_{t_1}(\tau) = 1(0+0,\ 0+0,\ 0+0,\ 0+0) = 1(0,0,0,0)$. There is no valid internal constraint on t_1. Then we can set the valid interval to be [0,0]. The fuzzy occurrence time of t_1 can be computed by $o_{t_1}(\tau) = MIN\{1(0+0, 0+0, 0+0, 0+0), 1(0+0,\ 0+0,\ 0+0,\ 0+0)\} = 1(0,0,0,0)$. The fuzzy delay of t_1 is 1(1,2,3,4). Because r_1, r_2 and r_3 are all output places of t_1, the fuzzy time stamp of r_1, r_2 and r_3 can be computed by $\pi_{r_1}(\tau) = \pi_{r_2}(\tau) = \pi_{r_3}(\tau) = o_{t_1}(\tau) \oplus$ 1(1,2,3,4) = $\min\{1,1\}(1+0, 2+0, 3+0, 4+0) = 1(1,2,3,4)$.

Let [0, 0] be the valid interval of the token in r_1 because there is no valid interval constraint on it and the valid interval of the token in r_3 is [0,3] according to Step 1. r_1 and r_3 are both input places of t_2, then $e_{t_2}(\tau) = latest\ (1(1+0, 2+0, 3+0,\ 4+0),\ 1(1+0,\ 2+0,\ 3+3,\ 4+3)) = \min\{1,1\}(\max\{1,1\},\max\{2,2\},\max\{3,6\}, \max\{4,7\}) = 1(1,2,6,7)$. We can also get $e_{t_3}(\tau) = 1(1,2,6,7)$. According to the state mentioned above, the valid interval of t_2 is $[0.1, 0.1+0.05] = [0.1, 0.15]$. Thus $o_{t_2}(\tau) = MIN\{1(1+0.1, 2+0.1, 6+0.15, 7+0.15), \max\{1,1\}\ (\min\{1+0.1,\ 1+0.1\}, \min\{2+0.1,\ 2+0.1\},\ \min\{6+0.15,\ 6+0.15\},\ \min\{7+0.15,\ 7+0.15\})\} = 1(1.1,2.1,6.15,7.15)$. Analogously. we can get $o_{t_3}(\tau) = 1(1.1,2.1,6.15,7.15)$.

The fuzzy delay of t_2 is 1(0.05,0.1,0.2,0.3) and r_4 is an output place of t_2. Thus, $\pi_{r_4}(\tau) = 1(1.1,2.1,6.15,7.15) \oplus 1(0.05,0.1,0.2,0.3) = 1(1.15,2.2,6.35,7.45)$.

Analogously, we can get $\pi_{r_4}(\tau) = 1\ (1.1,2.1,6.15,7.15) \oplus 1(0.1,0.2,0.3,0.5) = 1(1.2,2.3,6.45,7.65)$.

[0,0.5] is the valid interval of the token in r_4 according to Step 2, and [0,0.4] is one of the token in $r5$ according to Step 3. Then r_4 and r_5 are both input places of t_4. Thus, $e_{t_4}(\tau) = latest(1(1.15+0,\ 2.2+0,\ 6.35+0.5,\ 7.45+0.5),\ 1(1.2+0,\ 2.3+0,\ 6.45+0.4,\ 7.65+0.4)) = \min\{1,1\}(\max\{1.15,1.2\},\ \max\{2.2,2.3\},\ \max\{6.85,6.85\},\ \max\{7.95,8.05\}) = 1(1.2,2.3,6.85,8.05)$.

According to the state mentioned above, the valid interval of t_4 is [0.1,0.15]. Then $o_{t_4}(\tau) = \min\{1(1.2+0.1,\ 2.3+0.1,\ 6.85+0.15,\ 8.05+0.15),\ 1(1.2+0.1,\ 2.3+0.1,\ 6.85+0.15,\ 8.05+0.15)\} = 1(1.3,2.4,8,8.2)$.

The fuzzy delay of t_4 is 1(0.3,0.4,0.5,0.6) and place o is the output place for it. Thus $\pi_o(\tau) = 1(1.3,2.4,8,8.2) \oplus 1(0.3,0.4,0.5,0.6) = 1(1.6,2.8,8.5,8.8)$.

If the process should finish within 7 hours, the expected fuzzy time stamp $\pi_f(\tau)$ when the process arrives at the end place o is 1(7,7,7,7). As shown in Fig. 5.10, the area of the whole trapezoid is $(8.5-2.8+8.8-1.6)\times 1\div 2 = 6.45$, while the area of the left part of the trapezoid is $(7-2.8+7-1.6)\times 1\div 2 = 4.8$. Thus, $\Pi(\pi_o(\tau) \leqslant (\pi_f(\tau)) = 4.8\div 6.45 = 0.74$. It means that the time possibility that the process can finish within 7 hours is 0.74.

References

[1] Murata T (**1996**) Temporal uncertainty and fuzzy-timing high-level Petri nets. Application and Theory of Petri Nets, Lecture Notes in Computer Science 1091, New York: Springer-Verlag, pp 11 – 28

[2] Zhou Y, Murata T (**1999**) Petri net model with fuzzy-timing and fuzzy-metric. Special Issue on Fuzzy Petri Nets, Int J Intell Syst, 14(8): 719 – 746

[3] Li W, Fan Y (**2002a**) Overview on managing time in workflow systems. Journal of Software, 13(8): 1552 – 1558

[4] Ling S, Schmidt H (**2000**) Time Petri nets for workflow modeling and analysis. In: IEEE International Conference on Systems, Man and Cybernetics, Nashville, TN USA, 4: 3039 – 3044

[5] Du S, Tan J, Lu G (**2003**) An extended time workflow model based on TWF-net and its application. Journal of Computer Research and Development 40(4): 524 – 530

[6] Li W, Zheng G, Wang X (**2002**) A workflow model based on timed Petri net. Journal of Software 13(8): 1666 – 1671

[7] Li W, Fan Y (**2004**) Workflow model analysis based on time constraint Petri nets. Journal of Software 15(1): 17 – 26

[8] Li W, Fan Y (**2002b**) Schedulability analysis algorithm for timing constraint workflow models. Computer Integration Management System 8(7): 527 – 532

[9] van der Aalst WMP (**1998**) The application of Petri nets to workflow management. The Journal of Circuits, Systems and Computers 8(1): 21 – 66

[10] Zhou Y, Murata T, DeFanti A (**2000**) Modeling and performance analysis using extended fuzzy-timing Petri nets for networked virtual environment. IEEE Transactions on System, Man, and Cybernetics (Part B: Cybernetics) 30(5)
[11] Yu Y, Tang Y, Liang L, Feng Z (**2004**) Temporal extension of workflow meta-model and its application. In: Proceedings of the 8th International Conference on CSCW in Design. Xiamen, P. R. China, vol. 2, pp 293 – 297
[12] Dubios D, Prade H (**1989**) Processing fuzzy temporal knowledge. IEEE Transactions on System, Man, and Cybernetics 19(4): 729 – 744
[13] Tian F, Li R (**2003**) The CSCW analysis method based on Fuzzy-timing high-level Petri nets. In: Proceedings of the Second International Conference on Machine Learning and Cybernetics. Xi'an, P.R. China, pp 2547 – 2552
[14] Tian F, Li R, Zhang J (**2004**) Modeling and analysis collaborative design activities using fuzzy-timing high-level Petri nets. Journal of Computer-aided Design & Computer Graphics 16(3): 267 – 274
[15] Hollingsworth D (**1995**) The workflow reference model. Workflow Management Coalition
[16] Chiu D, Li Q, Karlapalem K (**1999**) A meta modeling approach to workflow management systems supporting exception handling. Information Systems 24(2): 159 – 184

16 Temporal Knowledge Representation and Reasoning

Na Tang[1,2], Yong Tang[1,2+], Lingkun Wu[2], and Hui Ma[2]

[1] Computer School, South China Normal University, Guangzhou 510631, P.R. China
[2] Department of Computer Science, Sun Yat-sen University, Guangzhou 512075, P.R. China

Abstract The importance of dealing with temporal properties in knowledge representation and reasoning has attracted researchers' attention for long. A number of temporal knowledge representation frameworks are proposed, however, few are implemented in practice. Since production system is a knowledge representation mechanism used by most knowledge based systems and expert systems in practice, in this chapter, temporal properties are integrated into traditional production system. In other words, both rules and facts process temporal attributes. Related definitions and reasoning algorithms are presented. A prototype implementation of a salary system is given to show the validity of the framework.

Keywords *temporal knowledge representation, temporal knowledge reasoning, temporal production system*

16.1 Introduction

Time is an important attribute in everything (Tang et al. 2004). With the prevalence and development of computer applications, **temporal information processing** has attracted a lot of attention. Temporal information processing becomes a novel technology in the new generation of databases and information systems, especially in the fields of **knowledge representation and reasoning**. In the domains like GIS, medical system, and marketing, temporal property even plays a critical role.

Researchers have been aware of the importance of temporal property for a long time. The study on this subject could date back to the 1950s. Starting from **Temporal Logic**, researchers have developed a series of formal logic frameworks, such as Modal Temporal Logic (Pnueli 1977), Propositional Temporal Logic (Sistla and Clark 1985). In the 1980s, McDermott (1982) proposed Event Calculus and

[+] Corresponding author: issty@mail.sysu.edu.cn

Allen (1983) introduced Interval Algebra, which are milestones in temporal logic researches. The work mentioned above show insights into temporal property from different facets. However, research on **temporal knowledge representation** has not been done systematically.

Previous work on temporal knowledge representation and reasoning was mainly based on temporal logic. However, sometimes the temporal logic is so expressive that some reasoning problems become impossible to decide, let alone the efficiency of solving problems in practice. In recent years, more frameworks like description logic (Baader et al. 2002) and ontology (Guarino 1998) are introduced into temporal knowledge representation.

Until now, the work mentioned above is mainly theoretical studies. Still there are a few applications based on temporal knowledge in practice. The reason is immaturity and high complexity of those ideas. When it comes to the practical applications of knowledge-based system, the framework of **production system** has taken up a great proportion. Besides, decades of experience in practice proves that production system has become one of the most mature, typical and implementation-friendly knowledge representation frameworks.

Three main components constitute the production system, namely Global Database, Production Rules and Control Strategies. The major operating mechanism of the production system is its control strategies, while the facts in the database are constantly taken out to match the selected production rules until the goal is achieved.

Traditional production systems assume that all facts and rules in database are currently valid, disregarding their temporal attributes. So far, research on temporal properties in production systems is inadequate. There are relatively very few researches in this area. (Kabakcloglu 1992a, 1992b) added relative valid time to facts, so did. Maloof and Kochut (1993), both of which contribute a lot to time sequential related reasoning. However, they have not yet taken into account the temporal constraints of production rules, which may prove to be of great importance. Some reasoning procedures involve not only currently valid rules, but also historical rules that have expired at present, whereas they were valid in history. For example, the question "Did Tom get a job promotion in 1990" needs historical facts and rules. Without temporal constraints on production rules, there have to be many different versions of the same production system and each handles the knowledge that belongs to its own valid time period. Obviously, this resolution is not satisfying enough, since all these production systems have the same kernel. Different versions of the same production systems are just replications and a waste of resource.

In this chapter, a new **temporal production system** is introduced, in which both rules and facts possess temporal attributes. Section 16.2 gives its definitions. Then the temporal reasoning algorithm is proposed. A prototype implementation of a salary system is illustrated in Section 16.3. Finally, the conclusion and ideas for future work are pointed out in Section 16.4.

16.2 Temporal Production System

16.2.1 Basic Definitions

In order to transplant the concept of temporal properties into production system, the traditional production system needs to be extended.

First, the definitions of **interval endpoint indicators** and **temporal expression** are given, which are used to define the valid time of preconditions in the definition of **temporal production rule**.

Definition 16.1 Interval endpoint indicators (IEI) are used as time points and time interval endpoints, and it can be positive integers. ∞ stands for the maximal time point. Its BNF definition is:

```
<IEI> ::= '∞' | < positive integer >.
```

In the definition above, "positive integer" is an ordinary positive integer in mathematics.

Example 16.1 5, 21, ∞ are all IEIs.

Definition 16.2 Time expression (TE) is used to express the valid time when something is true. Time expression is constructed by Boolean expressions of IEIs and a time unit quantifier, or a single IEI and a time unit quantifier. Existence quantifier ($\exists$) and universal quantifier ($\forall$) can be used in TE. Its BNF definition is:

```
<TE> ::= {<quantifier> <variable>} [<time variable interval>] ',' <variable
         expression> | <IEI> <unit>
<quantifier> ::= '∃' | '∀'
<time variable interval> ::= <left bracket> <IEI or variable> ','
                             <IEI or variable>  <right bracket > <unit>
<IEI or variable> ::= <IEI> | <variable>
<left bracket> ::= '[' | '('
<right bracket> ::= ']' | ')'
<variable expression> ::= <variable conjunction formula> '∨' <variable
                          expression> | <variable expression>
<variable conjunction formula>::= <variable Boolean formula> '∨'<variable
                                  conjunction formula> | <variable boolean
                                  formula>
<variable Boolean formula> ::= <variable arithmetic formula> <comparator> <IEI>
                               <unit>
<comparator> ::= '=' | '<=' | '>=' | '<' | '>' | '!='
<unit> ::= 'year' | 'month' | 'day' | 'hour' | 'minute' | 'second'
```

In the definition mentioned above, "variable arithmetic formula" is an ordinary arithmetic formula (only operation of plus, subtract, multiply, divide are allowed) of variables in mathematics, such as $x+3$, $y\times 5$. "year", "month", "day", "hour", "minute" and "second" are all ordinary time units.

Example 16.2 $\exists x, x>3$ day, $\exists x \exists y[x,y], y<5 \wedge x>1 \wedge x<y$ minute, $[5, \infty)$ year, 1999 year $\cdots$ are all TEs.

Definition 16.3 Temporal production rule has the following form:

RuleID RuleName ValidTime: IF PreCondition THEN PostCondition Remarks

"PreCondtion" and "PostCondition" are composed of facts and their own TEs. The scope of a variable, wherever it appears in "PreCondition" or "PostCondition", is throughout the rule, i.e., the variable x in PreCondition and variable x in PostCondition is the same one. Thus, any quantifier should be placed in front of where its variable first appears and its BNF definition is presented below:

```
<temporal production rule>::= <header> '::' 'IF' <precondition> 'THEN'
                              <postcondition> [<remarks>]
<header> ::= <rule ID> <rule name> <valid time>
<rule ID> ::= <character string>
<valid time> ::= <time interval union>
<time interval union> ::= <time interval union> '∨ ' <time interval> | <time
                          interval>
<time interval> ::= <left bracket> <time point> ',' <time point> <right bracket>
<time point> ::= <positive integer> 'year' <positive integer> 'month' <positive
                 integer> 'day' <positive integer> 'hour' <positive integer>
                 'minute' <positive integer> 'second' | 'NOW'
<precondition> ::= <precondition union>
<precondition union> ::= <precondition union> '∨ ' <condition intersection> |
                         <condition intersection> | NULL
<condition intersection>::= <condition intersection> '∧ ' <atomic condition> |
                            <atomic condition> | NULL
<atomic condition> ::= '{' <TE> '}' ':' <property> <comparator> <value>
<property> ::= <character string>
<value> ::= <const> | <property>
<const> ::= <integer> | <real number> | <character string>
<postcondition> ::= <condition intersection>
<remarks> ::= '//' <character string> | NULL
```

In the definition above, "character string" is a string of ordinary characters, "integer" and "real number" come from mathematics, "∞" are defined in Definition 16.1, "∧" is intersect operator, "∨" is unite operator, "∃" and "∀" are quantifiers, which are defined in Definition 16.2.

With reference to definition of the temporal production rule mentioned above, a rule can have its own valid time plus its preconditions and postconditions.

Example 16.3 Rule 1 item 1 running the red light [1995 year 1 month 1day 0 hour 0 minute 0 second, 1999 year 12 month 31 day 23 hour 59 minute 59 second] :: IF {∀ x, x>=0 hour ∧ x<=6 hour ∨ x>=18 hour ∧ x<=23 hour}: run red light THEN

```
{(x, ∞)}: penalty (100) ∧ subtract (5)
```

The rule mentioned above means: from 1995-1-1 00:00:00 to 1999-12-31 23:59:59, anyone who runs red light between 00:00:00 and 6:00:00 or between 18:00:00 and 24:00:00 would be punished with a $100 fine and 5 points would be subtracted from his/her license credit. Furthermore, the penalty becomes valid (i.e., it could be executed) after the time he/she runs the red light. Obviously, it can be used to denote union of time periods and cyclical time periods.

16.2.2 Temporal Reasoning

Roughly, reasoning framework in traditional production system could be divided into three categories: data-driven, goal-directed, bi-direction reasoning. All these kinds of reasoning frameworks could be applied to the temporal production system proposed in this chapter.

In this section, how to extend the reasoning algorithm of a traditional production system to a **temporal reasoning algorithm** in a goal-directed reasoning framework is demonstrated below. Other reasoning frameworks could be extended to the corresponding temporal one in a similar way.

A traditional goal-directed reasoning algorithm is listed below:

Step 1: User inputs expected goal.

Step 2: Put user expected goal into goal set.

Step 3: If the goal set is empty, then algorithm ends successfully and output recorded production rules. Otherwise, select a goal and delete it from the goal set.

Step 4: Select a new production rule *R* from the rule base (i.e., one that has not been selected before), whose postconditions contain the selected goal. If there is no such rule, ① if the selected goal is the user expected goal, algorithm ends unsuccessfully; ② otherwise, delete all goals in goal set and all recorded rules, go to Step 2.

Step 5: Record this rule (for further explanations), then match the preconditions of this rule with facts in fact tables to see if all preconditions are held. If all preconditions are held, go to Step 3. Otherwise, add those un-holding preconditions into goal set and go to Step 3.

The reasoning algorithm mentioned above is actually undecided, because no specific method for selecting a rule is given in Step 4. Different methods obviously cause different reasoning processes, and it is still an open problem to decide the best rule selecting strategy. It is beyond the scope of this chapter to present another new method. Therefore, all methods that already exist could be adopted here, such as irrevocable method, probing method, and graphic search method.

The main temporal-extension to the traditional one is give in Step 5. That is, extending the traditional matching process of preconditions and facts to a temporal matching process, in which both rule and its preconditions have valid time attributes. The matching process could be carried out in the way of resolution in predicate logic. Its main idea can be expressed as follows: negate the goal, which needs validating, add it to the already-known fact set, then process the resolution. If "NULL" clause is obtained, the original goal is true. If the resolution halts without a "NULL" clause, the original goal is false.

Likewise, the traditional resolution can be extended to handle temporal attributes as follows.

Let the valid time of the rule be $[Vs,Ve]$, precondition be $\{TF\}:P$. Then the valid time of this precondition is $TF \wedge [Vs,Ve]$. Hence, this precondition becomes $\{TF \wedge [Vs,Ve]\}:P$. If we negate it, we get $(\{TF \wedge [Vs,Ve]\}:(\neg P))$. Then go through the resolution process with $(\{TF \wedge [Vs,Ve]\}:(\neg P))$ and those known facts. If a "NULL" clause is obtained, the precondition $\{TF\}:P$ is true, else it is false. The valid time of preposition may change (to be more accurate, "shrink") along with the resolution process, and the final valid time of preposition may be used to decide the valid time of postcondition. This resolution process is illustrated as an example in Fig. 16.1.

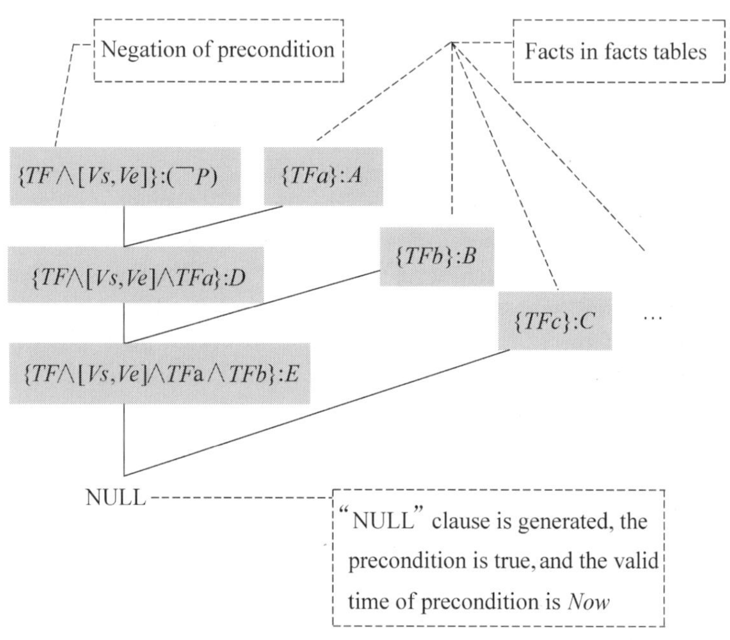

Figure 16.1 Resolution process

The foundation of the above extension is based on the following fact: nothing can be both true and false at the same time. To be more specific, if A is true in a specific time, $\neg A$ is false at the same time. Therefore, clause "if P is true in $[Vs,Ve]$ ($[Vs,Ve]:P$ is true), then $\neg P$ is false in $[Vs,Ve]$ ($[Vs,Ve]:(\neg P)$) is false)" holds at all times.

The temporal reasoning algorithm in temporal production system is as follows:

Step 1: User input expected goal.

Step 2: Put user expected goal into goal set.

Step 3: If the goal set is empty, algorithm ends successfully, and output recorded production rules. If the goal set is not empty, select a goal and delete it from the goal set.

Step 4: Select a new production rule R (its valid time is $[Vs,Ve]$) from the rule base, which has not been selected before, and whose postconditions contain the selected goal. If there is no such rule, ① if selected goal is the user expected goal, algorithm ends unsuccessfully, ② otherwise, delete all goals in goal set and all recorded rules and go to Step 2.

Step 5: Record this rule (for further explanations), then match the preconditions (suppose precondition $\{TF\}$:*Pre*) of this rule with facts in fact tables to see if all preconditions hold. The matching process is shown in the following steps:

Step 5.1: Negate the clause "*Pre*" and intersect valid time of precondition with valid time of rule: $TF \wedge [Vs,Ve]$, then we get "$\{[Vs,Ve] \wedge TF\}:(\neg Pre)$", put it in a set *S*.

Step 5.2: Get all the known facts in fact tables whose valid time has intersection (intersection is not NULL) with $[Vs,Ve]$ and put them into set *S*.

Step 5.3: Select two clauses from set *S*, which have not been chosen before. Let them be "$\{TFa\}$:*A*" and "$\{TFb\}$:*B*", respectively. If it fails, then precondition "$\{[Vs,Ve] \wedge TF\}:(\neg Pre)$" is false or unknown. Put all the conjunction sub-clauses of the precondition into set *S*. All sub-clauses inherit the precondition's valid time. Go to Step 3.

Step 5.4: Go through the resolution process of clause "$\{TFa\}$:*A*" and clause "$\{TFb\}$:*B*". The rule of resolution is — for clauses "$\{TFa\}$:*A*" and "$\{TFb\}$:*B*",

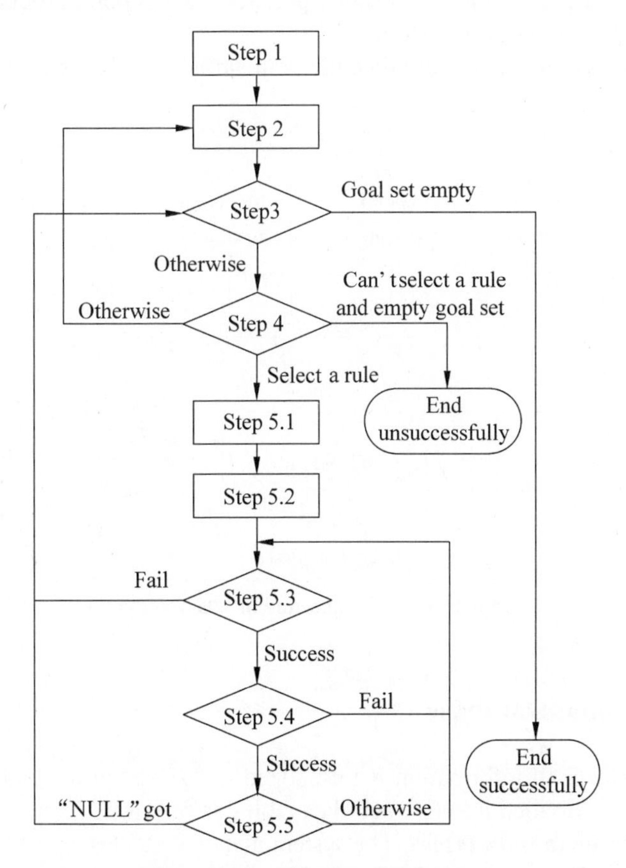

Figure 16.2 The reasoning process

if $A \wedge B = C$ and $TFa \wedge TFb$!= NULL, then those two clauses can process the resolution. The result is $\{TFa\}: A \wedge \{TFb\}:B = \{TFa \wedge TFb\}:C$. If A or B is "*Pre*" or evolved from "*Pre*", update its valid time to $\{TFa \wedge TFb\}$, otherwise, those two clauses cannot do the resolution process, then go to Step 5.3.

Step 5.5: If "NULL" clause is obtained, then precondition $\{TF'\}$:*Pre* holds (TF' is the final valid time as clause "*Pre*" evolves). Go to Step 3, otherwise go to Step 5.3.

The reasoning process flowchart is illustrated in Fig. 16.2.

A temporal production system is a system with temporal reasoning algorithm as shown above, temporal production rules as discussed in Section 16.2 and facts with temporal attributes.

16.3 Prototype Implementation in a Salary System

Salary policies possess strong temporal characteristics. Therefore, for better representation and reasoning over salary policies, the temporal production system can be applied to the salary system.

The salary system's entire architecture is illustrated in Fig. 16.3.

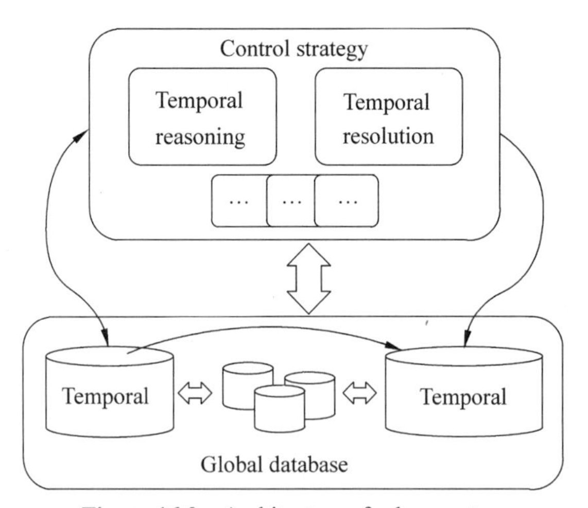

Figure 16.3 Architecture of salary system

16.3.1 Global Database

In the first place, all information is formalized and stored in the global database, which could be divided into five sub-data tables as follows:

(1) **System meta-data tables**: The system meta-data tables are used to keep the meta-data that the system needs.

(2) **User meta-data tables**: User meta-data tables keep the meta-data used for defining the practical system, for example, atomic actions (e.g., basic-salary-level-promotion-action), phrase links that map user-defined phrases to exact columns in the exact data table.

(3) **Rule base**: Rule base are used to store temporal production rules, which are defined by atomic concepts from system meta-data tables and user meta-data tables. Its detailed structure is discussed below.

(4) **Facts tables**: Facts tables preserve the facts in the system and they can be used for reasoning.

(5) **Temporary tables**: Temporary tables are the places where useful temporary information (generated in the process of reasoning) is stored.

Figure 16.4 shows the relationship between these tables.

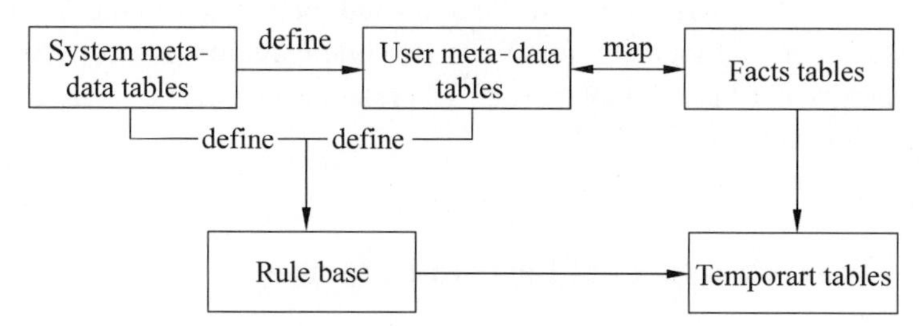

Figure 16.4 Relation between data tables

16.3.2 Data Structures of Temporal Production Rules in Database

In this part, we have discussed how to physically persist the temporal production rules better in database for reasoning convenience. Take the following rule for example.

Example 16.4 Salary policy of Mth piece in Nth item, performance annual check and salary rise (published at 2002-10-1, become valid from 2003-1-1 00:00:00 to *Now*). Those who have already acquired professorship before 1999 (1999 not included) can get one level basic salary promotion and one level office salary promotion, if he/she gets an "excellent" two consecutive times in annual performance check.

According to the rule definition in Section 16.2, the above salary policy can be transformed into the following:

Nth item Mth piece Performance annul check and salary rise [2003-1-1 00:00:00, *Now*) :: IF $\{\exists y, y<1999$ year$\}$: title = professor $\wedge$ $\{\forall x\ [x, x+1]$ year, $x>2002$ year$\}$: Performance check status = excellent THEN $\{x+2$ year$\}$: basic-salary-promotion(1) $\wedge$ office-salary-promotion(1) // published at 2002-10-1.

Because databases can only store structured data, the rule above needs to be further structured and inevitably divided into pieces for storage in different data tables. As in the prototype salary system, the rule mentioned above is divided and stored in 6 different data tables:

- **Rule-table** (ID, RuleID, RuleName, StartTime, EndTime, PreConditionID, PostConditionID, Remarks)
- **Precondition-table** (ID, PreConditionID, PreConditionIntersectionID)
- **Precondition-intersection-table** (ID, PreConditionIntersectionID, Atomic-ConditionID)
- **Atomic-condition-table** (AtomicConditionID, Property, MetaRelation, Value, TemporalConstraint)
- **Post-condition-table** (ID, PostConditionID, AtomicPostConditionID)
- **Atomic-postcondition-table** (AtomicPostConditionID, AtomicActionID, Porperty, MetaRelation, Value, TemporalConstraint)

Furthermore, the DBMS view technique could be applied to the 6 rule tables mentioned above and logically join them into one view for user convenience. In the view, one record is a complete rule. The relationship and structure of the rule tables are illustrated in Fig. 16.5.

16.3.3 Data Structures of Facts in Database

As noted, a single table is sufficient for simple facts. The fact of some products' stocks, for example, can be stored in a single table: stock (ProductID, ProductName, Quantity, StartTime, EndTime). As for complex facts like a person's whole information, the same mechanism as storing rules can still be adopted here. In this way, normal form requirements can be satisfied, and redundancy can be avoided. View technique still holds true here.

16.3.4 Details in Reasoning

There are two methods to activate the reasoning: automatic trigger and user event. Automatic trigger is based on predefined triggers. When certain condition is changed or some values are satisfied, reasoning is triggered. User event method depends directly on the user's action. If reasoning is needed, the user just fires it.

In this salary system prototype, goal-directed reasoning is adopted. Firstly, atomic post conditions (in atomic-post-condition-table) are found through expected goals (atomic actions in user meta-data table). Secondly, post conditions (in post-condition-table), which include those atomic post conditions, are obtained. Next, rules containing those post conditions are located. Therefore, up to now, all the rules that need scanning are sieved out and kept in a temporary data table. *Now*, the temporal reasoning algorithm can be applied.

Nth item Mth piece Performance annul check and salary rise [2003 - 1 - 1 00:00:00,Now) ::
IF{$\exists y, y$<1999 year}:title=professor $\wedge$ {$\forall x[x,x+1]$year, x>2002 year}:Performance check status =excellent THEN {x+2 year}: basic - salary - promotion(1) $\wedge$ office - salary - promotion (1) // published at 2002 - 10 - 1

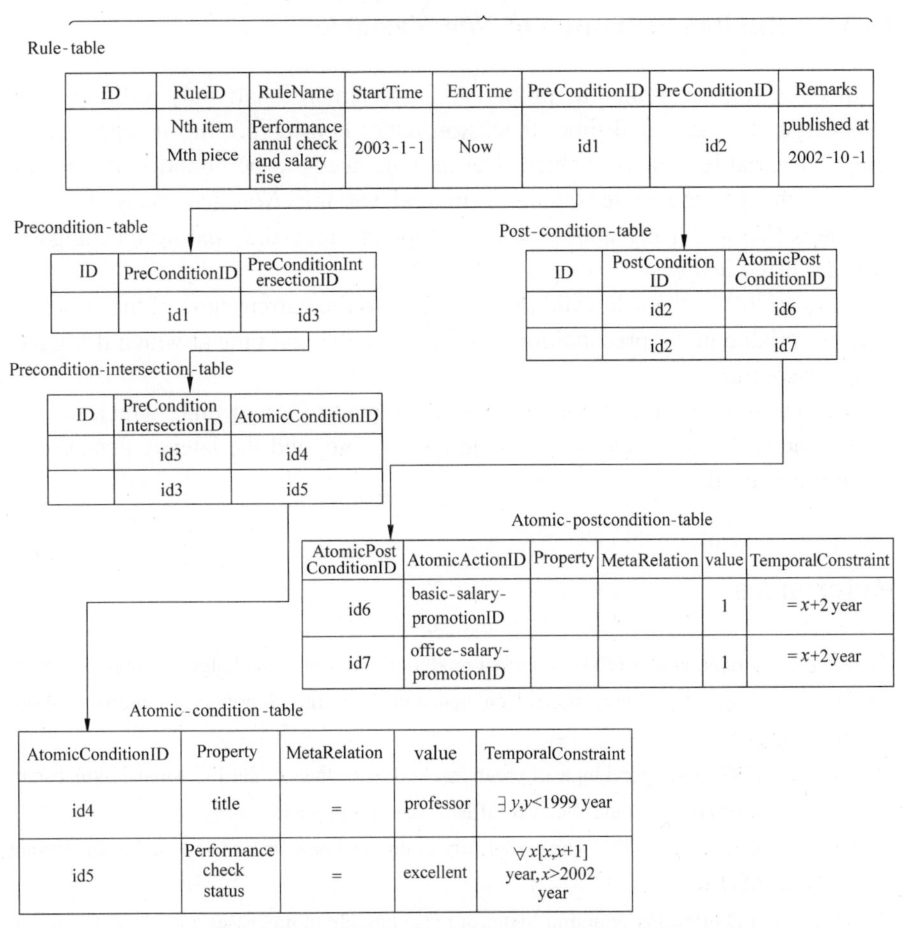

Figure 16.5 Physical structure of a rule in rule base

In the matching process of a rule, it is suggested that all the atomic conditions of a precondition (with the same PreConditionIntersectionID) load to main memory from disks at one time to reduce I/O time. Besides, if the matching of an atomic condition fails, there is no need to match the rest of the atomic conditions of this precondition. Instead, we can go straight to match the next precondition of the rule.

All the atomic conditions in the precondition are defined by phrases predefined in the user meta-data tables. The phases map can be found in specific data tables and specific columns.

If there is a temporal variable in a precondition, we can update it along with the reasoning process until it ends. If there is an identical temporal variable in the postcondition, assign its value with the one in the precondition.

16.3.5 Binding Semantics of *Now* Variable

In this system, the temporal property is one-dimensional: only valid time is considered. Transactional time dimension is not considered. In this application, temporal variable *Now* is enabled. Hence *Now* needs to be bound with specific value in the process of reasoning. Temporal variable *Now* has many different meanings (Ye and Tang 2005). In our sample system, the binding meanings of *Now* are listed below:

- *Now* variable in the rule valid time is bound to the current time of the system.
- *Now* variable in the precondition is assigned to the real time at which it triggers the reasoning.
- *Now* variable in the postcondition is also bound to a real time, which is later than the real time of the trigger of this reasoning, and the latency depends on the specific rule.

References

[1] Tang Y, Tang N, et al. (**2004**) A unified model of temporal knowledge and temporal data. In: Proceedings of 8th International Conference on Computer Supported Cooperative Work in Design, pp 711 – 713

[2] Pnueli A (**1997**) A temporal logic of programs. In: Proceedings of the 18th Annual Symposium on Foundations of Computer Science, IEEE, New York, pp 46 – 57

[3] Sistla AP, Clarke EM (**1985**) The complexity of propositional linear temporal logics. Journal of the ACM (JACM), 32(3)

[4] McDermott D (**1982**) A temporal logic for reasoning about processes and plans. Cognitive Science (6): 101

[5] Allen JF (**1983**) Maintaining knowledge about temporal intervals. Communications of ACM 26(11): 832 – 843

[6] Baader F, Calvanese D, et al. (**2002**) The description logic handbook: theory, implementation and applications. Cambridge University Press

[7] Guarino N (**1998**) Formal ontology and information systems. In: Proceedings of the 1st International Conference on Formal Ontology in Information Systems. Trento, Italy: IOS Press, pp 3 – 15

[8] Kabakcloglu AM (**1992a**) Temporal production systems. In: IEEE Proceedings of Southeastcon'92, vol. 2, pp 697 – 698

[9] Kabakcloglu AM (**1992b**) Artificial intelligence for medical knowledge representation/reasoning/acquisition. In: Proceedings of the 1992 International Biomedical Engineering Days, pp 186 – 191

[10] Maloof MA, Kochut KJ (**1993**) Modifying rete to reason temporally. In: Proceedings of the Fifth International Conference on Tools with Artificial Intelligence, pp 472 – 473

[11] Ye X, Tang Y (**2005**) Semantics on *Now* and calculus on temporal relations (in Chinese with English abstract). Journal of Software 16(5): 838 – 845

[12] Steiner A (**1998**) A generalisation approach to temporal data models and their implementations. PhD thesis, Swiss Federal Institute of Technology.

[13] Chen Z, Tang Y, et al. (**2006**) The design and development of temporal database middleware. Journal of Computer Research and Development 43(Suppl.)

17 Temporal Application Modes and Case Study

Jianguo Li[1], Yong Tang[1,2], Na Tang[1,2], and Hanjiang Lai[1]

[1] Computer School, South China Normal University, Guangzhou 510631, P.R. China
[2] Department of Computer Science, Sun Yat-sen University, Guangzhou 512075, P.R. China

Abstract There are various kinds of temporal application modes in current information systems. In this chapter, we divide the temporal application modes into three kinds—the pure temporal mode, embedding temporal mode and mixed temporal mode. The temporal information view that links the temporal data and temporal knowledge is proposed. The temporal technologies in cooperative software are discussed, such as temporal role management and temporal workflow. Finally, the cases showing the comprehensive utilization of temporal technologies are described.

Keywords *temporal application mode, temporal information, data view, cooperative software, middleware, temporal knowledge, temporal data*

With the development of database technology and web application, increasing importance has been attached to temporal technology. Though the temporal database technology has made great progress in recent years (Snodgrass 2007; Tansel and Imberman 2007), the application of temporal information processing technology still has problems due to lack of temporal software development tools. There are various kinds of application modes in temporal information systems (Tang et al. 2007). In this chapter, we divide the temporal application mode into three kinds—the pure temporal mode, embedding temporal mode and mix temporal mode. In Section 17.2, we propose a temporal information view that links the temporal data and temporal knowledge. In Section 17.3, we propose a cooperative software framework and then discuss the temporal technology in cooperative software, such as temporal role management, temporal workflow and temporal collaboration information. We also give an example of temporal application in collaboration software, which is a comprehensive utilization of temporal technologies. In Sections 17.4, as a typical illustration, a temporal knowledge based salary system developed by us is described. We discuss the temporal data model for employee information, the temporal representation and inference

issty@mail.sysu.edu.cn

mechanism of salary policies and the temporal event mechanism of this system. Finally, we give the summary of the chapter and propose some trends of temporal application.

17.1 Temporal Application Modes

Time is an objective attribute in the natural world but it is not supported by conventional information systems. Hence, we classify the systems into **non-temporal application** system and **temporal application** system according to whether the temporal attribute can be handled or not. Moreover, the latter is partitioned into three sub-classifications: Entire Temporal Application System, Embedding Temporal Application System and Mix Temporal Application System, as shown in the Fig. 17.1.

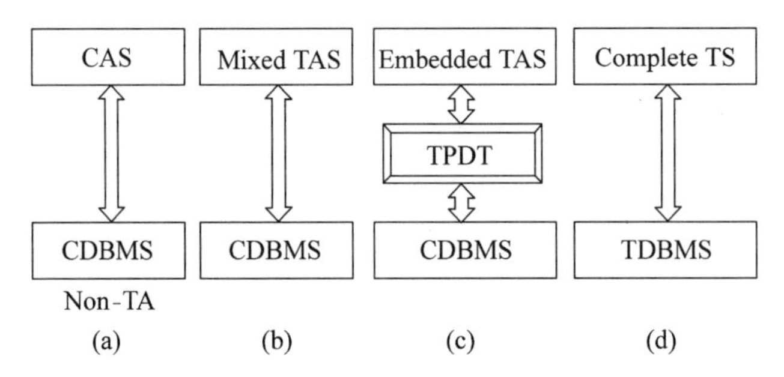

Figure 17.1 Various temporal application modes
C: conventional; T: temporal; A: application; S: system; TPDT: temporal processing development tools

17.1.1 Entire Temporal Application Mode

Entire Temporal Application Mode is also called **Pure Temporal Application Mode** or **complete temporal application**, which indicates that it requires entire temporal database support based on a temporal application system, and the basis of which is a temporal database management system (TDBMS), as shown in Fig. 17.1.

Complete temporal application mode relies on a real temporal database management system (TDBMS), which supports the temporal data type and its processing mechanism in the layer of DBMS. This mode provides the most convenient mode for the temporal application developer. However, no real TDBMS exists so far, as there are some problems of temporal model and processing mechanism, for example, the incomplete temporal calculative system, low efficiency of current temporal query languages. The so-called temporal database management systems at present include TimeDB and TempDB, which are not real TDBMS.

They are the temporal software development middleware supported by the conventional DBMS, such as Oracle, DB2.

17.1.2 Embedding Temporal Application Mode

Embedding Temporal Application Mode, the ground for which is conventional DBMS, implements the functionality of temporal information handling of application systems, as shown in Fig. 17.1. To develop an application by using a model of this type needs the support of temporal handling software, such as TimeDB of Time Consult Corporation and TempDB that we have developed.

Logically, embedding temporal application software is constructed from three layers as follows. This is also shown in Fig. 17.1.

(1) Temporal application layer: The layer is the window associated with users, which provides temporal preprocessed tools to define temporal data view, edit temporal knowledge library, call temporal middleware service and accomplish temporal information handling.

(2) Temporal middle layer: It provides temporal information handling service beyond database platform based on web, including temporal query, temporal deduction.

(3) Database layer: It implements temporal information representation based on the conventional database platform.

17.1.3 Mix Temporal Application Mode

Mix temporal application mode integrates temporal data handling techniques and application information handling techniques to implement the support of temporal attributes in the application system, the ground of which is also DBMS, as shown in Fig. 17.1. Currently, most of the temporal application systems belong to this type.

In this mode, the applications associated with temporal information are implemented by temporal data model and conventional technology. The interpretation of temporal part is operated by another application, but not the database itself.

17.2 Temporal Data/Knowledge View

17.2.1 Temporal Data View

Temporal information unit is the basic element for information library, such as one tuple of the temporal database, or one rule of the temporal knowledge base.

Generally, non-temporal information and temporal information are included in temporal application. While conventional database is not able to support temporal attribute, linking non-temporal information and temporal information into conventional database seamlessly is one of the key techniques.

To complete the seamless link of non-temporal information and temporal information, we propose implementing logical uniform temporal view based on temporal index. We adopt the concrete methods as follows. We split the data field into conventional data field and temporal data field, each of which is implemented by a separate database table. Then two data fields are combined by temporal index. When users need to operate the temporal database, a **temporal view** is created by the two tables to be the middleware for users' operation, as shown in Fig. 17.2.

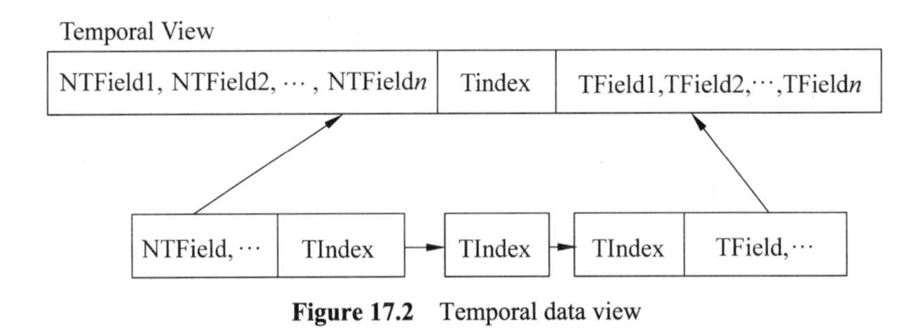

Figure 17.2 Temporal data view

17.2.2 Temporal Data/Knowledge Model

Since temporal information handling technique is the core technique of developing temporal application system, temporal middleware research has been the research tendency of temporal development tools. We propose a **temporal information model** (TIM) and are studying and developing temporal information processing middleware based on temporal information model. In short, all this is called TIM.

TIM is mainly composed of two parts: Temporal Information Processing Software Component and Temporal Knowledge/Data Model, the basic framework of which is shown in Fig. 17.3.

Here, *Temporal information* contains *Temporal data* and *Temporal knowledge*. Hence, TIM includes three stages: TDM (**Temporal Data Model**), TKM (**Temporal Knowledge Model**) and TKDM (Temporal Knowledge Data Model).

17.2.3 Links of Temporal Knowledge and Temporal Data

Temporal information contains temporal knowledge and temporal data. The basic method of temporal knowledge expression extends from knowledge expression

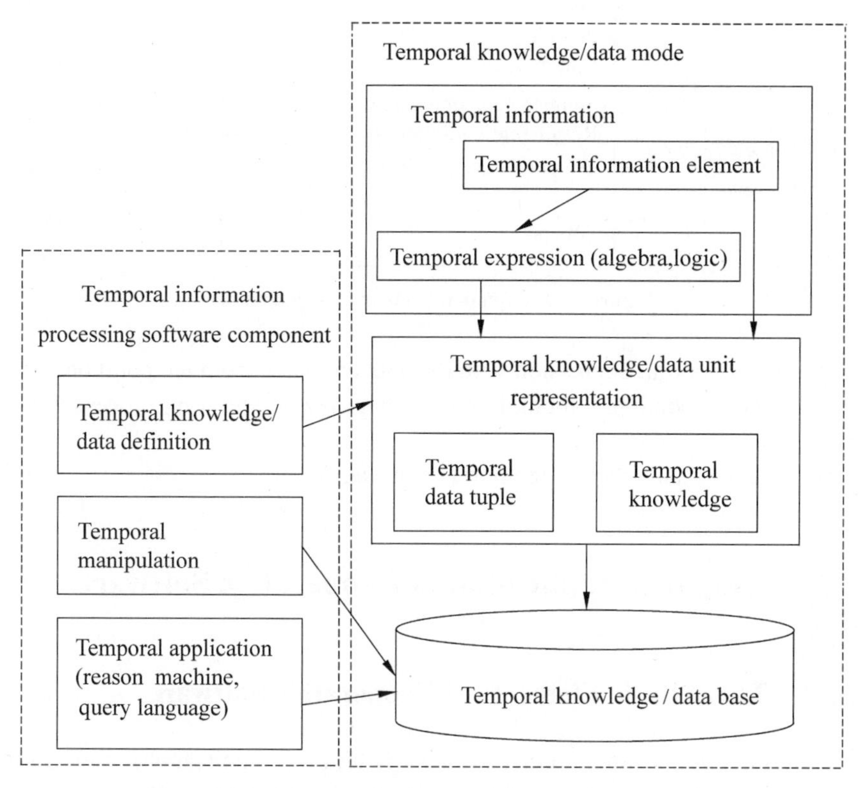

Figure 17.3 Framework of temporal information processing

mode. In the same way, we can inject temporal attribute into knowledge base and build the database expression mode of temporal knowledge through temporal attribute, thus implementing unified temporal knowledge data model from temporal knowledge and temporal data.

Figure 17.4 shows the temporal knowledge data model combined by rule and procedure. On the premise of rule, temporal expression, which is stored through temporal attribute in temporal database, is proposed to express temporal attribute of knowledge and implement the valid time of temporal knowledge. As temporal attribute is the temporal data type proposed by the project, it is accessed by temporal function and temporal query language. Hence, we implement the links of temporal knowledge and temporal data.

We will introduce the inference process. Before starting an inference rule, we choose the corresponding knowledge rule according to the temporal attribute value of temporal data. The match criteria of this is to perform temporal logic operation using temporal attributes of temporal data and temporal knowledge, and view the operation result as the true rule. The steps of temporal knowledge inference are as follows:

(1) Estimate the corresponding data according to the temporal expressions on the premise of rule.

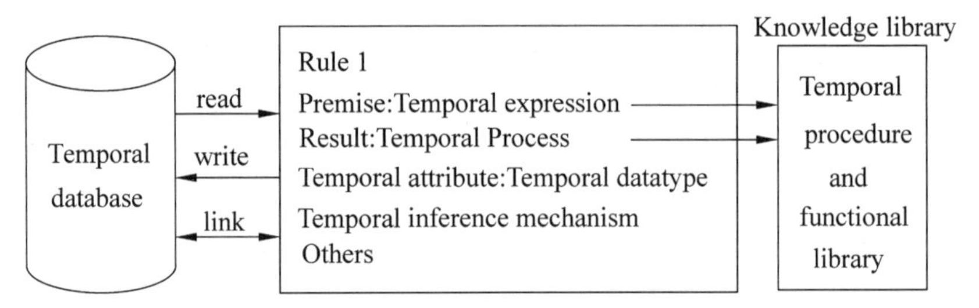

Figure 17.4 Temporal data knowledge model

(2) Match temporal expression with the data extracted from temporal database.

(3) The corresponding temporal process that may be a nesting process will be carried out.

(4) Write the result back to temporal database.

17.3 Temporal Application in Cooperative Software

17.3.1 Three Basic Elements of Cooperative Software

With the rapid growth of Internet technology and web applications, cooperative software is getting attention of increasing number of users and developers in business, industry and government.

What is cooperative software? Cooperative software, also called collaborative software, have many definitions in the business and academic with many different views. For example, in the web site of Wikipedia [http://en.wikipedia.org/wiki, 2009-04-03], the definition of **Collaborative software** is software that is designed to help people involved in a common task to achieve their goals. Collaborative software is the basis for computer supported cooperative work. Software systems such as email, calendar, text chat and wiki belong to this category.

A cooperative application has three basic elements including collaboration role (the collaboration objects), the collaboration information (the collaboration content) and software platform including collaboration tools, as shown in Fig. 17.5. Implementation of a cooperative application needs the software technology to handle role, information and collaboration procedure (Tang et al. 2007).

Every collaboration happens at a specific time and a specific space, so that the time and space information are two basic attributes.

Temporal information exists in all the components of cooperative software. Temporal technology has been applied in collaboration roles, collaboration message, collaboration procedure, collaboration tools and so on. This section introduces some temporal applications in cooperative software.

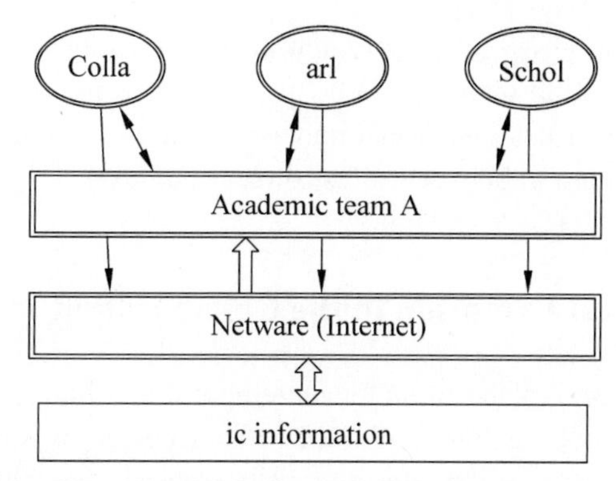

Figure 17.5 Three basic elements of cooperative application

17.3.2 Temporal Relation of Collaborative Roles

Collaboration roles are the core components of cooperative software. Temporal is the basic attribute of collaborative role. In recent years, there are increasing number of research reports on role collaboration, such as Role Based Access Control, role awareness, and role-based chatting. However, only a few papers are involved in temporal attribute of roles.

We have researched the collaborative relation describe model of role, expanded the temporal attribute to the collaboration role and their relation, and proposed a temporal role relationship diagram (TRR Diagram). Figure 17.6 shows the three aspects of TRR Diagram.

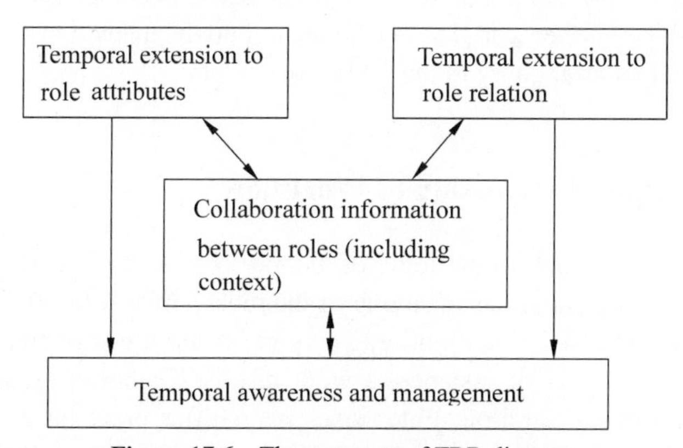

Figure 17.6 Three aspects of TRR diagram

(1) To extend temporal of the definition attribute of temporal role, for example, the position of a role.

(2) To extend temporal relation between collaborative roles. Every relation between collaborative roles has temporal attributes, for example, the one project.

(3) The role awareness based on temporal role relationship can provide more convenient collaboration manner. To be aware of a temporal role or a temporal relationship between roles also needs understanding of the temporal attributes of collaboration information between roles, such as context.

17.3.3 Temporal Extension in the Collaboration Information

The information for collaboration have temporal attributes, for example, the medical records of one patient is allowed to be accessed by some invited doctor only when he takes part in the consultation. In other words, the medical records have the valid time attribute for this **cooperative action**.

XML is the common technology for collaborate information processing. XML documents will be modified as time goes by. Compared to the traditional relational model, XML and XQuery can better support the expression of temporal information and temporal queries.

The successive versions of an XML document are expressed by temporal XML document in an incremental fashion, which provides us a high effective method for version management. Therefore, the topics of representing, querying and updating temporal XML documents have received increasing attention. In (Tang and Tang 2008), we present an XML data model for tracking historical information in an XML document. This model can present the temporally ungrouped information.

The **temporal information** is expressed in the XML documents. In this case, to query and update temporal information in XML documents efficiently is a difficult problem. Further work has to be done to find the method to extend XQuery to support the temporal query in the XML documents.

17.3.4 Temporal Extension of Workflow

Workflow is the basic technology of cooperative software. Recently, time management in workflow has been one of the most active research areas in both academic and industrial communities. There are many kinds of time constraints in business processes. For instance, a work task or a whole process should be finished in a limited duration. Enterprises may suffer great losses when these time constraints are violated. For example, in a customer claim handling process, claims that are not being handled in time may depress the customers. An enterprise has to pay penalty when a commercial contract has not been executed on time. Thus, it is of great importance to manage the time information in workflows effectively and to avoid the violation of time constraints.

Current researches on time constraints in workflow mainly focuses on time relevant process modeling and efficiency analysis based on the temporal attributes of processes and activities. Yu and others (Yu et al. 2004) presented the concept of **Temporal Workflow**. Through extending and modifying the WfMC's Basic Process Definition Meta-model, a modified workflow meta-model is presented, and the temporal attributes of elements and their relations in it are analyzed in detail. Based on this, the primary elements in workflow are formalized and time is introduced into workflow as a dimension.

Temporal information management in workflow has been recognized as one of the most significant tasks in workflow management. The temporal uncertainties and valid time constraints on resources and activities should be taken into consideration in workflow models. In Pan's paper (Pan et al. 2006), the authors proposed a new workflow model named Fuzzy Temporal Workflow Nets (FTWF-nets). The calculation of temporal elements in FTWF-nets is given. Time modeling and time possibility analysis of temporal phenomena in FTWF-Nets are also investigated. FTWF-Nets can be used to model temporal information in those workflows, which have temporal uncertainties and time constraints on resources and activities, and to analyze the time possibility on some typical constraints.

17.3.5 Case Study

The academic information is very important for academic exchanges. The scholar collaboration is always based on their academic information. We are developing a collaborative software platform based on scholar's academic information, called SCHOL@.

There are two kinds of roles: Scholar and academic team. Scholar is the basic role of SCHOL@ and their information is the source of academic team. Figure 17.7 shows the **role-tree** and **role relationship** of SCHOL@. In SCHOL@, both the role and the relationship between them have temporal attribute. For example, one scholar is the member of Team A during 2004-2008 and also the member of Team B from 2006 until now.

The temporal is also the important attribute of academic information. One tuple of the information may have different temporal attributes, including the scholars, for the teams and for a collaboration event. For example, to a paper information of one scholar, the valid-time to himself is from the data was published until forever, but the valid-time to the team who joined is the same as the valid-time of the relationship between him and the team.

In SCHOL@, we based on temporal database model and technology, we can realize the mechanism and handle the temporal roles relationship and academic temporal information.

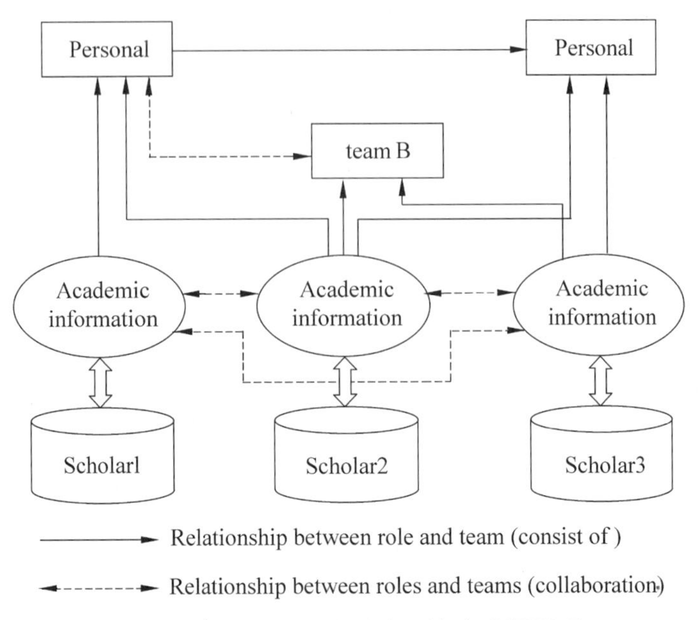

Figure 17.7 Roles and relationship in SCHOL@

17.4 SIDSS: A Typical Example of Temporal Application

17.4.1 Introduction

The intelligent decision support system of salary (SIDSS) is a temporal application, which was developed by our team during the period from 1998 to 2004. SIDSS has been used by about 10 thousand departments of local government of Guangdong Province, China since 2000.

The employee information is classified into two categories. One is the common information, which is not relevant to employee's salary, such as name, ID, birthday. The other is the information relevant to employee's salary regulation, such as educational qualification, position promotion, rewards, penalties and annual assessments. This kind of information is time varying. Salary policies have time-varying characteristic, which can usually be changed by their maker with the passage of time.

The correct method to determine an employee's salary is that based on the temporal information of employee by the time valid salary policies.

In a salary system, the key problems are to model the employee temporal data and to determine the salary of employee by temporal salary policies.

SIDSS is a typical temporal knowledge/data application, so we also called it temporal knowledge based salary system. Figure 17.8 shows the basic elements of SIDSS.

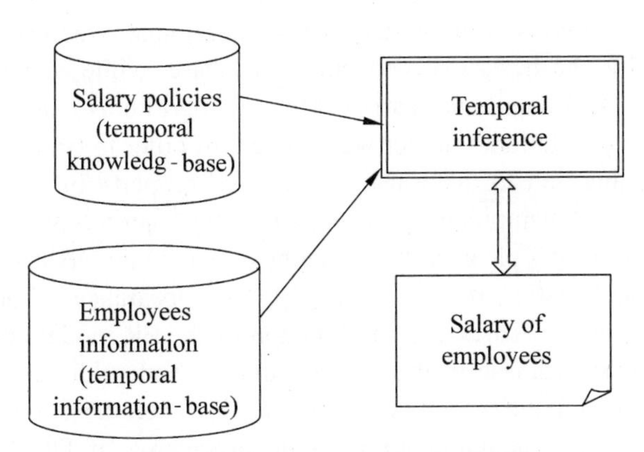

Figure 17.8 Basic elements in SIDSS

17.4.2 Temporal Data in SIDSS

The SIDSS contains three kinds of information: the first one does not influence salary, such as staff's number, name, and ID. The second one influences salary but is not influenced by time, such as the date of taking part in the job. The third one influences salary and is impacted by time, such as education, the promoting of position, and the standard of salary. Moreover, the historic records of staff's salary that were produced in the process of checking the salary are also temporary. One important characteristic of the intelligent decision support system of salary is that it can reappear at staff's history of salary changes since he took part in job according to the staff's information entered by the user, to ensure the correctness of salary checking. Therefore, SIDSS has valid time as well as transaction time. Therefore, SIDSS uses bitemporal database. However, in the case where the staff's information was not influenced by time, there will be duplicate records. This will produce too much of redundant data.

In order to reduce data redundancy, we put the attribute that does not depend on time in a table and put the time-changing attribute that depends on time in another table, and then index with the key word "staff ID".

From this, the object of temporal database is a 3D entity. It is reluctant to use traditional relational database to simulate it, because an object's historical expression needs some 1NF tuples. In recent years, not-1NF's theory, implementation and application have all made great progress. The essence of not-1NF is that it allows that a relation's attribute can be another relation, so it can describe great and complex architecture. Hence, it is convenient and intuitive to describe 3D objects.

Many semantic models have been raised since 1970s and ER model is one of them. Later, ER model was widely used in the design of information systems, and many deformations had also been generated. However, nearly all of these ER models were only involved in the database's diagrammatic representation, which were suitable for the design of conceptual database, while representative data mode uses relational data type. Various kinds of algorithms that were converted from ER model to relational model were raised. In order to prevent the problem while transforming and to make use of the good property of ER model, many scholars raised data manipulation language or DBMS prototype that were based on ER model. Some people even think that next generation DBMS would use ER model to replace relation model. Currently, the appearance of object-oriented databases can prove it. To extend the function of the ER model and to meet the needs of temporal database in the present time, Theodoulidis and other people raised an ER model that has temporal attribute. However, these models lack valid operation layer that is suitable for the data processing of DBMS, which can explain that its function is weak.

In SIDSS, we use a **Nested Temporal Entity-Relationship Model** (NTER) to describe the temporal salary data. NTER is a temporal ER model based on class. It extends ER model not only in diagrammatic representation but also in semantics.

Figure 17.9 is a basic metafile of NTER model. The meaning of metafile is as follows:

(a) Attribute
(b) Entity Set
(c) Relationship Set
(d) Temporal Relationship Set

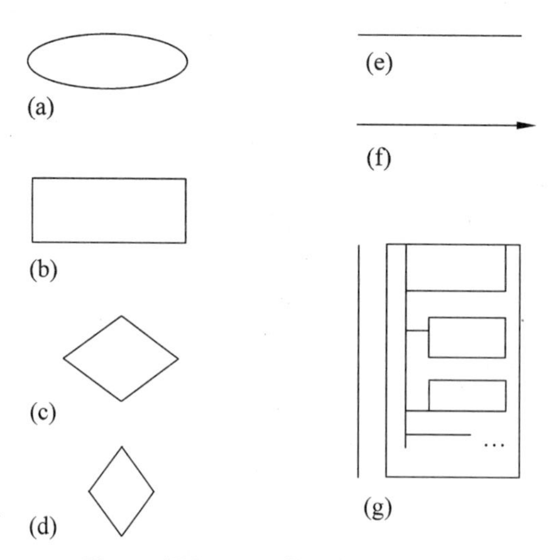

Figure 17.9 Metafile of NTER model

(e) Connect attribute to Entity Set or connect Entity Set to Relationship Set (when the relation between Entity Set and Relationship Set is "much")

(f) Connect Entity Set to Relationship Set (when the relation between Entity Set and Relationship Set is "one")

(g) Complex attribute with nest relation. The rectangular box at the top is the complex attribute and itself is also an entity set. It contains the following attributes in the rectangular box.

Figure17.10 is a simple example of NTER model in SIDSS.

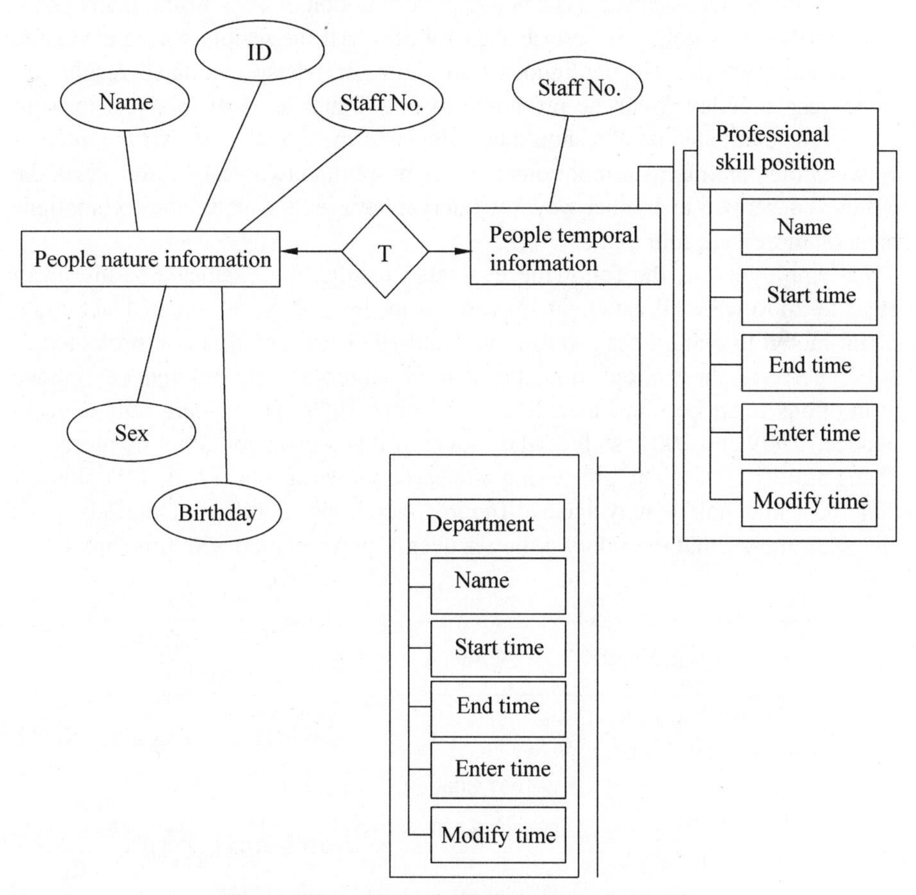

Figure17.10 Example of NTER model in SIDSS

17.4.3 Temporal Knowledge in SIDSS

Wage policy is a typical temporal knowledge and its effectiveness is closely related with the time. The most common saying is "×× policy begins to implement from ××year ××month ××date.” As a field of knowledge, it is often very complicated. The characteristics of the unstructured are very prominent. If we can find a better way to describe the knowledge, it will be an inspiration and reference to express other fields of knowledge.

As the most important knowledge in the SIDSS, the salary policy cannot be considered as good or bad, because every salary policy is right during some special period (Tang and Li 2000). When the policy is prohibited to be used, it does not mean that the policy is false, but only means it cannot be applied anymore. On the other hand, it is valid during the life cycle.

For example, let us consider the policy of “Promotion of Two-Years’ Excellence”. Before 1999, this policy is described as follows: All the people whose evaluated results are excellent for continuous two years, or whose evaluated results are “One year excellent”, can be promoted to one higher level of salary. The valid year of the evaluated results should be calculated from next year, and the interval between the continuous promotions must be more than two years. After 1999, the policy is described in another way: the interval between the continuous promotions must be more than four years.

An employee has the following evaluated results: the evaluated results from 1995 to 2000 were all excellent. According to this policy, his salary level could be promoted to one higher level in 1997 and 1999, but could not be promoted in 2001, because the policy stipulates that the interval between the continuous promotions must be more than four years after 1999. There were only 2 years between 1999 and 2001, so the salary level of this person could not be promoted (Tang and Li 2000). The processing is clearly shown in Fig. 17.11. It is obvious that the same policy may have different operations in different periods. This character means that the salary policy is deeply implemented with time brand.

Year	Evaluated result	Action	Description of the policy
1995	Excellent		
1996	Excellent		
		In 1997, allowed to be promoted with one level	The interval must be more than 2 years
1997	Excellent		
1998	Excellent		
		In 1999, allowed to be promoted with one level	
1999	Excellent		
2000	Excellent		
		In 2001, not allowed to be promoted with one level	The interval must be more than 4 years

Figure 17.11 Temporal character of salary policy

It can be concluded that the salary policy has the temporal characteristics, which is absolutely different from the normal knowledge. Relevant salary policy should be used in different periods. In this way, the salary may be accurately conformed.

As shown in the example, all the processes involving the salary policy must be judged first and handled specially for the temporal characters. In fact, almost all of the events in SIDSS are related to the temporal judgment caused by the mutability of salary policy and salary standard. If we can abstract this temporal judgment and form a model for formalization inference, it not only sets up a foundation and structure of the knowledge database, but is also convenient to the layout and the maintenance of the system. In this chapter, in order to take the events in SIDSS to the formalization inference, we put forward an inference model of temporal logic, and set up the search space. We still link the example discussed previously and illuminate it as follows.

All rules and their tables concerning the process of the policy of "Two-Year Excellence" promotion are as follows. We called the action as "Skipping Promotion".

Table 17.1 Rule table of promotion

Rule ID	Rule name	Rule description	Rule script	OCE	Rule life cycle
R1	Normal promotion	Handle the skipping promotion caused by evaluated results before 1998	Skipping promotion	C1	[1993/10, 1998/12]
R2	Skipping promotion	Handle the skipping promotion caused by evaluated results after	Skipping promotion	C2	[1991/1, *Now*]

Table 17.2 Condition expression table of skipping promotion

Condition ID	Condition	THEN	ELSE
C1	Evaluated results are all excellent in continuous two year	C3	Null
C2	Evaluated results are all excellent in continuous two year	C4	Null
C3	(Current year – Last skipping promoted year >= 2) OR (Never skipping promoted before)	A1	Null
C4	(Current year – Last skipping promoted year >= 4) OR (Never skipping promoted before)	A1	Null

Table 17.3 Skipping action table

Action ID	Action	Next
A1	Salary level + 1	Null

From Table 17.1 to Table 17.3, they are all the rules (Liu et al. 2002), conditions and actions of the Skipping Promotion. Both the "life cycle" in Table 17.1 and the conditions in Table 17.2 have the temporal characters. *Now*, we can run the example as follows.

Assuming that Person A's evaluated results from 1997 to 2000 are excellent, excellent, excellent and passing, the flow of execution is shown as follows:

(1) Reading the evaluated results at 1997 and 1998, which are excellent and excellent, oriented in rule list, result of 1998, which is in lifecycle of R1 and original condition ID, which is C1.

Finding out the condition of C1 and the results of 1997, 1998, which are all excellent and the ID of THEN, which is C3, sequentially judging that it is condition ID.

Finding out the condition of C3, which is never skip a grade and the ID of THEN, which is A1, sequentially judging that it is action ID.

According to A1 in action list, find out its action, the pay rank of person A skips promotion by one class. After finding that the value of NEXT is NULL, sequentially the rule execution ends.

(2) Then reading the evaluated results at 1998 and 1999, which are excellent and excellent, oriented in rule list, result of 1999, which is in lifecycle of R2 and originating condition ID, which is C2.

After finding the condition of C2 and the results of 1998, 1999, which are all excellent, and getting that the ID of THEN is C4, sequentially we judge that it is condition ID.

Finding out the condition of C4, the results of 1999-1998, which is less than 4, getting that the result of ELSE is NULL, sequentially the rule execution ends.

(3) Finally, reading the evaluated results at 1999 and 2000, which are excellent and passing, oriented in rule list, result of 2000, which is in lifecycle of R2 and originating condition ID, which is C2.

Finding out the condition of C2 and the results of 1999, 2000, which are not continually excellent and getting that the result of ELSE is Null, sequentially the rule execution ends.

17.4.4 Implementation of SIDSS

According to the particularity of the salary policy, the structure of SIDSS has three bases: rule base, method base and database. Figure 17.12 shows the system structure.

In this structure, the traditional four bases are simplified as three bases (database, rule base and method base). The facts in knowledge base migrate to database and knowledge base is simplified as rule base. The base of four-base model is equal to the mixture of rules and methods, so the model base is replaced by rule base and method base.

Event handler (Tang and Li 2000) consists of these components: event scheduler, event queue unit, event handling sub modules.

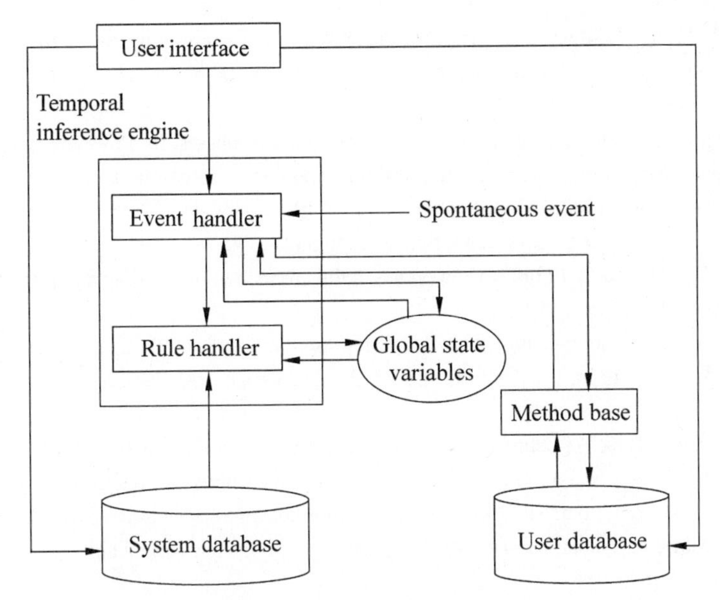

Figure 17.12 Structure of SIDSS

The function of event scheduler is to receive events sent by user interface and some spontaneous event, and to send these events to event queue unit.

The function of event queue unit is to put these events into corresponding queues and select the event that should be handled according to priority table. Then event scheduler will call relevant event handling sub modules. Event queue unit is a very important module for gathering information and regulating salary. Gathering information means that a number of events for salary adjustment will be triggered and these events must be queued and processed according to their triggering time and priority. All this work is done by event queue unit.

The function of event handling sub module is R/W global state variables, calling rule handler and method base to R/W user database.

Rule handler (Tang and Tang 2000) consists of rule matcher and rule base. The function of rule matcher is to receive events sent from event handler and global state variables, selecting relevant rules from rule base, processing the rules according to ECA sequence, modifying global state variables and sending modified global state variables to event scheduler.

References

[1] Liu Dongning, Tang Yong, Tang Na, Wei Wei, Deng Zhi (**2002**) An inference model of temporal logic in an intelligent decision support system of salary. In: Proceedings of the 7th International Conference on Computer Supported Cooperative Work in Design, pp 351 – 354

[2] Pan Y, Tang Y, Ma H and Tang N (**2006**) Workflow analysis based on fuzzy temporal workflow nets. Computer Supported Cooperative Work in Design Ⅱ, LNCS 3865, pp 545 – 553

[3] Snodgrass RT (**2007**) Towards a science of temporal databases. In: Proceedings of the 14th International Symposium on Temporal Representation and Reasoning, pp 4 – 5

[4] Tang N, Tang Y (**2008**) Mapping bitemporal XML data model to XML document. Lecture Notes in Computer Science 5236, Springer, pp 342 – 352

[5] Tang Y, Ji GF, Zhu J (**2007**) Cooperative software technology and its applications. China Machine Press, pp 3 – 10

[6] Tang Yong, Li Song (**2000**) Knowledge representation of an intelligent decision support system of salary. In: Proceedings of the Seventh International Conference on IE&EM, pp 519 – 523

[7] Tang Yong, Tang Na (**2000**) The design and implementation of intelligent decision support system of salary. In: Proceedings of the 7th International Conference on IE&EM, pp 385 – 389

[8] Tansel AU, Imberman SP (**2007**) Discovery of association rules in temporal databases. In: Proceedings of the 4th International Conference on Information Technology New Generations, pp 371 – 376

[9] Yu Y, Tang Y, Tang N, Ye XP (**2004**) A meta-model of temporal workflow and its formalization. In: Grid and Cooperative Computing, LNCS 3251, 987 – 992

Appendix

A.1 Extension ATSQL of TempDB 2.1

1. Statement composing

```
Statement ::= (query | ddl | dml | control) ';'
```

2. Query statements

```
timeFlag ::= [ 'nonsequenced' ] 'validtime' [ identifier | interval ]
coal      ::= '(' 'period' ')'
query           ::= [ timeFlag ] queryExp
queryExp ::= queryTerm { ('union' | 'except') queryTerm }
queryTerm       ::= queryFactor { 'intersect' queryFactor }
queryFactor     ::= '(' query ')' [ coal ] | sfw
sfw     ::= 'select' selectItemList
                    'from' tableRefList
                    [ 'where' condExp ]
                    [ 'group' 'by' groupByList ]
                    [ 'having' condExp ]
selectItemList ::= '*' | selectItem { ',' selectItem }
selectItem             ::= scalarExp [ alias ]
tableRefList    ::= tableRef { ',' tableRef }
tableRef        ::= '(' query ')' [ coal ] alias [ '(' colList ')' ] |
                          identifier [ coal ] [ alias ]
alias   ::= ['as'] identifier
condExp ::= condTerm { 'or' condTerm }
condTerm     ::= condFactor { 'and' condFactor }
condFactor   ::= [ 'not' ] simpleCondFactor
simpleCondFactor ::= '('condExp ')'                                |
                     'exists' '(' query ')'                        |
                     constScalarExp commonOp constScalarExp        |
                     constScalarExp commonOp('all'|'any'|'some')'('query ')'|
                     constScalarExp['not'] 'between' constScalarExp     |
                     'and' constScalarExp                               |
                     scalarExp [ 'not' ] 'in' '(' query ')'             |
                     tempScalarExp timeOp tempScalarExp                 |
                     tempScalarExp timeOp ('all'|'any'|'some')'('query')'|
                     eventTerm ['not']'between' eventTerm 'and' eventTerm
condOp   ::= commonOp | timeOp
commonOp ::= '<' | '>' | '<=' | '>=' | '<>' | '='
timeOp   ::= 'before'|'contains'|'overlaps'|'meets'|'starts'|'finishes'|
```

```
                  'equals'
groupByList  ::= colRef { ',' colRef }
scalarExp    ::= constScalarExp | tempScalarExp
constScalarExp ::= term { ('+' | '-') term }
term         ::= factor { ('*' | '/') factor }
factor       ::= [ ('+' | '-') ] simpleFactor
simpleFactor ::= colRef|const|'('constScalarExp')'|'abs''('constScalarExp')'
colRef       ::= identifier [ '.' identifier ]
const            ::= integer | float | ''' string '''
tempScalarExp ::= interval|eventTerm|span {('+'|'-') span}|colRef'-'event|
                   event '-' event
eventTerm     ::= event{('+'|'-')span}|colRef{('+'|'-') span}
interval      ::='validtime''('identifier')'|'period' intervalExp|'period''('
                  eventTerm ',' eventTerm ')'
intervalExp ::= '[' time '-' time ')'
time         ::= tempDBDate | eventExp
event        ::= ( 'begin' | 'end' ) '(' interval ')'          |
                 ('first'|'last')'('eventTerm','eventTerm')'|eventExp
EventExp     ::=        'now'|'beginning'|'forever'|'date' dateString|'date'|
              tempDBDat | 'timestamp' timestampString
dateString       ::= '"' YYYY '-' MM '-' DD '"'
timestampString ::= '"' YYYY '-' MM '-' DD ' ' HH ':' MM ':' SS '"'
tempDBDate   ::= '"' YYYY ['/' MM [ '/' DD [ '~' HH [ ':' MM [ ':' SS ]]]]] '"'
span         ::= 'interval' spanExp
spanExp      ::= integer qualifier { integer qualifier }
qualifier    ::= 'year' | 'month' | 'day' | 'hour' | 'minute' | 'second'
```

3. Data defination statements

```
ddl          ::= ddlTable | ddlView | dropTable | dropView
ddlTable     ::= 'create' 'table' identifier ( tableDef | ddlQuery )
ddlView      ::= 'create' 'view' identifier ddlQuery
tableDef     ::= '(' colDefList ')' [ 'as' 'validtime' ]
ddlQuery     ::= [ '(' colList ')' ] 'as' query
colDefList   ::= colDef { ',' (colDef | tableConstraint) }
colDef          ::= identifier dataType [ columnConstraint ]
columnConstraint::= primKeyCol | refIntegrity | checkConstraint
tableConstraint  ::= ['constraint' identifier] (primKeyTab | foreignKey |
                         checkConstraint)
primKeyCol   ::= 'primary' 'key'
primKeyTab   ::= 'primary' 'key' '(' colList ')'
refIntegrity     ::= 'references' identifier '(' identifier ')'
foreignKey       ::= 'foreign' 'key' '(' colList ')' 'references' identifier
                      '(' colList ')'
checkConstraint  ::= 'check' '(' condExp ')'
colList          ::= col { ',' col }
col        ::= identifier
dataType   ::= 'integer' | 'float' | char' typeLength |'varchar' typeLength |
                      interval'|'datetime'
typeLength     ::= '(' integer ')'
```

```
dropTable      ::= 'drop' 'table' identifier
dropView    ::= 'drop' 'view' identifier
```

4. Data table operation statements

```
dml          ::= insert | delete
insert           ::=insertByValues | insertByQuery
insertByValues ::= ['validtime' 'period' intervalExp] 'insert' 'into' identifier
                  'values' '(' valList ')'
insertByQuery ::= [timeFlag] 'insert' 'into' identifier queryExp
delete     ::= [ 'validtime' 'period' intervalExp ] 'delete' 'from' identifier
              [ 'where' condExp ]
valList    ::= val { ',' val }
val        ::= integer | float | ''' string ''' | intervalExp | dateString |
              timestampString | tempDBDate
```

5. Control statements

```
control          ::= 'commit' | 'rollback'
```

A.2 API of TempDB 2.1

1. initContext()

```
public static void main(String[] args){
    TDBStatement statement = new TDBStatement();
    statement.initContext();
}
```

Executing these codes, TempDB initializes runtime environment according to the setting.

2. initContext(String path)

```
public static void main(String[] args){
    TDBStatement statement = new TDBStatement();
    statement.initContext("F://config.xml");
}
```

Executing these codes, TempDB initializes runtime environment according to assigned configuration file “F://config.xml”.

3. executeSelect(String atsql2)

```
public static void main(String[] args){
    TDBStatement statement = new TDBStatement();
    statement.initContext();
    Object result = statement.executeSelect(args[0]);
}
```

Executing these codes, TempDB will perform the query statement assigned by parameter args[0], and the result will be stored in variable “result”.

4. executeInsert(String atsql2)

```
public static void main(String[] args){
    TDBStatement statement = new TDBStatement();
    statement.initContext();
    statement.executeInsert(args[0]);
}
```

TempDB will perform the insert statement assigned by parameter args[0].

5. executeDelete(String atsql2)

```
public static void main(String[] args){
    TDBStatement statement = new TDBStatement();
    statement.initContext();
    statement.executeDelete(args[0]);
  }
```

TempDB will perform the delete statement assigned by parameter args[0].

Index